Immortal River

Immortal River

The Upper Mississippi in Ancient and Modern Times

Calvin R. Fremling

THE UNIVERSITY OF WISCONSIN PRESS

The University of Wisconsin Press
1930 Monroe Street, 3rd floor
Madison, Wisconsin 53711-2059

3 Henrietta Street
London WC2E 8LU, England

www.wisc.edu\wisconsinpress

4 6 7 5 3

Library of Congress Cataloging-in-Publication Data
Fremling, Calvin R., 1929–
Immortal river : the upper Mississippi in ancient and modern times / Calvin R. Fremling.
p. cm.
Includes bibliographical references (p.) and index.
ISBN 0-299-20290-9 (cloth: alk. paper)
ISBN 0-299-20294-1 (pbk.: alk. paper)
1. Mississippi River—History.
2. Mississippi River Valley—History.
3. Mississippi River—Environmental conditions.
4. Mississippi River Valley—Environmental conditions.
I. Title.
F351.F84 2004
977—dc22 2004005381

ISBN-13: 978-0-299-20294-1

Contents

Illustrations

Abbreviations

BALMM	Basin Alliance for the Lower Mississippi in Minnesota
BOD	biochemical oxygen demand
BOWSR	Board of Water and Soil Resources
CWA	Clean Water Act
DNR	Department of Natural Resources
EIS	Environmental Impact Statement
EMP	Environmental Management Program
ESA	Endangered Species Act
FAO	United Nations Food and Agriculture Organization
FEMA	Federal Emergency Management Agency
FWPCA	Federal Water Pollution Control Administration
GLO	General Land Office
GREAT	Great River Environmental Action Team
HREP	Habitat Rehabilitation and Enhancement Program
LTRMP	Long Term Resource Monitoring Program
MICRA	Mississippi River Interstate Cooperative Resource Agreement
MNRRA	Mississippi National River and Recreation Area
MRC	Mississippi River Commission
MRRC	Mississippi River Research Consortium
MWBAC	Minnesota-Wisconsin Boundary Area Commission
NBS	National Biological Survey
NEPA	National Environmental Policy Act
NOAA	National Oceanic and Atmospheric Administration
NPDES	National Pollution Discharge Elimination System
NPS	National Park Service
NSF	National Science Foundation
OECD	Organization for Economic Cooperation and Development
OPEC	Organization of Petroleum Exporting Countries
RRAT	River Resources Action Team

RRCT	River Resources Coordinating Team
RRF	River Resource Forum
SBA	Small Business Administration
UMESC	Upper Midwest Environmental Sciences Center
UMRBA	Upper Mississippi River Basin Association
UMRBC	Upper Mississippi River Basin Commission
UMRCC	Upper Mississippi River Conservation Committee
USACE	U.S. Army Corps of Engineers
USCG	U.S. Coast Guard
USDA	U.S. Department of Agriculture
USEPA	U.S. Environmental Protection Agency
USFWS	U.S. Fish and Wildlife Service
USGS	U.S. Geological Survey
WRDA	Water Resources Development Act

Acknowledgments

Writing this book has been an immense undertaking, the labor of my life, but the research and writing have been enjoyable, stimulating me to walk new paths, see the river through new eyes, work with interesting people, and kindle new friendships. The Upper Mississippi River is so expansive and extraordinarily complex that no mortal can ever "know it all." Fortunately, I've had river-wise colleagues to assist me in subjects as diverse as geology, terrestrial ecology, aquatic ecology, river commerce, agriculture, economics, and politics. Equally important has been the wisdom offered by my hunting and fishing buddies, commercial fishermen, trappers, and farmers. The last two chapters are controversial; the contained material may be displeasing to some. In my effort to "tell it like it is," I hope that I have retained my objectivity and not editorialized excessively.

Jerome Christenson, reporter and columnist for the *Winona Daily News,* edited every draft of my manuscript and made valuable suggestions for organizing the book and making my scientific jargon intelligible. A gifted river historian in his own right, Jerome provided valuable insights in every chapter. His inspiration, encouragement, and sense of humor kept me on course.

I am grateful to the following reviewers, who graciously donated their time, for their corrections, suggestions, and encouragement: Thomas Bayer (chapters 4, 5, 6), Thomas Claflin (22), Barry Drazkowski (22), James Eddy (22), Frank Grether (4, 5, 6), Katherine Grulkowski (1, 2), Nancy Jannik (4, 5, 6), Carol Jefferson (6–11), James Meyers (4, 5, 6), Stephen Nagel (1, 2, 5, 8), Dennis N. Nielsen (4, 5, 6), David M. Pennington (12), Jerry L. Rasmussen (22), Robert Sloan (4, 5, 6), Catherine Summa (4, 5, 6), James L. Theler (8), Rory N. Vose (18, 22), Thomas Waters (1, 2, 5, 8).

The Department of Resource Analysis at St. Mary's University of Minnesota produced charts and maps, and digitized photographs.

The following people assisted me by providing materials, services, and vital information: Richard V. Anderson, Robert Bollant, Sid Bond, Mike Cichanowski, Ronald Deiss, Barry Drazkowski, Chris Ensminger, John Gabbert, John W. Gorman Inc., Jeffrey Janvrin, Jeffboat Co., Richard Karnath, Daniel Krumholz, Richard Lambert, Stanley Ledebuhr, Francis Losinski, Kenneth Lukaszewski, Robert H. Meade, Mathew Merchelwitz, Philip Moy, John C. Nelson, Doug Nopar, Kent Pehler, Mark Peterson, Bud Ramer, Larry Robinson, Martha Roldan, Robert Romic, Jerry Schneider, Richard E. Sparks, and Daryl Watson.

The inspiration for the title of my book was Lafayette Bunnel's 1897 classic *Winona and its Environs in Ancient and Modern Days.*

Last, and most important, I thank my wife, Arlayne, who has shared my adventures for over fifty years and kept me on task, but never to the point where book writing interfered with my hunting, fishing, and other outdoor foolishness.

Immortal River

1

Prologue

> The river chattered on to him, a babbling procession of the best stories in the world, sent from the heart of the earth to be told at last to the insatiable sea.
>
> Grahame, *The Wind in the Willows*

The air was clear on September 14, 1805, enabling First Lieutenant Zebulon Pike and three of his men to view a spectacular Mississippi River panorama from their lofty overlook. A brisk early fall wind cooled the perspiring explorers as they caught their breath atop a precipitous 560-foot bluff known as Wapasha's Cap.

They could see twenty miles up and down the vast valley and the labyrinth of channels and islands that lay before and behind them. High bluffs flanked the river, at that point separated by a two-mile expanse of channels, islands, wetlands, and the broad, treeless prairie that would one day be the city of Winona, Minnesota.

On the river, whitecaps raced the expedition's boats, working upstream against the current under full sail, giving their crews a rare and well-earned respite from the rowing, poling, and pulling that had muscled their heavily laden craft upstream from St. Louis for the past thirty-seven days (Coues 1965).

Two years before, the fledgling United States of America had managed one of the great land deals of all time. For fifteen million

Zebulon Pike viewed the grandeur of the Mississippi River valley from this bluff at Winona, Minnesota, in 1805. Rising over 560 feet above the valley floor, the rounded dome with a fringe of evergreens on its crown was a landmark for explorers and traders, and later for tourists and riverboat pilots. When Chief Wapasha II's band of Dakota moved to the sandy prairie beneath it in about 1807, the bluff was nicknamed "Wapasha's Cap" because of its likeness to the red voyageur's hat, which had been presented to Wapasha by a British army officer. In the mid-1800s pioneers began to call it "Sugar Loaf" because it looked like the familiar cone-shaped masses of crystallized sugar that predated granulated sugar. By today's standards it was hardly a mountain, but it was to settlers who had never seen the Rockies (courtesy of the Winona County Historical Society).

dollars—about three cents an acre—the new American nation purchased all the land drained by the western tributaries of the Mississippi River from France. Of immediate importance, the Louisiana Purchase ensured right of passage of American boats on the Mississippi River past the French fortifications that controlled access to the Mississippi River at New Orleans, but it also included the vast expanse of unexplored, but occupied, land that stretched all the way from the Mississippi River to the crests of the Rocky Mountains. With a stroke of a pen, the land area of the young upstart nation was doubled.

In order to find out just what it was he had bought, President Thomas Jefferson ordered expeditions sent into the western American wilderness.

The first and most famous expedition was approved by Congress even before the Louisiana Purchase was proposed. Led by Captain Meriwether Lewis and Lieutenant William Clark, it headed westward from St. Louis, up the Missouri River toward the Pacific Ocean in May 1804 (Ambrose 1996).

A year later, the second expedition, commanded by twenty-six-year-old Lieutenant Zebulon Pike, was ordered by General James Wilkinson to explore the Mississippi River northward from St. Louis to its source. Pike was also to select sites for military posts, negotiate with the Indians, try to make peace between the warring Sioux and Ojibwas, and establish American authority over British fur traders who still occupied posts in United States territory (Coues 1965).

As best he could, Pike was additionally to make zoological, botanical, geological, and meteorological observations. The son of an army major, Pike had enlisted in the army when he was only fifteen years old and had only a "common school" education obtained in the military posts commanded by his father. But Pike was a keen observer, enabling him to make profound observations about the Mississippi River and its valley. On the evening of September 14, probably by candlelight, he penned an account of his side trip to Wapasha's Cap into his journal:

> Mr. Frazer, Bradley, Sparks, and myself, went out to hunt. We crossed first a dry flat prairie; when we arrived at the hills we ascended them, from which we had a most sublime and beautiful prospect. On the right, we saw the mountains which we had passed in the morning and the prairie in the rear; like distant clouds, the mountains at Prairie Le Cross; on our left and under our feet, the valley between the two barren hills through which the Mississippi wound itself by numerous channels, forming many beautiful islands, as far as the eye could embrace the scene; and our four boats under full sail, their flags streaming before the wind. It was a prospect so variegated and romantic that a man may scarcely expect to enjoy such a one but twice or thrice in the course of his life. I proposed keeping the hills until they led to the river, encamping and waiting the next day for our boats; but Mr. Frazer's anxiety to get to the boats induced me to yield. After crossing a very thick bottom, fording and swimming three branches of the river, and crossing several morasses, we at twelve o'clock arrived opposite our boats, which were encamped on the east side. We were brought

> over. Saw great sign of elk, but had not the good fortune to come across any of them. My men saw three on shore. Distance (advanced) 21 miles.

As he viewed the raw grandeur of the Mississippi River and its vast valley, Pike could not have conceived of the *Virginia,* a wood-burning, steam-powered boat paddling past this site, carrying mail, on its way upstream just eighteen years later. He could not have dreamed that by 1836, steamboats would regularly take holiday cruises on the Mississippi, carrying sightseers 670 miles upstream from St. Louis to St. Anthony Falls, the future site of Minneapolis, Minnesota.

Perhaps Pike could have envisioned a river spanned by bridges and flanked by highways and even railroads, but he could not have foreseen nuclear power plants, diesel-driven towboats and their long strings of barges, or swarms of pleasure boats, water-skiers, and personal watercraft. The huge dams of the Nine-Foot Channel Project that created expansive, shallow reservoirs that are wider than the Mississippi is at New Orleans would have amazed him. He could not have imagined the incredible system of levees and flood walls that flank the modern river and separate it from its floodplain, running all the way from New Orleans northward to Iowa. He would have been impressed by the devastation wrought by the disastrous midsummer flood of 1993, when many flood works were breached, causing the flooding of cities and farms.

It is doubtful that the word "pollution" was in Pike's vocabulary, and the word "ecology" had yet to be coined. Indeed, Pike would have been surprised to know that most of the very bluff that he had climbed would be quarried, leaving only the vestige now called "Sugar Loaf."

My book documents the history of the Mississippi River and the dramatic changes that have transpired as humans have exploited and attempted to subdue the Father of Waters. The most sudden changes have occurred subsequent to the American-European invasion of the Mississippi Basin, and most have taken place in less than 150 years—a mere blink of an eye in geologic time. The chronicle is written as a Mississippi River primer, designed to present serious recreationalists, technicians, planners, and beginning river scientists with the basic natural and human history of one of the world's greatest, most famous rivers. How can we know where we're going if we don't know where we've been?

There is no shortage of information about the Mississippi River, but most of the "good stuff" is widely scattered. Much is highly technical, and most is buried in seldom-seen government documents and scientific publications. I have attempted to assemble and simplify some of this information, condensing it into a single volume, so that each new student of the river needn't "reinvent the wheel."

The book is written in a nontechnical style for the reader with little science background, and terms are defined the first time they are used. My hope is that increased knowledge of the river and awareness of environmental concerns will promote responsible, holistic management of Mississippi River resources, so that others, most yet unborn, can love and enjoy the river as I have.

No one can ever know the entire river. From the high vantage point of his wheelhouse, the towboat captain is intimately familiar with the navigable commercial channel but may never run the backwaters or hear the eerie cry of a loon on a Headwaters lake. The engineer has been accused of viewing the river through an eighteen-inch dredge pipe, disdaining ecology and aesthetics, and (like a beaver) being intolerant of the sound of running water. The duck hunter tends to see the river over his shotgun barrel, forgetting about economics and conflicting river uses. The floodplain dweller peers fearfully in the spring at the awesome, rising river from the top of his jerry-built levee or through the floodgates of a massive, government-subsidized floodwall. The environmental extremist would like to see all of the dams removed. The grain exporter wants the lock and dam system modernized and enlarged.

This account is not meant to be a travelogue. It concerns the natural and human-induced processes that have shaped the river and its ecology, and especially the basic underlying problem presently facing Old Man River: how to satisfy increasing demands on a diminishing resource. Hopefully, it presents a balanced view of the river as seen through the eyes of towboat captains, engineers, scientists, poets, and river rats.

In current scientific literature the metric system of measurement is nearly universal. However, the English system of measurement (miles, feet, pounds) is used throughout this primer because it is directed at a readership accustomed to the English system. Furthermore, for more than two hundred years most Mississippi River measurement data, including maps and charts, have been published in English units. All navigation aids on the river have their units of measure expressed in the English system.

For river scholars, references are cited throughout the text. Wherever possible, I have used references that are available in regional libraries. I write in the first person in some sections because I agree with Mark Twain who believed that, "One should only write in the third person if he is President of the United States or has a tape worm."

The reader will notice that "Upper" and "Lower" are capitalized when referring to sections of the Mississippi River. This is not only traditional but also proper because the adjectives describe distinctly different sections of the river that have been precisely defined geographically for more than 150 years. During the early years of our nation's history when rivers were our superhighways, most access to the Mississippi River was gained by boating down the Ohio River from the industrial cities of the eastern United States to meet the Mississippi at Cairo, Illinois. From there travelers could go downstream on the Lower Mississippi toward New Orleans or upstream on the Upper Mississippi toward the frontier. Other reasons for this important delineation will become apparent in subsequent chapters.

I feel almost qualified, after about sixty years of experience, to write about the Mississippi River. My mother, who was an amateur historian, kindled my interest early. Her great-grandfather, John Rollins, constructed and operated the first steamboat, the *Governor Ramsey,* on the Mississippi River above St. Anthony Falls. He also built and operated a flour mill powered by St. Anthony Falls. He and his son later became loggers in the St. Croix watershed, and my grandmother was a cook in their logging camps. I grew up in Brainerd, Minnesota, and learned to swim in the Mississippi's tea-colored waters at Lum Park.

The river has been like a magnet to me. Except for two years in the military, I have lived somewhere along the Mississippi River at least part of every year of my life. I'm on speaking terms with the entire river—from its source in northern Minnesota to its mouth in the Gulf of Mexico. I know some sections very well and have learned to run some of the labyrinthine backwaters near my home at Winona, Minnesota, by starlight.

As a university professor, I have learned from other biologists, sportsmen, commercial fishermen, chemists, engineers, geologists, hydrologists, and city planners. In my research I have worked closely with personnel of the U.S. Army Corps of Engineers, the U.S. Fish and Wildlife Service, the U.S. Geological Survey, and the Departments of Natural Resources of several states. I have been privileged to study navigation first hand on a U.S. Coast Guard buoy tender and aboard commercial towboats from Winona, Minnesota, to New Orleans, Louisiana. One of my most pleasurable duties was lecturer for the Smithsonian Institution aboard the commercial tour boat *Delta Queen*.

I have been fortunate to be able to study glaciers in Alaska, Iceland, Mount Rainier, and the Canadian Rockies. With SCUBA, I have examined ancient, submerged shorelines in ocean environments, documenting changes in ocean levels in response to continental glaciation. Similarly, I have used SCUBA to study the depths of the Mississippi River during the summer and through the ice in winter, concluding that Australia's Great Barrier Reef is a "piece of cake" compared to the tailwaters of a navigation dam on the Mississippi, where a diver must contend with zero visibility, strong currents, intense boat traffic, and tangles of monofilament fishing line festooned with lost fishing lures.

Unlike many adventurers, I have never launched a canoe at the Mississippi's source and paddled to New Orleans in a single journey, but I have researched, hunted, fished, or boated virtually every mile of it at some time. My buddies and I have fished the walleye lakes of the Headwaters during winter and summer for over half a century. Unforgettably, in late March of 1993, we augered through three feet of ice in a -30°F windchill to fish for yellow perch on northern Minnesota's Lake Winnibigoshish, one of several large reservoir lakes that the Mississippi Headwaters runs through. In mid-April my wife, Arlayne, and I followed the spring flood crest down the

Mississippi by car—through the nation's midsection and the great Louisiana delta. We watched the floodgates being closed at Cape Girardeau and Hannibal, Missouri, never dreaming that a second, unprecedented, midsummer crest would devastate low-lying communities and farmland from central Iowa to southern Illinois. We ate our way through the delta country of Louisiana, documenting the poignant demise of the Cajun lifestyle, and the inexorable loss of Louisiana's barrier islands and the nation's most expansive wetlands. Enjoyably, we had learned many years ago to catch and cook Louisiana redfish, blue crabs, and crawfish.

Through my avocations of hunting and fishing I think that I have gained insights that would have escaped me had I remained cloistered in my university ivory tower. My leisure activities have forced me to scrutinize Old Man River in all of his fickle moods—at sunrise, in the black of starless nights, during all seasons, during floods and droughts, and in fog, rain, and snow. Most important, my pursuits have enabled me to learn about the "real world" from an extensive cadre of "river rats" who have exploited, studied, and loved the river from within, rather than analytically studying and digitizing it from without—as desk-bound, computerized scientists and technicians are increasingly wont to do.

2

Introducing Old Man River

The Famous Mississippi

"Old Man River," "Father of Waters"—the names conjure up images of Indians in birch bark canoes, explorers, fur traders, loggers, cotton plantations, southern mansions, live oaks festooned with Spanish moss, slaves, Civil War, steamboats, showboats, riverboat gamblers, Mark Twain, Tom Sawyer, Huck Finn, pearl button factories, dams, towboats, mud, and giant catfish.

Immortalized in prose, poetry, and song, the river has been a defining theme in the cultural life of the nation. The Mississippi River and its tributaries played major roles in the exploration and colonization of the midcontinent and in the westward expansion of the United States.

Along with the Grand Canyon of the Colorado River and Yellowstone Park, the Mississippi is the natural feature that foreign tourists want to see most when they visit the United States. The Mississippi River is a national treasure, and a gateway to the world.

Vital Statistics

The Mississippi River—the name is an Ojibwa (Chippewa) word meaning "great river" or "gathering of waters"—is the largest and longest river in North America.

Of the world's rivers, the Mississippi is the third longest, has the third largest drainage basin, and is the eighth largest in average annual discharge (Nace 1970; Leopold 1994). With the Missouri, its longest tributary, it forms the world's longest river system (4,321 miles), even surpassing the Nile.

The Mississippi drainage basin, or watershed, is shaped like a huge funnel, stretching from the Allegheny Mountains in the east to the Rocky Mountains in the west. It drains an area of 1.24 million square miles, about 40 percent of the contiguous United States and about one-eighth the area of North America. The river's watershed includes all or parts of thirty-one states and two Canadian provinces (Meade 1995).

Flowing to the south, the Mississippi cuts through a cross section of America's heartland. It wanders from its headwaters for 535 miles as a young, often rocky stream within Minnesota. Below Minneapolis it flows 1,766 miles as a much older, sediment-choked river along the borders of Minnesota and nine other states, gathering the waters of its many tributaries and pouring them into the Gulf of Mexico.

> Down the Yellowstone, the Milk, the White and Cheyenne;
> The Cannonball, the Musselshell, the James and the Sioux;
> Down the Judith, the Grand, the Osage and the Platte,
> The Skunk, the Salt, the Black, and Minnesota;
> Down the Rock, the Illinois, and the Kankakee,
> The Allegheny, the Monongahela, Kanawha, and Muskingum;
> Down the Miami, the Wabash, the Licking and the Green,
> The Cumberland, the Kentucky, and the Tennessee;
> Down the Ouachita, the Witchita, the Red, and Yazoo—
> Down the Missouri, three thousand miles from the Rockies;
> Down the Ohio, a thousand miles from the Alleghenies;
> Down the Arkansas, fifteen hundred miles from the Great Divide;
> Down the Red, a thousand miles from Texas;
> Down the great Valley, twenty-five hundred miles from Minnesota,
> Carrying every rivulet and brook, creek, and rill,
> Carrying all the rivers that run down two-thirds the continent—
> The Mississippi runs to the Gulf.
>
> *The River,* Pare Lorentz

The Mississippi is not just any river; it is the "Mighty Mississippi," a busy, vital intracontinental water highway that connects North America's "breadbasket" with the rest of the world.

The Mississippi basin is a supermarket to the world. Crops grown on some of the planet's richest soils feed one in twelve of the world's people. But over the years, the Mississippi, sometimes running like chocolate, has dumped unnumbered tons of the nation's irreplaceable topsoil irretrievably into the Gulf of Mexico.

From its source at Lake Itasca, in Minnesota's North Woods, the Mississippi runs 2,301 miles through the midcontinent, across the Gulf Coastal Plain, and through the subtropical great Louisiana delta to the Head-of-Passes. There it splits like the toes of a bird's foot into several distributaries or outlet channels called "passes" that lead to the Gulf of Mexico. The major pass is about twenty miles long.

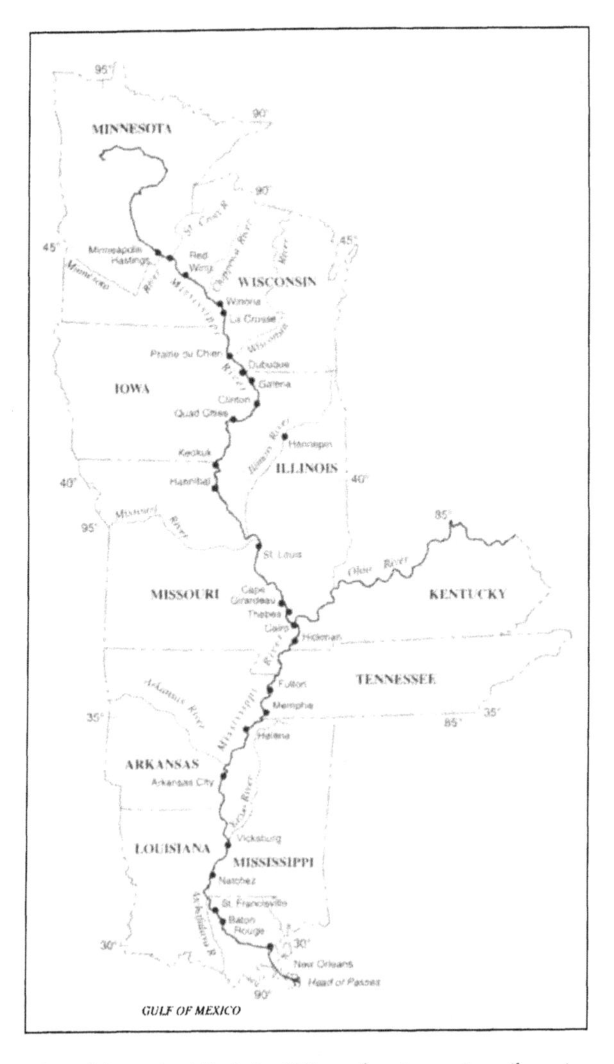

Map of the entire Mississippi River, showing major tributaries and selected cities. The Mississippi Headwaters includes the reach from the river's source at Lake Itasca in northern Minnesota downstream to St. Anthony Falls at Minneapolis, Minnesota. The Upper Mississippi includes the reach from Minneapolis downstream to Cairo, Illinois, at the mouth of the Ohio River. The Lower Mississippi includes the reach from Cairo to the Head of Passes in the Gulf of Mexico. The Atchafalaya River is the Mississippi's main distributary, diverting flow from the Mississippi directly into the Gulf of Mexico (adapted from Meade 1995).

The Mississippi River is customarily divided into three distinct segments: (1) the Headwaters—running 493 miles from the river's source (Lake Itasca in northern Minnesota) downstream to St. Anthony Falls in Minneapolis, Minnesota; (2) the Upper Mississippi River—running 854 miles from St. Anthony Falls downstream to the mouth of the Ohio River at Cairo (pronounced *kaye roe*), Illinois; and (3) the Lower Mississippi River—running 954 miles from Cairo to the Head-of-Passes in the Gulf of Mexico. Additionally, the 195-mile segment of the Upper Mississippi River from the mouth of the Missouri River (at St. Louis, Missouri) to Cairo is often referred to as the Middle Mississippi River because it is undammed and ecologically unique, mainly because of the influence of "Big Muddy"—the Missouri River.

If explorers had probed the Mississippi River from the Gulf of Mexico northward and seen the Missouri swollen with spring runoff, rushing huge trees along in its fast and dangerous current, they would have had good reason to think that the Missouri River was the real continuation of the main river or that the Upper Mississippi was only a large tributary.

The Missouri is more than sixteen hundred miles longer than the Upper Mississippi, and its drainage basin is about three times as large. Some have suggested that only that portion of the Mississippi from its source to the mouth of the Missouri should have been called "Mississippi" after its original Indian name. They have further suggested that the rest of the river that begins in Montana and ends in the Gulf of Mexico should have been called "Missouri River."

The literature contains many differing figures for the length of the Mississippi River. Its length changed naturally over the years as the Lower Mississippi meandered about in its broad valley, sometimes shortening itself suddenly as it cut off meander loops. In modern times the Lower Mississippi has also been shortened many times by engineers who cut off meander loops to speed navigation. My 2,301-mile figure was determined by adding the mileages shown on the newest official maps of the U.S. Army Corps of Engineers, the Mississippi River Commission, and the Minnesota Department of Natural Resources. It is unlikely the river's length will change significantly in the near future because the river is now constrained by massive channelization projects from the Twin Cities to the Gulf. Someday, however, it may spontaneously change its length a few miles in the Headwaters where it still meanders unfettered through flat glacial lakebeds. Because it occupies the bottom of a deep, relatively narrow trench, the Upper Mississippi from Minneapolis to St. Louis has not meandered significantly in the past ten thousand years.

Locations along the river's main channel are given in River Miles (abbreviated RM), starting with RM 0.0 at the Head-of-Passes in the Gulf of Mexico and proceeding upstream to the mouth of the Ohio River at Cairo, Illinois (RM 953.8). At Cairo, numbering starts at RM 0.0 again and continues up the Mississippi to its source. From the Gulf of Mexico to Minneapolis, mileages are printed on permanent, lighted buoys or shore markers that, along with unlighted buoys and markers, identify the edges of the commercial navigation channel. They provide an accurate means of locating sites along the river. The Headwaters segment of the river is not officially marked with buoys or other navigation aids, but mileages are accurately shown on free, detailed maps produced by the Minnesota Department of Natural Resources.

In contrast, RM 0.0 on the Ohio River is at Pittsburgh, Pennsylvania, at the confluence of the Allegheny and Monongahela Rivers. Mileages increase downstream to RM 981 at Cairo, Illinois, where the Ohio joins the Mississippi. This apparent anomaly is because exploration of the Ohio and the flow of manufactured goods were in a downstream direction. Mileages indicated how far downstream keelboaters were from homeport.

Nearly one half of the water discharged into the Gulf of Mexico is contributed by the Ohio River. The Mississippi and the Atchafalaya River, the Mississippi's main distributary, together discharge an average of about 420 billion gallons of fresh water per day to the Gulf (U.S. Geological Survey Circular 1133, 1995).

A Water Highway to the Sea

The Mississippi, its navigable tributaries, and man's engineering projects have created a water highway to the interior of the continent—and to the sea. For example, I could launch my sixteen-foot fishing boat just fifteen minutes from my home at Winona, Minnesota, and

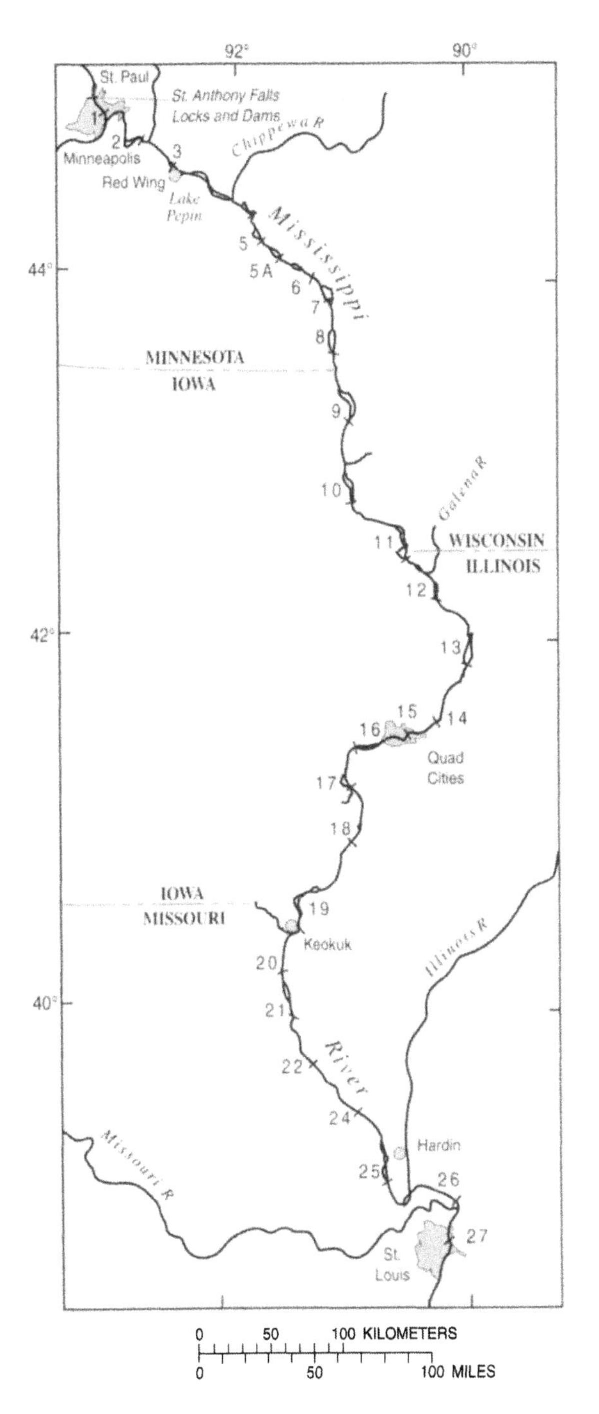

The Upper Mississippi River is partly impounded by locks and dams, built mostly for navigation, that control the depth and, to a lesser extent the flow of the river between Minneapolis and St. Louis. From Minneapolis downstream, the locks and dams are numbered from 1 to 27. There is no Lock and Dam 23, but there is an "extra one" at Winona, Minnesota (Lock and Dam 5A). With the Upper and Lower St. Anthony Locks and Dams in Minneapolis they total twenty-nine. There are no dams below St. Louis. The two largest impoundments are Lake Pepin, a natural lake formed by the partial damming of the Mississippi by the delta of the Chippewa River, and Pool 19, behind the hydroelectric dam at Keokuk, Iowa (adapted from Meade 1995).

A diesel-powered towboat pushes twenty barges of grain downstream below St. Louis. In this undammed reach, forty-barge tows are common, and most towboats produce about 10,000 horsepower. Upstream from St. Louis, where the river is dammed, tows seldom exceed fifteen barges due to the size of the navigation locks. Towboats in the upriver reach average about 5,400 horsepower.

travel 140 miles upstream to the Coon Rapids dam in north Minneapolis without ever taking it out of the water. Similarly, I could run downstream to New Orleans and continue into the Gulf of Mexico or enter the Gulf Intracoastal Waterway, and travel hundreds of miles along the Gulf and Atlantic coasts. Halfway down the Mississippi, I could detour up the Illinois River to Chicago, lock into Lake Michigan and the St. Lawrence Seaway, ultimately to enter the North Atlantic Ocean via the St. Lawrence River.

Commercial riverboats, once apparently doomed by railroads, now provide intense competition for all transporters of bulk commodities such as grain, coal, oil, gasoline, molasses, cement, and fertilizers. Modern towboats on the Upper Mississippi River push as many as fifteen barges and can carry the equivalent tonnage of approximately 225 rail cars.

The U.S. Army Corps of Engineers is responsible for maintaining

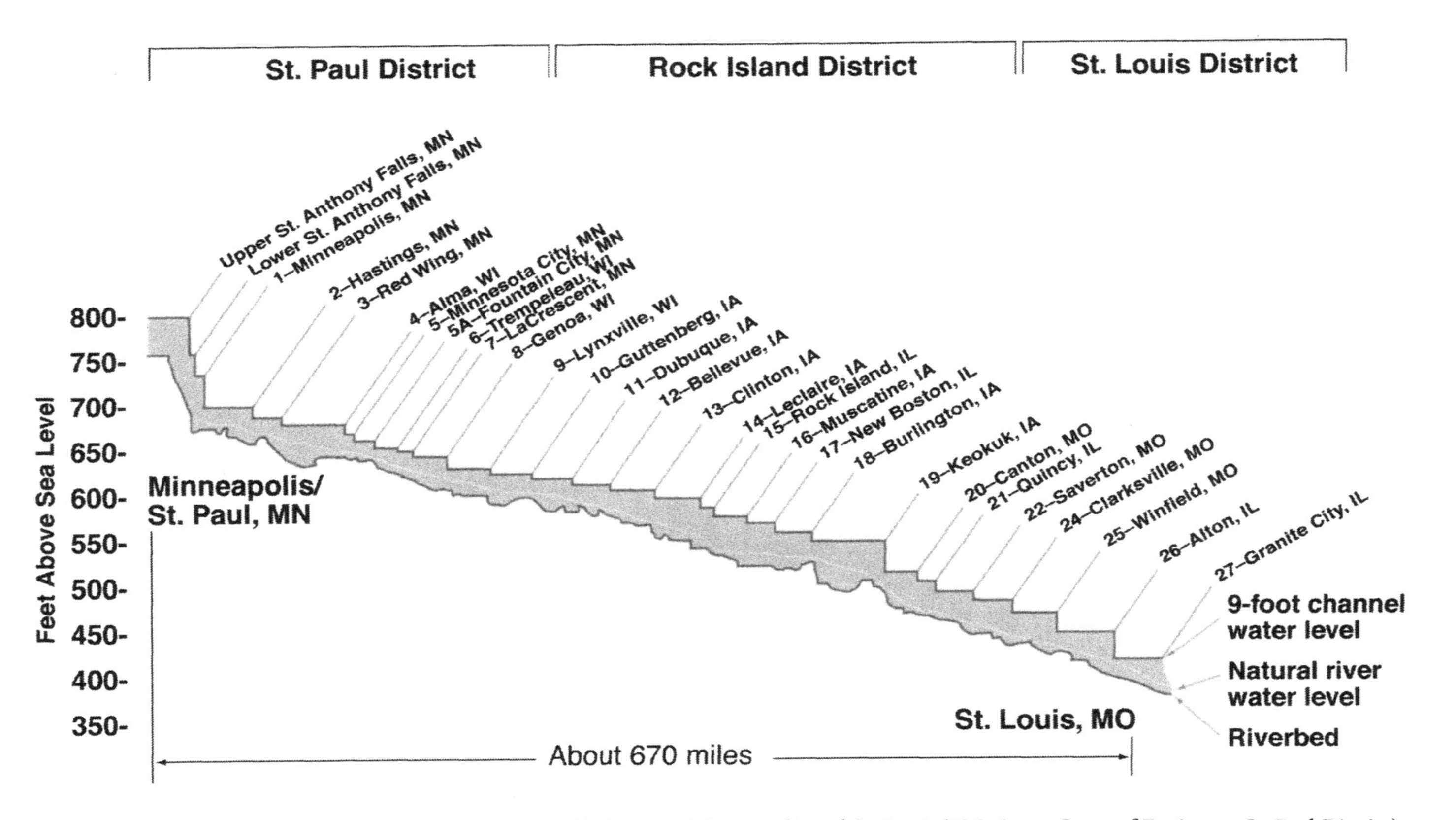

A "Stairway of Water" makes commercial navigation possible between Minneapolis and St. Louis (U.S. Army Corps of Engineers, St. Paul District).

federal dams in the Headwaters, federal flood levees system-wide, and the commercial nine-foot river channel from the Twin Cities southward. The twenty-nine navigation locks and dams are operated and maintained by the U.S. Army Corps of Engineers. The U.S. Coast Guard maintains the system of navigation aids that guides modern diesel-powered towboats (and thousands of pleasure boats) as they navigate the Upper Mississippi around the clock from early spring until early winter. Maintenance includes painting and repositioning buoys, and servicing lighted markers.

Recreation is a major use of the Mississippi River. Fishing and boating are among the most popular pastimes, but hunting, camping, swimming, birding, and visits to historic towns, archeological sites, and locks and dams are other common activities. On the Upper Mississippi alone, the national impact of recreation has been estimated to exceed $1.1 billion annually (U.S. Army Corps of Engineers 1993). This does not include the value of recreationally intense Headwaters and the hundreds of lakes that drain into it. The lakes are ringed with resorts, summer cabins, campgrounds, and year-round homes. Boating, fishing, and swimming are popular sports, and there is generally a high level of environmental awareness.

River Markers

Hundreds of red day markers denote the right side of the nine-foot-deep, commercial channel as boaters proceed upstream ("red right returning"). Similar green markers mark the left side of the channel. Numbering at the top of the marker tells the boater how far upstream she or he is from the mouth of the Ohio River at Cairo, Illinois. A solar-powered light (with storage battery) blinks twenty-four hours per day but is only visible at night or in dim light. A steady "on–off–on" signifies the left side of the channel. An intermittent "on–on–off–on–on" signifies the right. Lighted markers are usually located on bends, and successive markers are usually within sight of each other, making navigation possible at night. Floating red and green buoys also mark channel borders. By staying between the red navigation aids and green navigation aids, boaters are assured of at least nine feet of depth. Danger, in the form of rocks and stumps, lurks outside the marked channel. The U.S. Coast Guard maintains the navigation aids.

The river is more than a commercial or recreational resource. Its role in the natural ecology of the continent is unique and irreplaceable. The Mississippi River serves as a migration corridor for more than two hundred species of birds. Its backwaters are habitats to a vast array of mammals—such as beaver, otter, mink, muskrat, and raccoon. The 195 species of freshwater fish that occur in the Mississippi River comprise nearly one-third of the approximately six hundred freshwater fishes of North America. Because it is so ancient, the Mississippi is a virtual "Cretaceous Park," home to relict fishes about the same age as dinosaurs.

Today, most of the Mississippi River south of St. Paul, Minnesota, is a "working river," dominated by powerful, ponderous towboats. On their way downstream, the big ones wrestle six acres of grain-laden barges toward the deep-water ports of Baton Rouge and New Orleans. There the corn and soybeans are transferred to oceangoing freighters that distribute the grain worldwide. On their return trip upstream, the towboats may push barges of fertilizer for the farmers who raised the corn and soybeans, or fuel for cars and farm machinery.

Coal is shuttled upstream as well as downstream to supply the power plants that generate much of the electrical energy that runs the farms, cities, and industries. Coal goes both directions so that power plants can meet air pollution standards. Cheap high-sulfur coal from Kentucky may be blended with more expensive low-sulfur coal from Wyoming. The "cowboy coal" may be shipped directly to the power plants by rail or off-loaded to river barges near the Twin Cities and then barged downstream. Thus, the Mississippi is one of the world's most important commercial rivers and one of the most severely regulated. "Regulated river" is a recent euphemism describing rivers that are dammed and constrained.

Human influence on the river begins virtually at its source. The Headwaters segment of the Mississippi flows through nine natural lakes and the impoundments created by eleven dams. Some of the natural lakes have been increased in size by dams at their outlets. The dams have no locks, and there is no commercial boat traffic (except for charter fishing boats on the largest lakes).

Recreational boaters are advised that their boats must be properly equipped with personal floatation devices, running lights, and other safety gear because they may be inspected by law enforcement personnel of the U.S. Coast Guard, conservation officers of the U.S. Fish and Wildlife Service, conservation officers of states that border the river, and the river patrols of many counties and cities.

The Upper Mississippi River has been intensively channelized for navigation beginning in 1878. Broad, shallow impoundments were created on the Upper Mississippi when twenty-nine navigation dams were constructed, mainly during the 1930s, to create a slack-water navigation channel nine feet deep between St. Louis and Minneapolis. River travelers are usually surprised at the width of the Upper Mississippi in its impounded reaches where it is much wider (but much shallower) than it is at St. Louis or New Orleans where the river is undammed.

Because the impoundments alone are insufficient to maintain the nine-foot commercial channel, the river's main channel is routinely dredged in some reaches. Almost all sand islands along the main channel have been placed there as result of dredging. In recent years, attempts have been made to minimize the adverse environmental impacts of this practice.

The Lower Mississippi River has been channelized and shortened 142 miles, and its velocity has been increased commensurately. It remains undammed, but its floodplain has been reduced about 90 percent by levee construction begun in 1727 near New Orleans. About 40 percent of the runoff from the continental United States is carried by the Mississippi River system (Leopold 1994).

Downstream from Minneapolis, the Mississippi River has been intensively managed for the transport of commercial cargoes and for flood control for more than two hundred years. The U.S. Army Corps of Engineers, as mandated by the U.S. Congress, maintains a navigable channel nine feet deep from Minneapolis, Minnesota, to Vicksburg, Mississippi; a twelve-foot channel from Vicksburg to Baton Rouge, Louisiana; and a ship channel forty feet deep from there to the open waters of the Gulf.

First-time visitors to New Orleans are usually surprised to learn that the Mississippi is only about 2,200 feet wide at the foot of Canal Street. Equally surprising is that the river is 240 feet deep in the sharp bend just off the Esplanade Wharf at the edge of New Orleans'

French Quarter. For the last 450 miles of its run to the sea, the Mississippi's bed lies below sea level. Protected by massive levees, much of New Orleans lies below sea level, and virtually all of it lies below river level.

From New Orleans upstream to Dubuque, Iowa, the Mississippi is flanked continuously by massive flood-control levees or steep natural banks. Upstream from Dubuque, additional flood levees protect most low-lying cities.

The mainline flood levees are usually armored with rock; and along the Lower Mississippi even the riverbanks and portions of the riverbed are armored, usually with concrete or asphalt. With minor exceptions (St. Anthony Falls, Rock Island Rapids, Keokuk Rapids, and Chain of Rocks at St. Louis), virtually every rock larger than a cantaloupe—from Minneapolis to the Gulf of Mexico—has been placed there by the U.S. Army Corps of Engineers or the Corps' contractors.

A River in Trouble

The Mississippi River has suffered at the hand of man. Dams and levees, which aid navigation and floodplain agriculture, have reduced the river's natural ability to create habitat for fish and wildlife during periods of high flow. Floods have increased in frequency and severity. Navigation impoundments, side channels, and sloughs are filling with sediment—and the rate of filling may be exacerbated by proposed increases in commercial traffic.

Some river reaches are severely polluted. Exotic plants and animals are competing with native species, and whole ecosystems seem to be unraveling. Yet, we are exponentially increasing our demands on this diminishing resource. While the myriad manmade problems affecting the Mississippi are of recent origin, they have their foundation in the natural forces that shaped the river and its enormous watershed. A basic understanding of that fascinating geological history is necessary to appreciate today's magnificent river and its ills.

This book tells the story of the evolution of the modern Upper Mississippi River. The story is a long one, spanning about six hundred million years of earth's history, and it melds information from geology, ecology, and human history.

Rivers versus Lakes

> If you were to guess which last longer—rivers, mountains, or lakes—which would you choose? Lakes, especially those that are products of the ice age, disappear with astonishing speed in geologic terms. Mountains, often taken as symbols of rugged endurance, are thrown down by the relentless forces of erosion a few million years after their uplift has ceased. . . . Rivers, on the other hand, can last scores or even hundreds of millions of years.
>
> Wiggers 1997

Lakes are temporary features on the land; rivers are virtually immortal. They are relentless shapers of the land. The source of the Mississippi has not long been at Lake Itasca in northern Minnesota. The modern Headwaters is only about ten thousand years old—a mere toddler in geologic time—its present course set when the last glacier retreated across Minnesota. Within the last twenty thousand years, alternate sources have included two vast glacial lakes—Lake Agassiz and Lake Duluth.

Most of the Mississippi's course, from Minneapolis, Minnesota, to Cairo, Illinois, was determined over a million years ago as meltwaters etched a path along the eastern edge of the great Nebraskan glacier. The river's gorge was enlarged by meltwaters of subsequent glaciers.

The general course of the Lower Mississippi may be as old as the Atlantic Ocean. It has probably flowed into the Mississippi Embayment—the troughlike structure that reaches northward from the Gulf of Mexico to Cairo, Illinois—for more than 250 million years. "Rivers are born traveling, always wanting to move on, intolerant of restraint and interference—itinerant workers always rambling down the line to see what's around the next bend, growling or singing songs, depending on how things suit them. Now, a lake never goes anywhere or does much. It just sort of lies there, slowly dying in the same bed in which it was born. The lake is a set of more or less predictable conditions—at least, compared to the swiftly changing stream of physical, chemical, and biological variables that constitute a living river" (Madson 1985, 8).

I

The River Primeval

3

Majestic River Bluffs and Beautiful Valleys

Introduction

The immensity of the gorge of the Upper Mississippi River awed early explorers. Viewing the precipitous bluffs and cliffs that flanked the river, many thought that they were seeing low mountain ranges. They marveled at the rugged headlands and the layered rocks filled with the shapes and shells of animals they had never seen before.

To this day, travelers and people who have lived for years along the river wonder about these same things, unaware of how the river came to occupy this deep valley. Most are surprised when they ascend the "hills" and find themselves in the midst of beautiful, flat, highly productive agricultural land that is dissected by deep valleys. There are no mountains or hills, just valleys—valleys carved through a vast plateau of soft, incredibly old, sedimentary rocks.

This chapter and the next four chapters tell how this magnificent landscape came to be, offering a "big picture" of the geologic forces that shaped this landscape from the dawn of time to the present. A basic understanding of the geology that underlies the river and the land it flows through illustrates how natural forces have dictated the locations of our cities, highways, dams, and bridges. It is the story of the natural barriers to man's navigation of the river—St. Anthony Falls, the Rock Island Rapids, the Des Moines Rapids at Keokuk, Iowa, and the Chain of Rocks at St. Louis. It explains why some sections of the river must be dredged and redredged to maintain commercial navigation, and ultimately, why the Mississippi River is here at all.

To tell of the almost six hundred million years of extraordinary earth and ecological history pertinent to the story of the Upper Mississippi River I have stood on the shoulders of giants, drawing upon the written works of legions of geologists, biologists, anthropologists, and intellectually curious river folks. As is customary in geologic literature, I denote prehistoric times in years B.P. (Before Present). To avoid confusion, times just prior and subsequent to the birth of Christ will be denoted as B.C. and A.D., the system familiar to most readers.

Ancient Seas: Formers of Sedimentary Rocks

The land was born of rock, fire, and water. The infant earth was a glowing ball of molten rock whirling around the primeval sun. Over cosmic stretches of time the earth began to cool and rain began to fall, producing the feature that makes our planet unique in all the known solar system—liquid water. Of all the earth's naturally occurring compounds water is unique, the only substance found as a liquid, solid and vapor at naturally occurring surface temperatures and pressures. In each of these physical states water carves and shapes the earth's surface, wearing away, first at bare rock, later cutting through sand and soil to create much of the landscape that surrounds us.

The entire surface of the earth began as hot rock, gradually cooling under the first rains falling on the newly formed planet. These rocks formed in fire—igneous rocks—are the ancestors of all other rocks, sand, and soil. Igneous rock is still being formed when hot magma from deep in the planetary interior rises to the surface as volcanic lava or cools and solidifies quietly in the earth's crust. We are familiar with igneous rock as granite and basalt, hard and dense.

Water—exploding into steam as it contacts molten magma; irresistibly expanding in cracks and crevasses as it freezes; swirling and polishing as it flows as a liquid—is the great destroyer of igneous rock. It fractures the rock, breaking off fragments, which are carried in its currents in an abrasive stew that wears away the rock it flows across, weathering it into ever smaller pebbles, sand grains, and flecks of clay. As the current slows or the stream dries up, these particles are deposited as sediment.

Granite, familiar to most as the stone cut and carved into cemetery markers, consists mainly of the minerals quartz and feldspar. With complete weathering quartz usually produces sand-sized sediments, while feldspar weathers to silt and clay. Silt and clay particles are so fine they can only be distinctly seen though a high-powered microscope. Silt, the coarser of the two can be distinguished from clay by a simple test—when ground between the teeth, silt is gritty, clay is not. Fine sand, predominantly quartz, is hard and stable—the most common mineral not easily dissolved in water.

Larger sediment particles are often bits of rock containing more than one mineral. Larger still are granules, pebbles, cobbles, and boulders consisting of rock fragments, usually rounded by abrasion during their transport by water or ice (Leopold 1994). Particles larger than sand are generally called gravel. Because quartz is very stable, it tends to hang around longer and to be the most common mineral at earth's surface. Most of the Mississippi's sand is made of quartz.

Sediment transported by water is described as either suspended load or bed load. Suspended load consists of clay and silt and is what makes water muddy. Due to its size it ultimately settles out in quiet areas, but it may be resuspended again and again by river currents or wave action. Bed load, in contrast, consists of larger particles that are dragged along near the streambed (Leopold 1994). Oceans are the ultimate sinks for sediments. The heavier and coarser sand particles settle out near river mouths or where there are currents produced by the pounding of the surf, but silts and clays settle out in the calm areas, either in protected shallow areas or at great depths.

Left undisturbed, deposits of sediment may be transformed into sedimentary rock. Deposits of sand may ultimately form sandstones.

Silts and clays may form siltstones and shales. All sedimentary rock is formed in layers called strata. Strata can easily be seen and identified in sedimentary rock formations; therefore, sedimentary rock formations are often described as stratified, or layered. Most of our knowledge of our planet's history is derived from studies of the sedimentary rocks that blanket most of the earth's crust—the oldest rocks on the bottom and the youngest on the top.

The particles that compose sedimentary rocks such as sandstone or shale have been compacted by the immense weight of overlying strata, and are usually cemented together by minerals precipitated from ground water. Other sedimentary rocks, such as limestone and dolomite (dolostone), are composed largely of the calcareous skeletons of corals and seashells that may have been reduced to sand-sized particles. Limestone is simply calcium carbonate, but dolomite is calcium-magnesium carbonate. Technically, dolomite is the mineral and dolostone is the rock that contains it. However, the rock is still referred to as dolomite in most circles, especially in older literature.

With time, sedimentary rocks may be buried under an ever-thickening series of layers that may reach thicknesses of tens of thousands of feet. The earth becomes hotter at increasingly deeper levels, and the weight of overlying rocks exerts incredible pressure. In concert, the heat and pressure cause the deeper strata to form metamorphic ("changed form") rocks. The metamorphic rocks are a complex group that includes common rocks like slate (from shale), marble (from limestone and dolomite), and quartzite (from sandstone).

Sedimentary rocks that were formed in the depths of the sea may be uplifted by geologic forces to become mountains thousands of feet high. In the American West, the Rocky Mountains are the contorted sedimentary strata of an ocean bed that has been elevated over two miles. I was awed many years ago when I climbed one of the mountains, sat down to rest at the summit, and found it littered with fossil trilobites, oceanic invertebrates that have been extinct for 250 million years! Impossible as it seems, the crest of the world's tallest mountain, Mount Everest, is composed of limestone formed from the floor of an ancient ocean.

All of the rock formations visible along the Mississippi from St. Paul, Minnesota, to the Gulf of Mexico are sedimentary rocks.

Sedimentary rocks cover a much larger area of North America and its continental shelves than igneous rocks. Shale covers 52 percent, sandstone 15 percent, limestone and dolomite 7 percent, while granite and basalt only account for 18 percent (Leopold 1994).

The Rock Cycle

The earth's crust is restless. Supported by a highly plastic mantle, plates of the earth's crust, with their raised continents, have drifted and wrinkled. Relative to sea level, their surfaces have risen and fallen. The ocean basins have changed shape, and massive continental glaciers have repeatedly scoured large areas of the Northern Hemisphere.

The continual process of destruction and rebirth of the earth's crust is called the "rock cycle." The concept of the rock cycle was probably first stated in the late eighteenth century by geologist James Hutton: "We are thus led to see a circulation in the matter of this globe, and a system of beautiful economy in the works of nature. This earth, like the body of an animal, is wasted at the same time that it is repaired. It has a state of growth and augmentation; it has another state, which is that of diminution and decay. This world is thus destroyed in one part, but is renewed in another; and the operations by which this world is thus constantly renewed are as evident to the scientific eye, as are those in which it is necessarily destroyed" (Hutton 1795, 562).

Formation and Uplifting of the Mississippi Valley's Ancient Blufflands

There is general consensus among scientists that our planet is about 4.6 billion years old. About 4 billion years ago primitive life developed in the ocean. For the next 3.5 billion years, all life on earth was confined to the ocean waters.

About seven hundred million years ago soft-bodied creatures such as jellyfish and worms appeared in the Precambrian fossil record, setting the stage for the great explosion of life that was to come in the Cambrian period. Fossils found in rocks formed about 544 million years ago document the first explosion of hard-bodied invertebrate animals in the seas. That momentous event defines the beginning of the Cambrian period of the Paleozoic Era.

Era	Period	Significant Developments	Millions of years ago
Cenozoic	Quaternary	Upper Mississippi River drainage system develops. Pleistocene and Holocene Ice Age. Rise of man.	1.6
	Tertiary	Grasses abundant. Rise of mammals. Geological pressures push up entire North American continent. Continued deposition of sediments in southernmost Illinois. Continents assume present shapes and locations.	15 25 66
Mesozoic	Cretaceous	Rocky Mountains begin to form. Disappearance of dinosaurs. Rise of flowering plants. Renewed marine deposition in southernmost Illinois and Missouri. Beginning of Mississippi Embayment.	144
	Jurassic	Heyday of dinosaurs. Earliest birds.	208
	Triassic	Appearance of dinosaurs and mammals. Pangaea begins to break up. Atlantic ocean begins to form.	245
Paleozoic	Permian	Terrain above sea level. Early Appalachians develop. Pangaea in place. Trilobites become extinct.	286
	Pennsylvanian	Great coal-forming tropical swamp forests. Formation of supercontinent Pangaea.	320
	Mississippian	Shifting coastlines and depositional patterns.	360
	Devonian	Earliest amphibians. Fishes abound. First insects.	408
	Silurian	First jawed fishes. First land animals and plants. Midwest is equatorial. Coral reefs.	438
	Ordovician	First marine vertebrates. Diverse marine invertebrates. Oneota dolomite, St. Peter sandstone deposited.	505
	Cambrian	Jordan sandstone deposited. Mt. Simon sandstone deposited. Earliest abundant sea invertebrates. Trilobites predominate. Subsidence of most of Mississippi River Basin. Seas encroach.	570
	Precambrian Time (almost nine-tenths of earth history)	Evidence of organisms with advanced cell structure.	1,400
		Evidence of primitive organisms.	3,600
		Formation of earth's crust	4,600

A geological time scale noting significant events in the development of the Mississippi River Basin (Wiggers 1997; Ojakangas and Matsch 1982; Anderson 1983).

At the beginning of the Cambrian period, the North American continent was smaller than it is now. As new forms of life developed, geologic forces caused the earth's crust to subside (sink) throughout much of the interior of the continent. As the land sank, there was a general worldwide rise in the level of the sea, which flooded the low-lying, bleak, barren, surface of the land now drained by the Mississippi River and its tributaries.

The burst of evolution known as the Cambrian explosion began around 530 million years ago when sea levels rose dramatically, probably due to melting of vast glaciers, creating expansive shallow seas, giving evolution a tremendous boost.

The oceans did not advance at a uniform rate. Forces deep within the earth caused mild subsidence or down warping in some areas and broad, gentle uplifts in adjacent areas, causing shorelines to advance and retreat. The seas continued to rise until the late Ordovician period, about four hundred million years ago.

As the sea advanced, pounding surf attacked the uplands, stripping rock debris from still barren, severely weathered land. As in modern times, most sediment was carried to the sea by rivers—then reworked, sorted, and cleaned by the surf.

Beach zones were high-energy environments where wave action and currents continued the disintegration of the rock debris, winnowing it, and depositing the coarsest particles in the surf areas as clean, well-sorted beds of sand that ultimately formed sandstones. Silt and clay were wafted out into quiet, deeper waters where they settled and were compressed to form shales. Abundant lime-secreting organisms produced deposits that formed limestones and dolomites in warm shallow water, with little input of sand, silt, or clay.

During the ensuing five hundred million years the shallow epicontinental sea served as a collection basin for sediments eroded and washed outward from primordial uplands and mountain ranges. As ancient rivers entered the sea, which was seldom deeper than 150 feet, they lost their velocity and their ability to carry sediments, causing the larger or heavier sediments to be deposited near sea level. Even though the sediments were of relatively low density, their great weight facilitated a commensurate subsidence of the earth's crust. Subsidence and sedimentation were concurrent. The rate of accumulation was slow, perhaps averaging only an inch every two thousand years or so.

This average is misleading, however, because it portrays scenes of tranquility. The early earth was a geologically active and often violent place. Most of the sandstones and shales seen in Mississippi River bluffs were deposited when catastrophic earthquakes and volcanic eruptions sent avalanches and mud slides crashing into swollen rivers.

A large drop in sea level took place at the end of the Ordovician when vast quantities of water were distilled from the world's oceans and bound up as ice in continental glaciers (Mossler 1999). Thus, the shallow sea did not advance at a uniform rate over the land; numerous advances were interrupted by minor withdrawals causing the shoreline to alternately advance inland (transgress) and to retreat (regress). The changes in sediment deposition that coincided with the changing shoreline caused distinctive cyclic patterns in the sedimentary rocks. A sandstone stratum laid down as an ancient beach, for example, may be bounded above and below by shale or limestone formed in deeper calmer waters. These layers of sedimentary rocks are now hundreds of feet thick in southern Minnesota and thousands of feet thick in the Far West and Deep South.

It is generally accepted that during this interval of inundation the North American Plate straddled the equator. Nearly all of the marine fossils found in central North America are of animals that flourished in warm, tropical seas. Throughout most of the Paleozoic era, which ended about 245 million years ago, North America drifted northward, to become part of the supercontinent Pangaea *(all earth)*.

Pangaea included all of the earth's present continents. The eastern edge of the North American Plate once pushed against the African and Eurasian Plates, causing the rise of the Appalachian Mountains, while the eastern edge of the South American Plate abutted the African Plate. During the first hundred million years of this interval, land surfaces remained lifeless, barren, and easily eroded, but the seas teemed with an abundance and diversity of plants and animals (Ojakangas and Matsch 1982). Animals, plants, and fungi first came ashore about 450 million years ago, very recently in geologic time.

About two hundred million years ago, Pangaea began to break up into its constituent plates. The North American Plate, whose exposed land surface was the North American protocontinent, began to drift away from Eurasian and African Plates at about the speed that fingernails grow. This initiated the opening of the Atlantic Ocean.

As recently as 95 million years B.P., our planet was still a world of shallow seas. The landmasses no longer formed a supercontinent, but they remained close together. Sea levels were higher than now, primarily because there were no large glaciers to store water, and

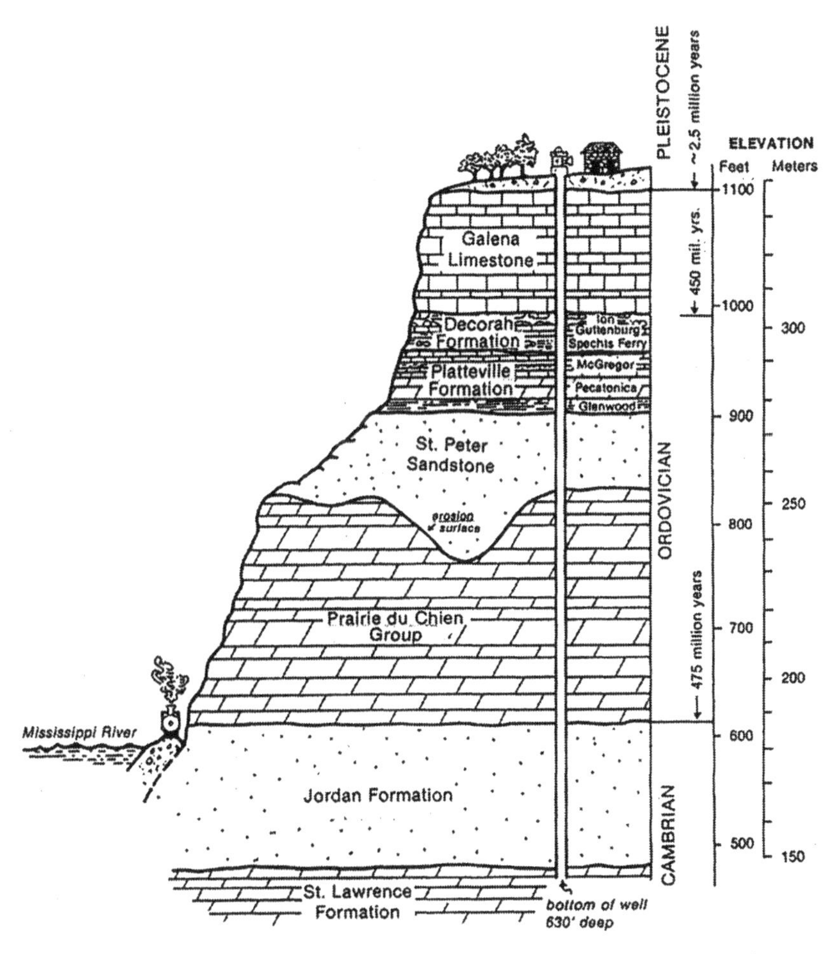

Generalized cross section of sedimentary rock strata at Pikes Peak State Park near McGregor, Iowa. Note that the upper surface of the Jordan Sandstone that occurs about six hundred feet above river level at Dresbach, Minnesota, is just above river level at McGregor (Koch, Prior, and Tuthill 1973).

oceans flooded the interiors of most continents during the Cretaceous period.

Geologic forces caused the general uplifting of the North American continent from the Mississippi River to the Pacific Ocean during the westward drift of the North American plate. The Rockies, some of the continent's tallest and youngest mountains, were thrust upward as the Earth's crust, near the plate's western margin, uplifted and folded.

Subsequently, to the east of the Rockies, a great sedimentary rock plateau rose from the sea, constructing a stable platform of relatively soft sedimentary rocks, bounded on the west by the youthful Rocky Mountains and on the east by the much older southern Appalachian Mountains.

Stripped of its sedimentary rock covering by repeated glacial advances, the remaining stable nucleus of the primordial North American continent is exposed in the northern United States and Canada as the Canadian Shield, an expanse of igneous and metamorphic rocks that include some of Earth's oldest rocks.

Jordan Sandstone

Jordan Sandstone, one of the most easily recognized rock strata along the Upper Mississippi, can be seen as a series of layers 75 to almost 175 feet thick extending from Hastings, Minnesota, southward to the Iowa state line. It is medium- to coarse-grained and white, turning yellow, brown, or orange by oxidation where it is exposed to the air. Spectacular outcrops of Jordan Sandstone can be seen, illuminated by the afternoon sun, in Granddad Bluff at La Crosse and in the bluffs at Trempealeau, Wisconsin.

Like other Cambrian sandstones, it contains 90 to 99 percent quartz grains whose ultimate origin was the igneous and metamorphic rocks of the Canadian Shield. The grains are uniform in size, rounded, and frosted—indicating a long complex history of being tumbled by river currents, ocean surf, and wind before they came to rest on the floor of the shallow Cambrian sea. As is typical of all sedimentary rocks, sandstone is layered. Wind, water, and humans dislodge sand grains from the outcrops. Thus freed, the grains are on their way to the sea again—perhaps to make more sandstone.

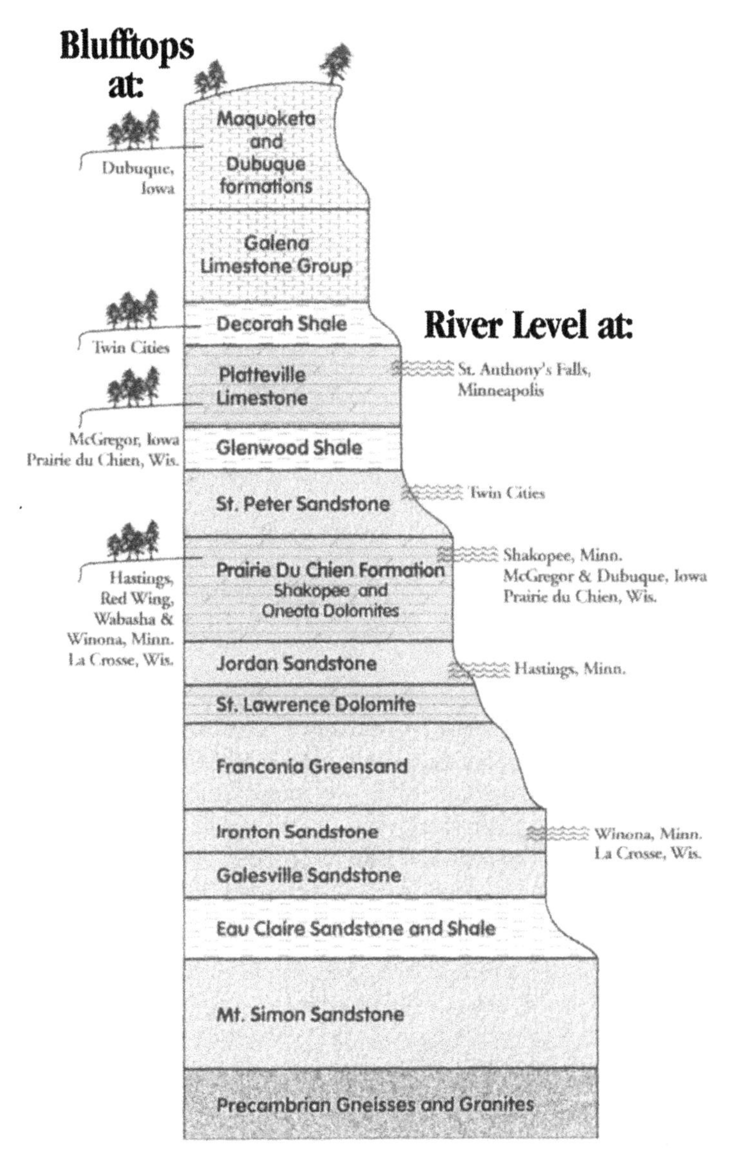

Sedimentary rock formations that can be seen along the Mississippi in northeastern Iowa, southeastern Minnesota, and southwestern Wisconsin. Where the right-side lines are vertical, cliffs are likely. Slopes are most probable where there are sloping lines. Town names along the right edge show where a particular formation lies at river level; town names along the left edge indicate which types form the crest of the bluffs nearby. The St. Lawrence Dolomite and Franconia Greensand layers in highway cuts can often be identified because trees usually take root in these soft stones (Sloan 1998b).

The sedimentary rock strata of the Upper Mississippi River were originally laid flat and for the most part remain that way. Only occasionally have they been significantly displaced by shifts in the earth's underlying crust. They have never been twisted, turned, folded, or melted, but they do bulge upward.

In the Mississippi Valley they reach their highest elevations near La Crosse, Wisconsin, then tilt downward to the north, west, and south. The distinctive, orange Jordan Sandstone is an excellent marker. At Brady's Bluff near Trempealeau, Wisconsin, it rises about 1,160 feet above sea level and is covered by a layer of Oneota Dolomite. In the Twin Cities its upper surface lies only about six hundred feet above sea level. At Pike's Peak, near Clayton, Iowa, it occurs at river level (about six hundred feet above sea level). Upstream and downstream from their highest levels in the La Crosse area, the Jordan Sandstone and Oneota Dolomite are covered by younger strata. Near the river, some of the youngest strata have eroded away.

The Twin Cities Basin

Rock strata that loom six hundred feet above the Mississippi River at La Crosse tip downward to the north and lie far beneath the ground surface in the Twin Cities. The Jordan Sandstone, for example, lies from 350 to 450 feet below ground. It averages ninety feet in thickness and forms an artesian basin that supplies most suburban wells. Atop the Jordan Sandstone lies the Prairie du Chien Formation that includes a stratum of Oneota Dolomite about eighty feet thick, followed by a 155-foot-thick stratum of white, sugarlike St. Peter Sandstone, overlain by about thirty feet of Platteville Limestone. Both the St. Peter Sandstone and Platteville Limestone are easily seen from river level in the walls of the river gorge in the St. Paul area. The latter two strata were crucial in the development of the area's waterfalls (Staff of the Department of Geology, University of Minnesota 1956).

The basin was the reason for the confluence of the Mississippi, St. Croix, and Minnesota Rivers—and the consequent birth and growth of the Twin Cities.

During the Tertiary period of geological time, which extended from about 70 million years B.P. to the time of Pleistocene glaciation that began about 2 million years B.P., the central portion of the United States was continually free from inland seas. Therefore, in a land that was generally flat, complex drainage systems had time to become well developed with wide, deep river valleys (Bray 1985).

Winona Travertine

Known to geologists as Oneota Dolostone, "Winona Travertine" is a building stone that, when sawed, shows an attractive, mottled, buff appearance, resembling long-famous Italian Travertine. Over one hundred tons of it are carved from a bluff near Winona every week, and precut to exacting specifications to be used for interior and exterior construction nationwide. The principal difference between Winona Travertine and Italian Travertine is that the imported stone is quite soft and easily worked when it is first quarried, but it hardens on exposure. Winona Travertine, on the other hand, is an extremely hard rock when it is quarried and remains hard, practically unaffected by climate changes, and is impervious to stain.

4

Glaciers

Sculptors of the Upper Mississippi River Basin

Glaciers have sculpted the landscape of the entire Upper Mississippi Basin, directly or indirectly. Those areas not subjected to the earth-moving activities of glaciers have been shaped by their meltwaters and by violent glacial winds.

The Blue Planet

Since the planet's surface cooled, water has been Earth's principal agent of erosion. This vital liquid is abundant on Earth, making our blue planet unique in our solar system. The total supply of planetary water remains virtually the same from year to year, as it has for the last five billion years or so, but the relative amounts occurring as seawater and ice have varied greatly through the ages. At present, the oceans contain roughly 97 percent of Earth's water. Ice and snow account for almost 2 percent, more than is contained in all lakes, rivers, streams, and groundwater combined.

Over geologic time, water has been periodically distilled from the seas to form continental glaciers (derived from the Latin *glacialis*—"frozen") that covered much of the land. As the ice masses increased in volume, reaching thicknesses of more than two miles, sea levels dropped as much as five hundred feet, and the incredible weight of the ice caused the Earth's crust to subside beneath them.

There is no modern analog of continental glaciers at forty-three degrees north latitude (the southern limit of Pleistocene glaciers), but continental glaciers may be studied today in Antarctica and Greenland. Large glaciers still exist as ice caps in Iceland and as rivers of ice called *valley glaciers* in mountain ranges worldwide. Only Australia lacks glacial ice presently.

Where Did the Boulders Come From?

Mysterious boulders, some of great size, and often scratched and worn, litter much of the landscape of the Northern Hemisphere.

These granite boulders in a pasture near the Mississippi River at St. Cloud, Minnesota, have not traveled very far from their source in the granite bedrock that outcrops in the area. However, boulders of the same composition are common in Iowa where there are no exposed granite formations. Because such boulders are foreign to the Iowa landscape, they are called *erratics* and are evidence that the area has been glaciated. The boulders in this photo were transported by the Wisconsin glacier, but they were also rounded and polished by erosion in a glacial river, probably one that roared through caves within the glacier.

They are often composed of rock types different from the local bedrock and to those who stopped to think about it, it was obvious they had traveled from faraway places. People speculated on their origin for many years, but what had carried them from so far away was, for centuries, a mystery. Equally mysterious were the forces that shaped the wide, U-shaped river valleys, valleys far too wide and deep to have been carved by the rivers flowing through them.

The biblical account of Noah's flood was the basis for the Diluvial Theory, widely held in Europe and North America, to explain these out-of-place boulders, called *erratics* by geologists, and various flood features they observed in the landscape. A competing view held by catastrophists maintained that the earth had been subjected to a series of destructive cataclysms, each followed by a new creation. The reality would prove to be even more challenging to the imagination.

Mountain people have long known that glaciers can carve valleys and carry rocks. As early as 1795, geologists realized that features known to have been caused by movements of glaciers in the Swiss Alps were distributed widely over the surface of northern Europe. Soon, similar features observed throughout Canada and the northern United States were also attributed to the work of glaciers, but it was not until the early 1800s in Europe that the process was recognized as occurring on a continental scale.

In 1837, a young Swiss naturalist, Louis Agassiz, first proposed the Glacial Theory to explain the presence of erratics and scratched bedrock in Europe. He argued that in the recent past the earth had been gripped by a colder climate, resulting in great sheets of glacial ice covering much of the northern hemisphere. Later, this period of continental glaciation was named the Pleistocene epoch.

During the period from 1837 to 1840, Agassiz, supported by geologists William Buckland and Charles Lyell, firmly established the Glacial Theory in Europe. By 1883 American geologists N. H. Winchell and Warren Upham had mapped North America's major glacial features and plotted its ice borders. Upham's major work was his study of Glacial Lake Agassiz, published in 1896 by the U.S. Geological Survey.

Nature's Earth Movers

Glaciers are awesome earth-moving machines, fueled by snowflakes and frost. Their basic mechanics are straightforward. When the climate is cold enough, more snow falls in winter than melts the subsequent summer. Incredibly, much of the "snowfall" may consist of "diamond dust," minute frost crystals formed from atmospheric moisture on frigid, clear days. As the snowfalls of successive years accumulate, the compressed snow at the bottom melts, recrystallizes, and turns to ice (much like a child packing an ice ball on a school playground).

The ice at the bottom of a continental glacier may be as hard as steel due to the immense pressure exerted by the weight of the ice above, but it behaves like a plastic. Valley glaciers, pulled by gravity, flow downhill like rivers of ice. The movement of continental glaciers results from a different set of forces. As snow and ice accumulates, the center of the ice sheet builds up to form a dome that may be a mile or more high. The unimaginable weight squeezes the underlying ice, giving it the characteristics of a viscous fluid, causing it to spread outward like a lump of clay pressed under the potter's hand.

The edge of the ice sheet represents a line of dynamic equilibrium between the rate of melting and the rate of advance. If the glacier moves forward faster than the leading edge melts away, the glacier advances. If forward movement is balanced by melting, the leading edge remains stationary. The ice retreats if climate change reduces snowfall in the area of accumulation or if the rate of melting exceeds forward motion, or both. But even when it is in apparent retreat, the glacial ice is still moving forward, carrying with it the rock and debris it has picked up during its long, slow advance.

As it flows across the surface of the land, the ice sheet plucks up pieces of bedrock, grinds them up, uses them as abrasives to scour Earth's crust, and may transport them for hundreds of miles. In summer, torrents of melt water roar through the bowels of the glacier, carving icy caves, transporting rocks and sediment, and often causing sinkholes to form in the glacier's surface.

The Mississippi's most spectacular loess formations are seen along the east side of the Lower Mississippi River at Vicksburg and Natchez, Mississippi, where the loess forms high, erosion-resistant bluffs with nearly vertical cliff faces. For scale, note the person standing at the foot of the bluff. During the Civil War, elaborate redoubts and caves were dug into the loess during the siege of Vicksburg. The famous "Natchez Under the Hill" was carved into a loess cliff along the Natchez waterfront.

As a glacier retreats, a sheet of rock debris called *drift* is left behind. The unsorted, unstratified drift deposited directly from the ice is called *till,* a mixture of boulders, cobbles, pebbles, granules, sand, silt, and clay. Till, carried by the ever moving ice, is deposited in ridges, called *moraines,* when the glacier is stationary for long periods of time. *Terminal moraines* mark the position of greatest glacial advance. *Recessional moraines* mark the glacier's position during periodic standstills along the path of retreat. Plotting the locations of moraines provides graphic evidence, almost like footprints, of where the glacier has (or has not) been.

At the edge of the ice sheet, beyond the moraines, streams of meltwater carry the finer portions of the till outward, sorting and distributing them as layers of gravel, sand, silt, and clay. If the meltwater is confined to a valley, the deposits form a *valley train,* but where the glacier retreats on a continental scale on relatively level terrain, the deposits may stretch out laterally for miles to form an *outwash plain.*

In an outwash plain, the largest, heaviest particles settle out first from glacial streams and rivers. Clay, composed of microscopic particles, travels farthest. Called *glacial flour,* it makes glacial streams look like milk and finally settles out in a quiet lake, estuary, or in the depths of an ocean. In Switzerland, glacier water is called *gletscher milch* (glacier milk).

Occasionally, as the great ice sheets retreated, immense chunks of the glacier would break off to be left behind. When they were covered by sediment and buried in a moraine or beneath the outwash plain, the blocks would melt slowly, insulated by the overlying sediment. When the ice blocks finally melted, the insulating sediments collapsed, leaving pits, which then filled with water to become kettle lakes and potholes.

Like most lakes, ice-block lakes are temporary features; their geologic life is very short. They usually fill with sediments, and they may disappear if the water table is lowered. If their outlet is eroded, they may drain away.

A classic example of a kettle moraine is the St. Croix terminal moraine just south and east of the confluence of the Mississippi and Minnesota Rivers. It encompasses the Minnesota cities of Burnsville, Mendota Heights, and Inver Grove. The moraine, which includes more than one hundred large potholes and small lakes, was created when isolated ice blocks melted. To the south of the St. Croix terminal moraine lays the Rosemount Outwash plain, transected by State Highways 52 and 56. The Pine Bend Refinery on Highway 52 lies on the north edge of the outwash plain.

Pleistocene ice sheets produced brutal *katabatic* (downward climbing) winds that were caused by the flow of cold, dense air down the long sloping surface of glaciers that were a mile or so high over northern Minnesota (Pielou 1991). The ferocity of the winds can only be imagined, but such winds in Antarctica have been clocked at two hundred miles per hour. The stinging assault of the katabatic winds matched those of tornados and hurricanes but lasted much longer, shrieking for days on end. Yet, the wind depth was just a hundred or so feet.

As the cold air plummeted downward it was compressed, making it warmer and drier. Especially in summer, gales roared over the barren, recently deglaciated land, picking up dried glacial drift to create windswept deserts where great sand and dust storms darkened the sky for weeks at a time. The winds further sorted and redeposited the sediments, with sand dunes forming close to the till, and thick beds of wind-blown dust called *loess* (pronounced *luss*) being deposited over wide areas downwind.

Loess is a medium- to coarse-grained silt. It contains some clay and may be calcareous. It was probably formed along the Mississippi River when silt from the river floodplains was blown to higher levels by strong winds during periods when the plains were exposed and dried. The soils of North America's Great Plains are composed mainly of loess. These soils, some of the most fertile in the world, comprise "the nation's breadbasket." They support the production of corn and soybeans that are barged downstream on the Mississippi River and its tributaries.

Pleistocene Glaciation in North America

Soon after discovery of Pleistocene glaciation in the last century, geologists learned that it included at least four periods of glaciation. Initial evidence for continental glaciation came from studying features of the continents themselves, but in recent years oceanographers have amassed a detailed glacial chronology by analyzing cores of deep-sea sediments.

The glacial record preserved in ocean basins is very complete because it is not subject to erosion by glaciers or by weathering. The record places the beginning of the Pleistocene epoch at 1.6 million years B.P., and shows that there have been at least eighteen glacials during the Pleistocene—one every hundred thousand years or so—with the intensity of the glaciations increasing toward the latter part of the epoch.

Our discussion will be limited to the four major glaciations that shaped the Mississippi River. These are the "classic" glaciations referred to in older geologic literature, and they occurred during the Pleistocene epoch in North America. The Nebraskan, Kansan,

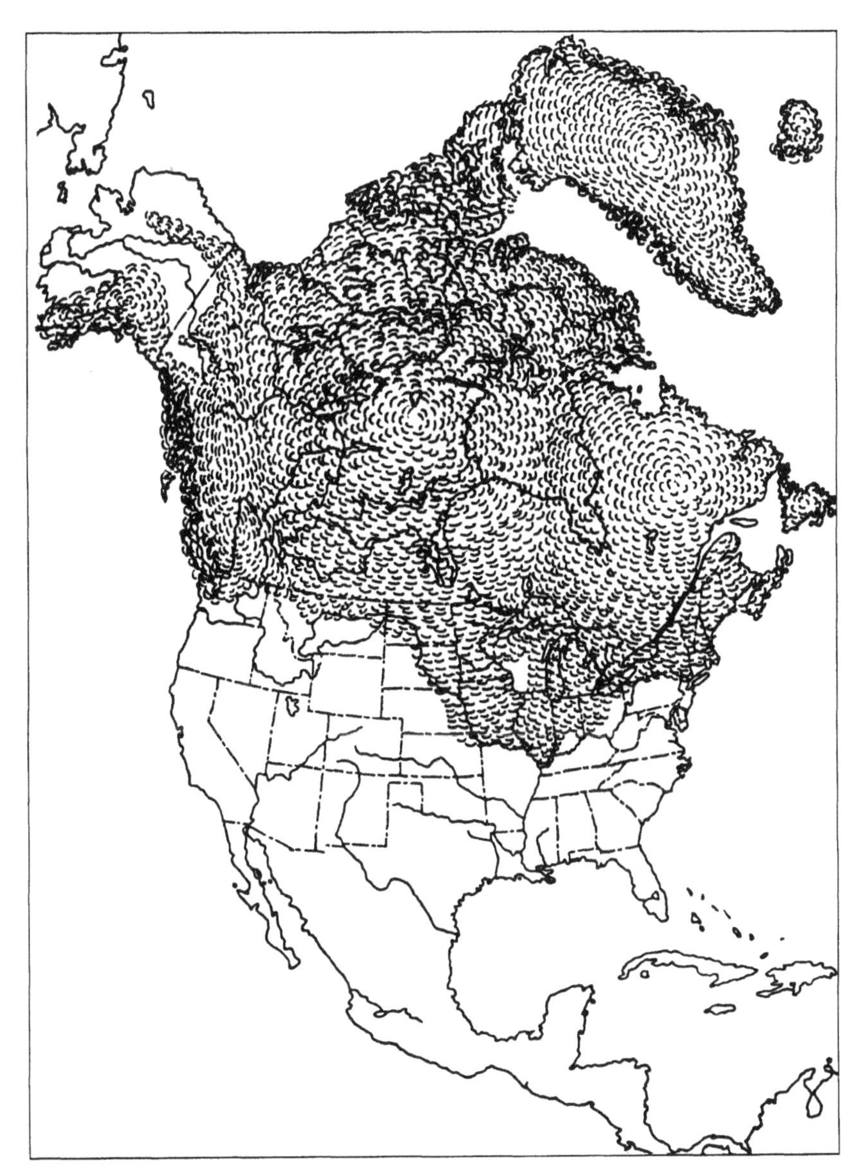

During the Great Ice Age, major ice sheets extended as far south as the present positions of the Missouri and Ohio Rivers whose valleys were carved by torrents of melt water flowing along the ice margins. Note the unique unglaciated area, mainly in southwestern Wisconsin (courtesy of the Minnesota Geological Survey).

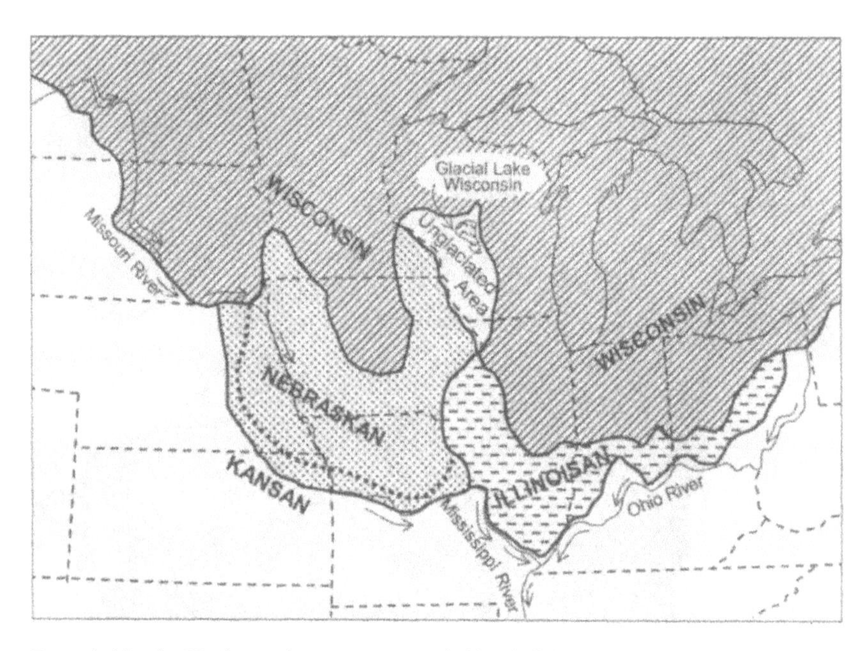

Remarkably, the Unglaciated Area is surrounded by drift left by four major ice sheets. Although glaciers encroached from many directions at various times, the area was never completely surrounded by ice at any one time. The area repeatedly served as a refugium for many species of plants and animals during glacial onslaughts (from "Glacial Map of the United States East of the Rocky Mountains," National Research Council, Division of Earth Sciences 1959).

Illinoisan, and Wisconsin glaciations are each named for the state where evidence of their maximum development is found. Each glaciation was followed by an interglacial interval in which the climate was similar to today's. It should be noted that most geologists today use the term *pre-Illinoian* (or *pre-Illinoisan*) for all glacial events that took place between two million and five hundred thousand years ago, prior to the Illinoisan (Mossler 1999).

The Nebraskan glaciation is the oldest named glacial. Its drift is entirely buried by younger drift and is known from wells. Its maximum advance, beginning about 1.6 million years B.P., was west of the Mississippi River. It extended as far south as Nebraska. It is generally believed that the present general course of the Upper Mississippi River south of the Twin Cities was determined as the river flowed as an ice-border stream along the east edge of the Nebraskan glacier. The *Aftonian interglacial* was a major warm period that resulted in almost complete melting of the Nebraskan ice sheet. It resulted in extensive loess deposits that formed the backbone of the fertile soils that developed atop the Nebraskan drift.

The Kansan glacial occurred between 900,000 and 600,000 years B.P. Its major advance was also west of the Mississippi River, and was the maximum advance of ice down the Great Plains. Its deposits overlie Nebraskan drift and the Aftonian soils of Kansas and Nebraska. The warm *Yarmouthian interglacial* followed the Kansan glacial. It lasted for thousands of years and allowed the development of soils on the Kansan drift.

The Illinoisan glacial represents the most southerly advance of ice east of the Mississippi River. It lasted from about 400,000 to 300,000 years B.P. Its drift is best preserved in Illinois. Warming of the climate during the *Sangamon interglacial* melted the Illinoisan glaciers and allowed extensive soil development on Illinoisan drift in Illinois and Iowa.

The Wisconsin glacial began about 100,000 years B.P. and ended about 10,000 years B.P. It was the last major glaciation in North America, and is the best understood because its deposits are widely exposed and have not been disturbed by subsequent glacial erosion and deposition. The Wisconsin glacial consisted of several advances and retreats of the ice, each with smaller advances. During at least one of these minor advances the ice overrode forests that had become established in front of the retreating ice sheet. The huge Des Moines Lobe moved southeast through western Minnesota into central and eastern Iowa, progressing as far south as Des Moines, Iowa. The Iowa State Capitol Building stands on the tip of this lobe.

The Wisconsin glacial was dominated by the periodic growth and decay of two great ice sheets. The Laurentide ice sheet, the larger of the two, was centered over Hudson Bay. The smaller Cordilleran complex of glaciers covered the mountain ranges of the northwest coast from Montana northward into the Aleutian Islands.

The formation of the Laurentide ice sheet was the most spectacular glacial event in the sixty-five million years of the Cenozoic era. It alone covered about four million square miles of North America and contained about 6.4 million cubic miles of ice, representing enough water to lower the level of the sea almost 250 feet (Pielou 1991). Concurrently, northern Europe and Asia were also glaciated, but the ice sheets of the Wisconsin glacial were at least 50 percent larger than those in Europe and Asia combined. At their maximum they formed an unbroken expanse about the size of Antarctica.

The combined effect of all of the earth's glaciers caused sea levels to drop as much as 426 feet during the Late Wisconsin stage low stand from 22,000 to 18,000 years B.P. (Saucier 1994).

The center of the Laurentide ice sheet was over Hudson Bay, and at its maximum about 20,000 to 18,000 years B.P., it may have been as thick as three miles (Pielou 1991). Amoeba-like, it advanced outward from its centers in Canada, covering over one third of North America. Its weight was incredible. For each three hundred feet of ice thickness the continental crust was depressed about one hundred feet. Areas depressed during the Wisconsin glacial are still slowly rebounding because they have been relieved of their incredible burden.

A Glacier in Summer

Before I studied glaciers first hand, I always imagined them in winter, snow-covered, pristine, and serene. Exploring Iceland's vast Vatnajökll Glacier in late summer dispelled those notions. The glacial surface had been melting all summer, exposing and concentrating the contained volcanic ash and sand. Walking was easy and quite safe in the fog of early morning because the frozen surface was like sandpaper, but as the fog lifted and the sunny day wore on, melting made it slippery and treacherous.

Streams of meltwater ran across the glacial surface and plunged into deep crevasses that crisscrossed the glacier. Far below, in the depths of the glacier, we could hear streams roaring through ice caves that ran for miles. The streams finally disgorged their water and sediment onto the outwash plain at the foot of the glacier. Uppermost in our minds was being sure that we didn't slip into a crevasse. Unforgettably, my companions and I flew kites in the stiff wind at the foot of Iceland's Vatnajökll Glacier.

North America's Pleistocene glaciers would have dwarfed Icelandic glaciers.

5

The Upper Mississippi River

Child of the Pleistocene

Worldwide Episodes of River Cutting and Filling

Worldwide, about twenty million square miles of the earth's surface were covered during Pleistocene glacial maximums. During the Pleistocene as much as 30 percent of Earth's land surface was ice-covered, compared with about 10 percent today. The average thickness of the ice sheets was about one mile. Combined, they stored so much planetary water that expansive tracts of the continental shelves of North America were then dry land. The Atlantic Coast of the United States, for example, was located more than sixty miles to the east of New York City (Lutgens and Tarbuck 2000).

Continental glaciation and commensurate changes in ocean levels affected erosional processes on the entire planet. Falling ocean levels caused river gradients to become steeper. Consequently, the rivers ran faster and were able to *degrade* or down cut through previously deposited sediments, carving the submarine canyons that exist today on continental shelves. Rising ocean levels, on the other hand, reduced the gradient of rivers, decreased their sediment carrying capacity, and caused valley floors to rise or *aggrade* as they became choked with sediment. This complex interplay of glaciation and fluctuating ocean levels alternately caused master valleys and tributary valleys to flush and to fill.

When glaciers were at their maximums and sea levels were lowest, all of the world's rivers cut their channels deeper, beginning at their mouths and working all the way back to their sources. Retreating waterfalls cut the Mississippi Valley northward, and they did the same thing in the Mississippi's tributaries (Sloan 1999).

In the case of the Mississippi River, the story is more complex because the rapid draining of glacial lakes, impounded by retreating glaciers late in the Wisconsin glacial, caused torrents of sediment-free water to entrench the Upper Mississippi Valley while the Lower Mississippi Valley was aggrading. As glaciers wasted away, sea level rose, and rivers filled in their valleys. For example, at Winona, Minnesota, the river fill from 300 feet above sea level to city level (540 feet above sea level) was all deposited in the last ten thousand years (since the end of Wisconsin glaciation).

Water: Master Carver of River Valleys

As the main-stem valley deepened, the gradients of tributary streams steepened accordingly. Because they frequently contained large volumes of glacial meltwater, their headwaters advanced rapidly into the relatively soft sedimentary rock strata of the plateau, creating the rugged topography and complex drainage patterns of the watershed. Such drainage systems are described as *dendritic* because they are branched like a tree. The master stream made the initial incision of the main-stem valley, but the steep-sloped tributaries with their great erosive capacity were mainly responsible for broadening the gorge.

As the channels of the glacial Mississippi changed their courses over time, their flow, especially on the outsides of bends, was directed along the bases of the bluffs that flank the river. This eroded the soft sandstone and undercut the harder, overlying limestones and dolomites. This natural quarrying action caused blocks of rock to break off, thus widening the river valley and forming precipitous cliffs and the characteristic steep *talus* (rock rubble) slopes that lie at their feet. Tributaries in these areas usually are significant contributors of sand eroded from the sandstones.

The variation in the width of the Mississippi Valley is due mainly to differences in hardness of the rocks intersected by the river and its tributaries. The valley is broadest where sandstone strata occur at river level. Conversely, the valley is narrowest where erosion-resistant limestones and dolomites occur at river level. As the river flowed over resistant rock strata and plunged into softer, underlying sandstones it created the St. Anthony Falls, the Rock Island Rapids, the Des Moines Rapids at Keokuk, and the Chain of Rocks Rapids at St. Louis.

Episodes of Cutting and Filling of the Upper Mississippi River during Wisconsin Glaciation

From about 14,000 to 9,500 years ago, the terminus of the Wisconsin ice sheet fluctuated back and forth across the continental divide, resulting in multiple cut-and-fill episodes in the Mississippi main-stem and tributary river valleys. Large main-stem floods interrupted periods of downcutting in tributaries within the Mississippi system, usually the result of rapid draining of glacial lakes. During these episodes the Mississippi back-flooded into the tributary valleys, forming lakelike environments where silt and clay were deposited, thus elevating valley floors.

As the Wisconsin ice sheet retreated northward, it stood across the valley of the Mississippi at St. Paul and discharged great quantities of water, gravel, sand, silt, and clay down the valley. This produced a massive valley train in the main stem, and as the floor of the main stem rose, the gradients of tributaries decreased commensurately, causing them to drop their sediment loads. This, in turn, additionally elevated the floors of tributary valleys, causing them to be flat and continuous with main-stem terraces, with eroded bedrock lying deep beneath the heavy alluvium that partially fills the valleys.

Mighty Glacial Lake Agassiz and the Awesome Glacial River Warren

Today, the Minnesota and St. Croix Rivers are important tributaries of the Mississippi, but their flows were much greater in the past.

Both are termed *underfit streams* because they now flow through valleys that are much too large to have been carved by the stream now flowing through them. Their huge valleys were carved by ancestral rivers that were so immense that they can scarcely be imagined. The rivers, in turn, were fed by overflow from huge glacial lakes.

The Minnesota River Valley is evident on most colored relief maps of the United States as a huge, green, V-shaped scar running across the lower third of Minnesota, from the west border to the east border. It is much more conspicuous, and obviously much larger, than the valley of the Mississippi Headwaters. The Minnesota River Valley includes Lake Traverse and Big Stone Lake on the Minnesota–North Dakota border and joins the Mississippi south of St. Paul at Fort Snelling. Thousands of motorists cross the Minnesota River Valley daily on Highways 35W, 494, and 55 just south of the Twin Cities, but most are probably too busy fighting traffic to wonder how the sluggish little Minnesota River could have carved such a magnificent valley.

In 1868, just sixty-three years after Zebulon Pike began his exploration of the Upper Mississippi River, General G. K. Warren published his explanation for the immensity of the Minnesota River Valley. As a result of remarkable geological sleuthing and insight, he was able to attribute it to the torrential draining of a vast proglacial lake, which he named "Lake Agassiz" in honor of the great Swiss proponent of the glacial theory, Louis Agassiz. The glacial River Warren, which drained the lake and carved the valley, was named in honor and memoriam of General Warren by another prominent geologist, Warren Upham, in 1882.

Glacial Lake Agassiz served as the source of the Mississippi River for almost three thousand years. At various times, this massive lake covered parts of Ontario, Manitoba, Saskatchewan, Minnesota, North Dakota, and South Dakota (Pielou 1991). Lake Agassiz was born as glacial ice retreated northward into the Red River Basin, blocking its normal drainage northward into Hudson Bay. Meltwater from the retreating glacier was ponded between the southern edge of the ice and a recessional moraine that created the lake's southern shore.

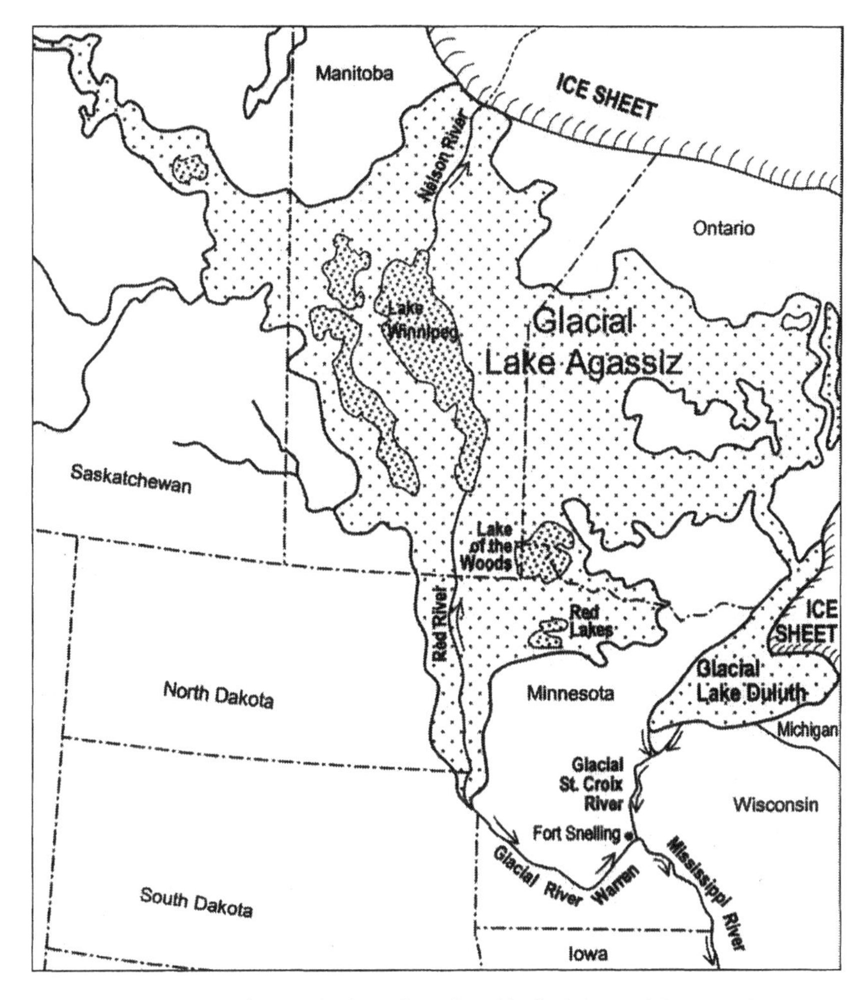

As the Wisconsin ice sheet melted northward, it blocked the Red River–Nelson River system that normally drains northward into Hudson Bay. Impounded meltwater formed Glacial Lake Agassiz, the largest lake to ever have existed in North America. It served as the source of the Mississippi River for almost three thousand years. Draining Lake Agassiz southward, the Glacial River Warren carved the great valley now occupied by the Minnesota River. It also scoured the Mississippi River Valley downstream from the Minnesota River's mouth at Fort Snelling. Although the lake's entire basin is shown, it was never completely inundated at any one time. Depending on lake level, there were other temporary outlets as well—like the one into Glacial Lake Duluth, ancestor of modern Lake Superior. Lake Duluth formed because ice prevented eastward flow of meltwater into the Atlantic Ocean via the St. Lawrence River drainage system. Lake Duluth overflowed via the Glacial St. Croix River into the Mississippi until eastern outlets were free of ice. With the disappearance of the ice sheet, Lake Agassiz drained to the north but left the scattered remnant lakes shown on the map (Elson 1967; Ojakangas and Matsch 1982).

Of the extinct glacial lakes of North America, Lake Agassiz was by far the largest. It fluctuated in size and depth, covering a total area of about 123,500 square miles and had a maximum depth of almost four hundred feet (Elson 1983). It was more than four times larger but much shallower than modern Lake Superior. Tall, precipitous cliffs of ice loomed upward on its windswept northern shore. The pounding surf undercut the ice cliffs, causing the calving of icebergs that drifted on the lake's surface, dropping their loads of sand, gravel, and boulders as they melted.

Depending on the position of the melting ice sheet, Lake Agassiz shifted its drainage four times; first into the Mississippi River drainage, then eastward via Lake Superior, and then into the Mississippi again. At one time its waters were even continuous with those of Alaska via the Mackenzie River. Consequently, Lake Agassiz was the hub of migration for cold-water fishes and many other species of aquatic life that now live in the interior of Canada, the northern United States, and much of Alaska.

The greatest of all Upper Mississippi floods began about 12,700 years ago when Glacial Lake Agassiz spilled over its southern rim, forming the Glacial River Warren. Thus, Lake Agassiz served as the source of the Mississippi River for about the next 2,700 years. Although the River Warren had a great volume of flow, it did not contribute much sediment to the Mississippi because its gradient was low, and also because most glacial sediment had settled out in Lake Agassiz.

The Falls of the River Warren: Ancestor of St. Anthony Falls

For eight miles downstream from St. Anthony Falls in Minneapolis, the valley of the Mississippi is not much larger than it is above the falls, but at Fort Snelling, where it is joined by the huge Minnesota River Valley, the Mississippi Valley suddenly becomes immense. At this location, it is obvious that the Minnesota River was once the master stream, and that the Mississippi River was its tributary.

Upstream from its junction with the River Warren at Fort Snelling, the Mississippi Headwaters was a modest river, a mere tributary of the massive River Warren, running through raw landscapes recently sculpted by retreating glaciers and their meltwaters—but downstream from its junction with the River Warren the Mississippi was an awesome torrent that thundered over an armored streambed of hard Platteville Limestone. About eight miles below Fort Snelling it plunged almost two hundred feet into a preglacial gorge cut into the underlying sandstone, forming the falls of the River Warren, then much larger than today's Niagara Falls. Below the Falls, the combined flows of the Glacial River Warren and

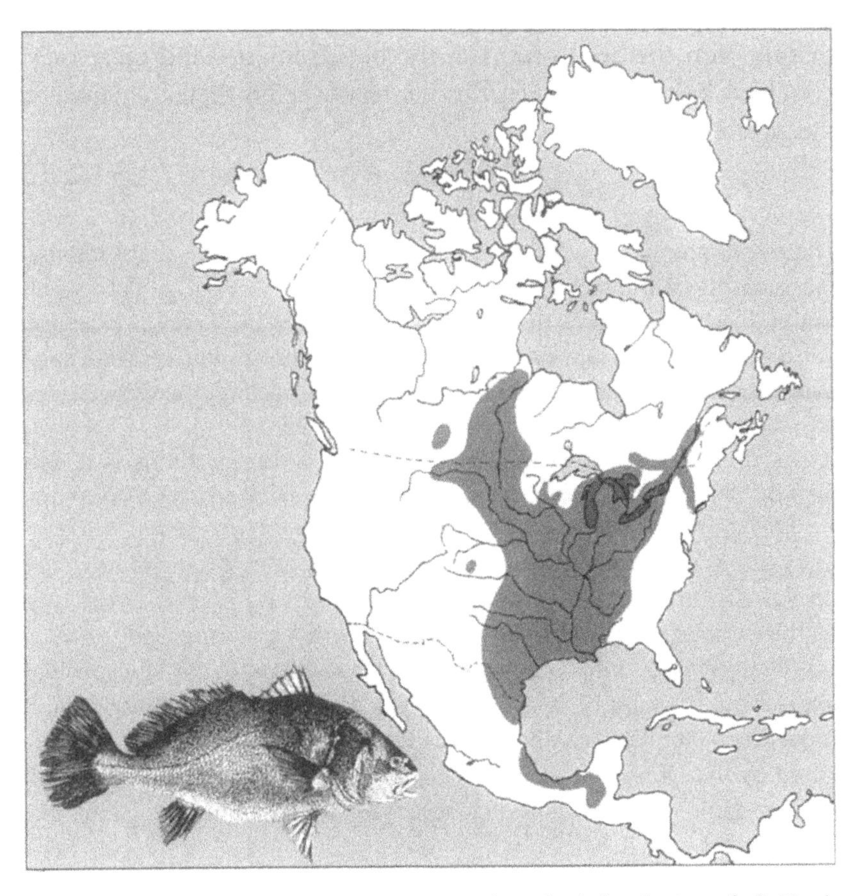

The versatile freshwater drum has the greatest latitudinal distribution of all North American freshwater fishes. It inhabits three continental drainage systems due to its migrations via interconnected glacial lakes and rivers during Wisconsin glaciation. The species is generally found in large rivers and lakes, but commonly strays into the Gulf of Mexico (Fremling 1980; drawing by Kent Parks, courtesy of Nasco International).

the Glacial St. Croix River rapidly scoured out much of the previously deposited drift and terrace sediments, and sliced downward into the underlying sandstones to a depth 175 feet below the present floodplain.

The Mississippi must have been spectacular at that time—a turbulent river rampaging through a gorge scoured over eight hundred feet deeper than the uplands. Its many tributary rivers and streams, swollen with spring meltwaters, also formed cataracts as they thundered into the gorge, enshrouding the balsam firs, spruces, and white cedars with spray. Rainbows of the mist must have contrasted sharply with the deep greens of the boreal forest—and there may even have been Native American hunter-gatherers there to marvel at the awesome scene.

The Freshwater Drum: Ice Age Migrant

The freshwater drum, commonly known as sheepshead, apparently has the greatest latitudinal distribution of any North American freshwater fish. It owes that broad distribution to changing continental drainage patterns during waning Wisconsin glaciation. Although this versatile species does best in large, silty lakes and rivers, it occurs in a wide variety of habitats—even straying into the Atlantic Ocean.

Its distribution indicates that the drum's ancestral home was the Gulf of Mexico. Freshwater drum apparently spent the years of Wisconsin glaciation in the Mississippi River and its tributaries south of the glaciers, but as Wisconsin glaciers retreated, the species extended its range up the Glacial Illinois River into the Great Lakes watershed all the way to the Atlantic Ocean via the Gulf of St. Lawrence. It ascended the Glacial River Warren into Glacial Lake Agassiz, and into the Hudson Bay watershed. The Falls of St. Anthony were an insurmountable barrier for movement into the Mississippi Headwaters until the falls were abrogated by the Upper St. Anthony Lock completed in 1963. Other Headwaters dams have prevented widespread distribution throughout the Headwaters.

In my opinion, the freshwater drum is the most underrated sport fish in the Mississippi River. The loins of these tenacious, easily caught fighters produce tasty fillets. Drum will take a variety of baits and artificial lures, but are most readily caught on the bottom with night crawlers and a slip sinker (Fremling 1978, 1980).

Characteristically, a waterfall retreats upstream as the turbulent waters of its plunge pool carve a shallow cave behind the falls, causing pieces of rock to break off from the precipice and fall into the plunge pool. This process is especially rapid if the cataract flows over hard caprock underlain by softer rock. Such was the case as the River Warren plunged over its streambed of hard Platteville Limestone into a gorge cut into the underlying, soft St. Peter Sandstone, forming the Falls of the River Warren. The boxcar-sized slabs of limestone that broke off from the lip of the falls can still be seen on the sides of the gorge formed by the falls' retreat. As the falls chewed its way up the River Warren, it first joined the much smaller falls of the Mississippi Headwaters at Fort Snelling and then divided into two falls. The larger of the two falls (the Falls of the River Warren) continued up the River Warren, and the other (St. Anthony Falls) started up the much smaller Mississippi Headwaters.

Due to the great flow of the master stream, the Falls of the River Warren raced two miles up the River Warren, only to die when it ran out of Platteville Limestone caprock. In its death throes, the Falls of the River Warren was reduced to rapids that could be negotiated by Mississippi River fishes.

Historic St. Anthony Falls

An early French missionary-explorer, Father Louis Hennepin, "discovered" St. Anthony Falls in 1680, mapped their position, exaggerated their height as forty to fifty feet, and named them the Falls of St. Anthony of Padua after his patron saint. Over the years, subsequent explorers made increasingly realistic estimates of the falls' height. George Catlin said it all in 1823, "The Falls of St. Anthony is about nine miles above this Fort (Snelling), and the junction of the two rivers; and, although a picturesque and spirited scene, is but a pygmy in size to Niagara, and other cataracts in our country—the actual perpendicular fall being but eighteen feet, though of half a mile or so in extent, which is the width of the river; with brisk and leaping rapids above and below, giving life and spirit to the scene" (Catlin 1841, 131).

After it separated from the Falls of the River Warren, St. Anthony Falls retreated an additional eight miles up the Mississippi from Fort

Snelling to its present position above Fort Snelling, with the height of the falls decreasing from seventy-five feet to eighteen feet. On its way upstream, St. Anthony Falls spawned Minnehaha Falls, which, in turn, has retreated over a half mile up Minnehaha Creek.

As the falls retreated upstream it left a trail of limestone rubble that prevented the passage of steamboats upstream from St. Paul during normal river flows. Consequently, St. Paul became the head of navigation, giving the city an economic "jump start" that contributed to it becoming Minnesota's capital rather than its now larger twin, Minneapolis.

No sooner was St. Anthony Falls born than it began its upstream death march. In its retreat, it traveled almost eight miles, leaving a deep gorge and a trail of limestone rubble. It was approaching extinction as it moved upriver to within less than a mile of Nicollet Island, the place where the limestone caprock would end. To preserve the falls, the Corps of Engineers in 1876 built concrete dikes and aprons to protect what was left of the limestone caprock, adding two low dams to keep the limestone under water, protecting it from freezing and splitting.

St. Anthony Falls is now located near the Third Avenue Bridge in Minneapolis in a stretch of river that drops about sixty-five feet in three-quarters of a mile. Now smothered by concrete and steel, St. Anthony Falls is hardly discernible to the casual observer. From Fort Snelling to St. Anthony Falls the river occupies a narrow gorge. Above St. Anthony Falls the valley is much smaller, flowing in a comparatively shallow trench, and flanked by wide terraces of glacial alluvium along most of its course. St. Anthony Falls marks the biological and geological division between the Headwaters Reach and the Upper Mississippi River. Before the Upper and Lower St. Anthony Falls locks were constructed, St. Anthony Falls was the head of navigation.

Glacial Lake Duluth and the Glacial St. Croix River

East of Lake Agassiz the retreating Wisconsin ice sheet created another great glacial lake, Glacial Lake Duluth. Because the eastern outlets of Glacial Lake Duluth into the St. Lawrence River system were blocked by ice, the water level of the lake rose until it was four or five hundred feet higher than today's Lake Superior. It overflowed down the valley of today's Brule River and greatly increased the flow of the Glacial St. Croix River, which was faster and bigger than any river in North America today. Overflow from Lake Duluth was facilitated because its southern shore was still being depressed by the weight of the retreating Wisconsin glacier. This flood carved the St. Croix River's mile-wide gorge and the spectacular St. Croix Dells and its potholes in hard basalt rock. The deluge augmented the flow of the Glacial River Warren, and the combined torrents scoured the Mississippi valley downstream from Prescott, Wisconsin.

Interstate State Park, located on the St. Croix River at Taylor's Falls, Minnesota, contains some of the world's most outstanding and unique geological features. Lake Superior's southern outlet via the St. Croix River was abandoned just 9,200 years ago. Today, the St. Croix River enters the Mississippi River at Prescott, and the modern Brule River now occupies the northern end of the Glacial St. Croix valley, flowing northward into Lake Superior, and running in the opposite direction as the Glacial St. Croix.

Flowing water tends to transport as much sediment as it can carry. Sediment-poor water is sometimes called *hungry water* due to its great erosive capacity. The hungry waters of the River Warren and the Glacial St. Croix greatly increased the erosive capacity of the Upper Mississippi River, enabling it to export sediments faster than they could be supplied by tributaries. This resulted in the entrenching of the Mississippi channel over two hundred feet in some reaches, and the formation of the terraces (remnants of the former floodplain) that presently flank the valley.

In Memoriam: Glacial Lake Agassiz, Glacial River Warren, and Glacial St. Croix River

As the Wisconsin glacier retreated into Canada, Glacial Lake Agassiz finally drained to the north and east, reducing the water supply of the River Warren, which was transformed into the much smaller Minnesota River about 9,000 years B.P. As ice melted from the Great Lakes basin, Lake Superior and the other Great Lakes resumed their drainage via the St. Lawrence River into the Atlantic Ocean. Relieved of their massive burdens of ice, the glacial outlet channels of both Lake Agassiz and Lake Superior began to rebound, completing the beheading of the River Warren and the Glacial St. Croix. Today, the Minnesota and St. Croix Rivers are underfit streams, "rattling around" in the large valleys carved by their mighty ancestral rivers.

By 8,000 years B.P., Lake Agassiz was a vestige of its former self. When it finally drained away less than six thousand years ago, Lake Agassiz left as remnants the large modern Lakes Winnipeg, Manitoba, and Winnipegosis of Canada as well as many smaller lakes such as Minnesota's Upper and Lower Red Lakes.

Today, part of the bed of Lake Agassiz forms the Red River Valley with its distinctive beach lines, incredibly flat plains, rolling uplands, and the world's most fertile, black, fine-grained soils. Today, 66 percent of the Red River Valley is in productive, geometric fields of canola, sunflowers, sugar beets, wheat, and potatoes.

Drainage was poor on the eastern side of Glacial Lake Agassiz, resulting in high water tables that inhibited the decomposition of dead vegetation, resulting in the formation of *peat*. The great peatland that formed north and east of Minnesota's Upper and Lower Red Lakes is one of the largest continuous peat deposits in the world (Tester 1995). Lakes and rivers within the ancient lakebed are stained red with tannins and humic acids, giving rise to names like "Red River," "Red Lake," and "Lake Vermilion." When the staining by this *bog tea* is extra dark, descriptive names like "Black River" and "Black Bay" are used.

The Big Picture

Throughout the Pleistocene the Twin Cities metropolitan area was the junction of three major drainages. The largest was that of the Minnesota River or Glacial River Warren, which drained Glacial Lake Agassiz. Next in size was the St. Croix River draining Glacial Lake Duluth (Glacial Lake Superior). Least in size was the Headwaters of the Mississippi River draining only the glacial interlobe area of central Minnesota (Sloan 1985).

The slope of the Mississippi River from Fort Snelling southward is low compared to the slope of its valley north of Fort Snelling. This is because the valley southward of the junction of the Minnesota and Mississippi Rivers was eroded more than two hundred feet deeper than the present valley floor and then filled to its present level with glacial outwash and sediments from steep-sloped tributaries. The riverbed is destined to continue its aggradation, and man-made obstructions, such as dams, exacerbate the rate of rise.

A Summary of the Major Course Changes of the Upper Mississippi and Its Tributaries during the Pleistocene

During pre-Pleistocene time (late Tertiary), the Central Lowlands of the northern United States had been drained principally by streams flowing northward into Canada. The northern tributaries of the Missouri River drained into the Arctic Ocean via Hudson Bay. The northern tributaries of the present Ohio River flowed northward across Pennsylvania, Ohio, and Indiana. They were tributaries of the St. Lawrence River system that flows into the North Atlantic Ocean via the Gulf of St. Lawrence (Fisk 1944; Thornbury 1965).

Nebraskan, Kansan, and Illinoisan glaciers advanced as far south as the approximate present position of the Missouri and Ohio Rivers. The modern courses of these rivers were determined as vast quantities of meltwater collected along the leading edge of the glaciers, forming rivers that had to flow in a general southerly direction because their northward flow was restricted by ice. They merged with the ancient Lower Mississippi River, which ran southward through the Mississippi Embayment (a sediment-filled northward extension of the Gulf of Mexico), and into the Gulf of Mexico (Fisk 1944, Pielou 1991).

Throughout the Pleistocene, the Upper Mississippi River was repeatedly displaced laterally by glaciers. The earliest glaciers (the Nebraskan and Kansan) pushed it eastward. Illinoisan ice pushed the Upper Mississippi westward, as did Wisconsin ice that forced it into its modern valley and obliterated the evidence of previously existing valleys by filling them with sediments.

Before the Nebraskan ice sheet had advanced as far south as central Minnesota, the Mississippi River had apparently carved a deep, broad channel across southern Minnesota and Iowa, and entered the present valley of the Mississippi just south of Muscatine, Iowa. From that junction southward the valley of the Mississippi is wide and old looking. It is believed to have been the mainline of discharge in pre-Nebraskan times as it is now.

Evolution of the modern Upper Mississippi River is generally believed to have begun about 1.6 million years ago when Nebraskan glacial ice that had approached from the west and northwest displaced the Mississippi River eastward from its northwest to southeast course through central Iowa to its present location as far south as Clinton, Iowa. As it flowed along the eastern edge of the Nebraskan glacier as an ice-border stream, it was forced to flow over rock strata rather than seek natural valleys (Trowbridge 1959; Sloan 1998a).

South of Clinton, the Mississippi occupied the lower two-thirds of the Illinois River Valley during much of the Pleistocene. Flowing around the east edge of the Nebraskan glacier, the Mississippi carved a valley southeastward from Clinton to Hennepin, Illinois, where it entered the valley now occupied by the modern Illinois River. The Mississippi flowed through this channel during the Aftonian Interglacial, and the channel was enlarged during the ensuing Kansan glaciation. This channel, now abandoned, is called the Princeton Channel. It was blocked temporarily by Illinoisan ice but reopened after the glacier had retreated. The final advance of the Wisconsin glaciers filled the Princeton Channel with sediments and forced the Mississippi westward into its present valley.

The Illinois River, now draining a much-reduced area, occupies the huge valley that was carved mainly by the ancient Mississippi (Trowbridge 1959; Fry et al. 1965; Simons et al. 1975). Flowing through this ancient channel, the modern Lower Illinois River has one of the lowest gradients in the world (0.64 inch per mile). The Mississippi River hydraulically dams it, as does the aggraded bed of the sediment-filled Mississippi (Hajic 1985). In glacial times, massive flows of the Mississippi episodically caused backward flows in the Illinois and Kaskaskia Rivers, causing the deposition of exotic sediments in the American Bottom (Young and Fowler 2000).

In sharp contrast with the broad river valley elsewhere, at Rock Island, Illinois, and Keokuk, Iowa, the Mississippi flows through a narrow gorge sliced through limestone, filling the whole valley floor from wall to wall. The Des Moines Rapids and gorge at Keokuk were formed during the Yarmouth Interglacial period. The subsequent

Illinoisan glacier forced the Mississippi back west, blocked its channel and those of the Cedar and Iowa Rivers, and formed Lake Calvin that covered much of central Iowa. Lake Calvin drained south over the Des Moines Rapids, enlarging them. They are thousands of years older than the Rock Island Rapids, which were formed during Wisconsin glaciation (Trowbridge 1959).

The major period of downcutting of the modern Upper Mississippi Valley occurred within the past fifteen thousand years as the Wisconsin glacier melted, causing increased river flow that incised the Mississippi valley below Fort Snelling, Minnesota, until it was almost one thousand feet deep. Cutting of the valley was not gradual, but episodic as a result of hundred-year or thousand-year floods caused mainly by the draining of glacial lakes.

In addition to Lake Agassiz and Lake Duluth, there were many more proglacial lakes throughout the drainage basin during Wisconsin glaciation. They were formed whenever ice blocked a drainage route, causing meltwater to be ponded, and they ranged greatly in size and life span. Because all of them did not exist at the same time, they could not drain simultaneously or drain at the same rate.

With the Great Lakes' outlet to the North Atlantic Ocean via the St. Lawrence River blocked by ice during Wisconsin glaciation, Lake Chicago (Glacial Lake Michigan) drained into the Mississippi via the Illinois River. Similarly, the Ohio River drained overflow from Lake Maume (Glacial Lake Erie) into the Mississippi (Pielou 1991).

Terraces: Bleacher Seats Overlooking the River

The floodplain is the flat area adjacent to the river channel, built by the present river in the present climate and subject to overflow. If the river channel deepens, what was formerly floodplain is abandoned and remains as a terrace—an abandoned floodplain. Multiple terraces can be observed along the entire Mississippi River, evidence of many episodes of filling and flushing (Leopold 1994).

The Mississippi tended to entrench itself during the floods caused by the draining of glacial lakes, but between floods the valley floor aggraded as tributaries brought in more glacial outwash than the Mississippi could carry away. The result was a succession of

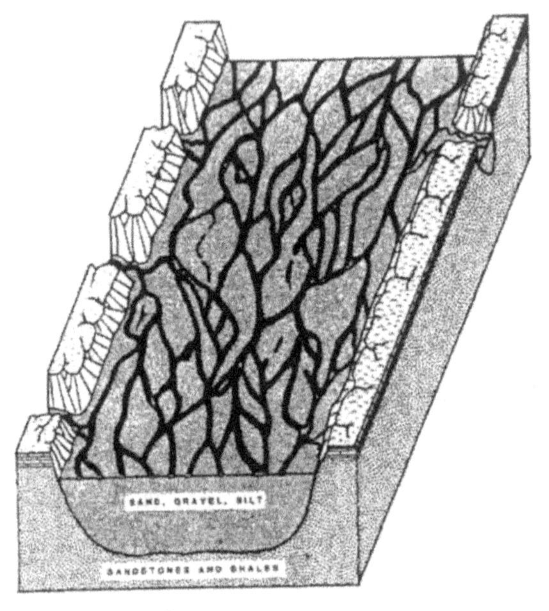

a

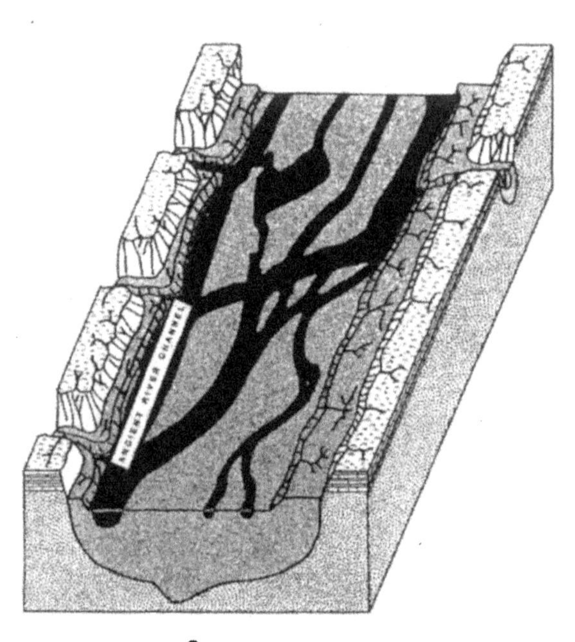

c

Conceptual drawings of the evolution of the Mississippi River Valley at Winona, Minnesota, which lies along the western edge of the unique Unglaciated Area. Because the most recent glaciers passed to the east and west of the Winona area, its rugged bluff land was left unscathed. River flow is from top to bottom, with Minnesota on the left and Wisconsin on the right. a) During Wisconsin glaciation, tributaries brought in sediment faster than the Mississippi could carry it away, causing the valley floor to rise about fifty feet higher than its present level. b) The valley was scoured about three hundred feet deeper by flows from the

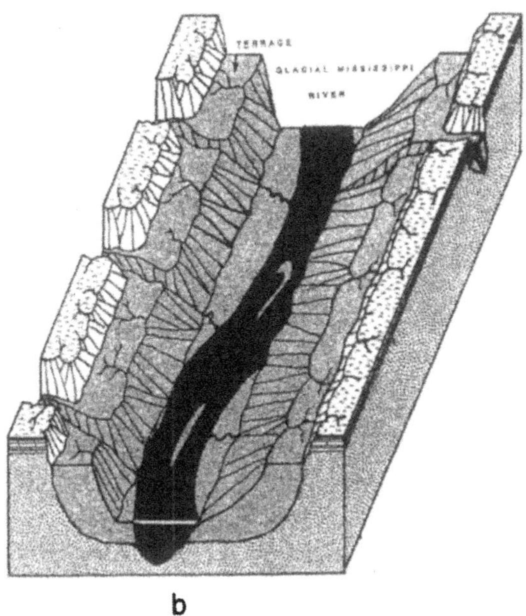

b

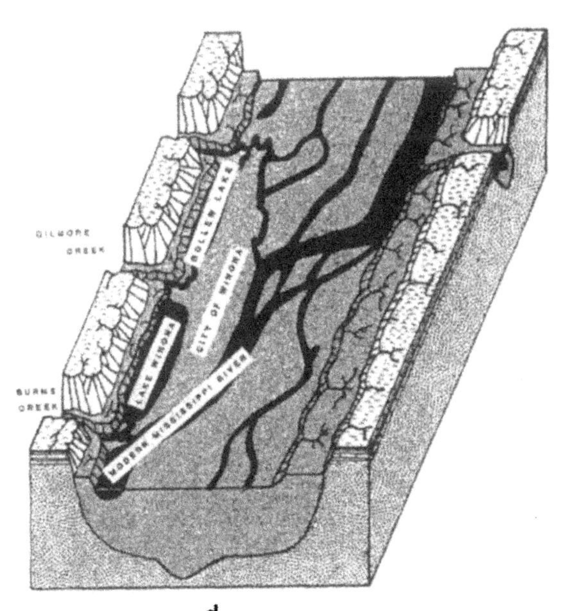

d

Glacial River Warren and Glacial St. Croix River. Note the terraces (remnants of the original valley floor) that flank the river. c) Subsequent to the glacial river episode, the valley partially refilled with sediment from steep-sloped tributaries. Note the channel along the Minnesota side of the valley. d) The side channel was segmented by deltas of tributary streams to form Lake Winona and Boller Lake. The City of Winona is situated on a huge sand bar formed in a braided river.

prominent, benchlike terraces. These fragments of old floodplains flank the river from St. Anthony Falls to the mouth of the Ohio River. They mark the elevations that the valley floor had attained during aggradation.

The highest terraces show that the valley had aggraded to more than fifty feet above its present level before being scoured by flows from glacial rivers, which entrenched the valley and secondarily caused the entrenchment of flat tributary valley floors. Native Americans used the terraces for summer encampments, especially if they occurred at the mouths of tributaries. Because the terraces are nearly level, and therefore less subject to flooding, they have been used as locations for communities. They are also used for agriculture, roads, railroads, and as home-building sites. Houses and farmsteads in rural tributary valleys are often located atop small scraps of terrace. Do the owners contemplate how their flat, perched lots developed in such a precipitous, rugged landscape?

Digging in the terraces reveals thick sequences of sand and gravel lying atop the bedrock sides and floor of an older valley, demonstrating that the valley has been entrenched, partially filled, and retrenched several times. Most terraces and most of the drainage basin are mantled with loess, resulting from wind transport of fine glacial materials during the postglacial period, before the glacial drift was stabilized by vegetation. Many broad terraces are covered with sand dunes.

Extensive terraces are obvious on the Minnesota side of the river near Lake City and Wabasha. They are also conspicuous on the Wisconsin side of the river at Cochrane and Buffalo City. Large areas of sand dunes, blowouts, and deflation basins can be seen in the Weaver Dunes Natural Area near Kellogg, Minnesota, and in the Trempealeau National Wildlife Refuge near Trempealeau, Wisconsin. Interestingly, many of the deflation basins, created during arid, windy conditions, are now wetlands due to a rising water table associated with the creation of the navigation pools of the Nine-Foot Channel Project during the 1930s.

The most conspicuous terrace is the Mankato Terrace that runs southward from the Twin Cities, flanking the Mississippi about eight hundred feet above sea level. Many river cities in Minnesota and Wisconsin are built, in part, upon it, including Inver Grove, Richfield, South St. Paul, Hastings, Red Wing, Wabasha, Minnesota City, Trempealeau, La Crosse, and Prairie du Chien. Winona, on the other hand, is built upon a huge sandbar that was formed by meltwaters of the Wisconsin glacier.

The Savanna Terrace is a distinct feature along the Upper Mississippi River and within its tributary valleys in northeastern Iowa. Its surface stands about sixty feet above modern river level and decreases in elevation, relative to present stream level, down the Upper Mississippi Valley, until it is only about ten feet above stream level in central Illinois. It is the highest terrace along the Upper Mississippi without a loess cover. A distinctive feature of the terrace are alternating bands of red and gray slack-water clays in the upper part of the terraces that occupy the mouths of streams tributary to the Mississippi River. The

Determining Rate of Retreat of St. Anthony Falls

With remarkable deductive reasoning, General G. K. Warren described the origin and history of St. Anthony Falls more than a century ago. "The valley of the Mississippi below the junction, and of the Minnesota above it, is wide and beautiful, and is continuous in direction and of nearly the same breadth, varying from one to two miles. In marked contrast is the valley of the Mississippi above their junction, it being only about one quarter of a mile wide and nearly at right angles with the other. It is a mere gorge, whose bottom is almost completely filled by the river, and evidently had its origin in the waterfall now at St. Anthony. The fall in recent times must have been where the river now joins the main valley, and has since receded to its present position seven miles above the junction" (Winchell and Upham 1884).

Using observations dating back to Hennepin's map of 1680, Minnesota's first state geologist, N. H. Winchell, calculated that it took 7,800 years for the falls to reach its present location. Later corrections to account for changes in bedrock thickness upstream and for a decrease in height of the falls due to valley filling produced a final estimate of twelve thousand years, very close to the presently accepted estimate of 11,700 years, based on carbon dating.

source of the gray silty clay was Glacial Lake Agassiz, and the source of the red clay was Glacial Lake Duluth (Bettis and Halberg 1985a).

The brickyards of yesteryear were often located at the mouths of tributary valleys where clay had been deposited in an estuarine environment. Clay and silt strata, which are almost impervious to water, retard its downward percolation, causing some residents located on the floors of tributary valleys to have wet basements and failed home sewage-treatment systems. Clay and the finest fractions of loess were made into pots by Native Americans and bricks by early settlers.

Because clay and silt are poor aquifers, wells must be drilled through such deposits into sandstone or other permeable, water-bearing, rock or sand strata. Before the tributary valleys were filled to their present level with alluvium, they tended to be V-shaped in cross section. Consequently, wells in the center of an alluvium-filled valley must often be drilled deeper to penetrate an aquifer than those nearer the sides of the valley.

6

The Land Missed by the Glaciers

Introduction

Chapters 3 through 5 told the story of how ancient seas and glaciers created the land forms typical of the valley of the Upper Mississippi River. This chapter discusses one of North America's most unique geological-ecological areas. Missed by glaciers, but carved by their meltwaters, the *Unglaciated Area* served as a *refugium,* a hideout where plant and animal species escaped glacial onslaughts, waiting for the climate to warm. The Mississippi and some of its tributaries flow through this dramatic area, providing opportunities for aesthetic, cultural, and recreational enjoyment. The last half of the chapter takes the reader on a downstream tour through the Unglaciated Area by boat or car, mainly discussing its natural features, but also some of its modern river-related features.

The Remarkable Unglaciated Area

Near Red Wing, Minnesota, the Mississippi begins a grand 250-mile route running southward to Savanna, Illinois, along the western fringe of the remarkable Unglaciated Area, a beautiful, stream-dissected landscape that contrasts strikingly with the glaciated lands that surround it.

Within the Unglaciated Area, the Mississippi flows through a gorge carved deeply into a great, flat-topped, sedimentary rock plateau. Because the Mississippi valley was even deeper in the past, having been scoured by successive glacial torrents, steep-gradient tributary streams cut their valleys correspondingly deep. They dissected the ancient Paleozoic Plateau, creating a complex, extensively branched drainage pattern and precipitous blufflands.

It is easy to see how early explorers believed that the Mississippi River, in this reach, was flanked on each side by a low range of mountains. Weary from the toil of their arduous trip up the river and impatient to get upstream, they seldom took time to leave the relative safety of the river and to climb the steep, six-hundred-foot bluffs. If they had, they may have realized that, once atop the bluffs, they were standing on flat land (the upper surface of the ancient plateau), and that the area was not hilly or mountainous, but "valley-y" (if there is such a word).

Much of the United States is also unglaciated, as in the southern states and a large part of the West, but the Upper Mississippi's Unglaciated Area is unique because it is completely surrounded by glaciated territory. The unglaciated landscape has been forming for many thousands of years, whereas the surrounding glacial landscapes, products of Wisconsin glaciation, are only about fifteen thousand years old or younger.

Travelers through the Unglaciated Area may not be aware of its geological history, but most notice that the spectacular area is remarkably different from the ordinary country that surrounds it. Cliffs and headlands loom above highways. Layers of sedimentary rocks exposed in the bluffs record the advance and retreat of ancient semitropical seas, representing more than five hundred million years of geologic history. Farm fields are devoid of the glacial erratic boulders so characteristic of fields in glaciated areas. Because the area is so rugged and well drained, lakes and bogs are rare. Even the trees look different. People who stereotypically think of Iowa as being flat and agricultural are amazed at the ruggedness of the extreme northeast corner of the Hawkeye State.

The Unglaciated Area is bounded approximately by Red Wing, Minnesota, on the north; by St. Charles, Minnesota, on the west; by Wausau and Madison, Wisconsin, on the east; and by Savanna, Illinois, on the south. Glacial erratic boulders are first commonly seen in farm fields near these cities. Technically speaking, only a narrow strip of extreme eastern Minnesota and Iowa along the Mississippi River is included in the Unglaciated Area. Although southeastern Minnesota and eastern Iowa were not glaciated during Wisconsin glaciation, they were during pre-Illinoisan glaciation (Mossler 1999).

The Unglaciated Area encompasses nearly fifteen thousand square miles, an area twice as large as the state of New Jersey or about as large as Denmark (Martin 1965). Lying within it is the historic lead-zinc mining district that includes Pitosi, Wisconsin; Galena, Illinois; and Dubuque, Iowa.

Protected by a dome of hard rock in northern Wisconsin, glacier after glacier approached this unusual area but left it virtually unscathed. Tongues of ice were diverted westward through the bed of Lake Superior and southward through the bed of Lake Michigan.

For many years it was called the Driftless Area because no evidence of glaciation, such as glacial drift, could be found in the region. In recent years, patchy deposits of glacial drift and scattered glacial erratics have been found in the Unglaciated Area, but most drift has either been lost through erosion or has been mantled with late-Wisconsin-age loess.

Some geologists prefer to call the area the Paleozoic Plateau because its bedrock is exposed, in marked contrast to adjacent areas where the terrain is more subdued and dominated by thick deposits of glacial drift that mantle and mask the underlying bedrock.

The topography of the Unglaciated Area is basically the product of erosion that took place during earliest Pleistocene time. The area was probably glaciated repeatedly in the pre-Illinoisan, with the last glaciation occurring about five hundred thousand years ago. Major stream incision probably began about 160,000 years ago. The area appears to have been free of ice and permafrost during the Wisconsin glaciation maximum (Lively and Alexander 1985), but the tributary stream valleys underwent a complex history of erosion and aggradation between seventeen thousand and ten thousand years ago in response to glacial drainage in the Mississippi River main stem (Halberg and Bettis 1985a, 1985b).

A Refugium for Plants and Animals

Although the Unglaciated Area is surrounded by glacial deposits, it was never completely surrounded by glaciers at any one time. This is known because the surrounding deposits are of very different ages, ranging from Nebraskan drift that may be as old as two million years, to Wisconsin drift that is only fifteen thousand years old. The area may have been an "embayment" into the ice margin that appears to have been nearly closed during both Kansan and Illinoisan advances.

The Unglaciated Area was probably at least partially covered with vegetation during all Pleistocene glacial advances. It was a refugium; a small isolated area that escaped the extreme changes undergone by surrounding areas. It was a safe haven where many species of plants and animals survived the devastation of ice sheets and then repopulated the surrounding areas as the glaciers retreated. At the time of glacial maximums, both temperature and precipitation may have been only slightly lower than today (Curtis 1959).

During periods of glacial melt, atmospheric moisture supplies were probably at their lowest, creating desertlike conditions on the high plains, destroying vegetation, resulting in production and transport of loess (wind-blown silt). During that arid time, the Unglaciated Area was like an oasis. Its steep valleys provided protection and a warm, moist growing season. During Wisconsin glaciation, maple-basswood forests may have dominated the landscape (Curtis 1959). It probably was also a refugium for two-needle pines (red and jack), spruce, balsam fir, and white cedar, but not for oak, hickory, or red cedar (Carol Jefferson, personal communiqué 2001).

The most recent glaciers did not scour the Unglaciated Area, but their meltwaters did, deepening existing valleys. Wind was also a powerful erosive force, especially the fierce katabatic winds (discussed in chapter 4) that roared down the slopes of nearby glaciers.

Early settlers and river travelers gave fanciful names—Chimney Rock, Paradise Point, Inspiration Point, and Castle Rock—to the beautiful pillars, palisades, and ramparts carved out of the soft sandstones by wind and water.

Most of the blufflands within the unglaciated area and along both sides of the river from the Twin Cities all the way downriver to Cairo, Illinois, are marked by *karst* landscape—characterized by sinkholes, caves, subterranean rivers, springs, and disappearing streams. These features developed when limestone and dolomite were slowly dissolved by rainwater mixed with carbon dioxide or soil acids.

Cave abundance is relative, depending on the size of the explorer. There are millions if you are a mouse, hundreds if you are a human, but none if you are an elephant. These passages have been carrying water intermittently, depending on climate, for the past hundred million years. Many streams in the karst area disappear into sinkholes, flow through caves, and reappear as springs (Sloan 1999). Another feature of karst is fractured bedrock covered by a thin mantle of soil. Consequently, the groundwaters of the karst region are extremely susceptible to pollution from farm fields, feedlot runoff, and failed sewage lagoons.

Forests

Throughout the Unglaciated Area sugar maple-basswood forests now flourish on moist north- and east-facing slopes, and oak-hickory forests dominate the drier, west- and south-facing slopes. The difference in forest types can be striking in valleys that trend east and west where one slope faces south and the other north. The south-facing slope is dry and relatively barren, but the north-facing slope is lush and heavily forested. In winter, the south-facing slope may be snow free while its counterpart is snow covered.

A floodplain forest of ash and silver maple is usually found along streams. American elms were dominants in these forests until the 1970s when Dutch elm disease destroyed them. Forests of the Unglaciated Area include species not familiar to most Minnesotans—shagbark hickory, bitternut hickory, butternut, black walnut, black oak, chinquapin oak, Hill's oak, swamp white oak, black cherry, bladdernut, Kentucky coffee tree, sycamore, blue beech, and wahoo.

Goat Prairies

Steep, rocky, and exposed, these steep south- and west-facing slopes appear to be places only goats might love—hence the name *goat prairie*. Faced and sloped like giant solar collectors, they lose their winter snow cover early and bake under the summer sun.

The steep slopes ensure that the runoff of rainwater is extreme on goat prairies, causing continual surface movement of soil particles. As a result the soil blanket is very thin, averaging less than four inches in depth, except where pockets have formed in the underlying limestone. The water-holding properties and nutrient supplies of the soil are good, but the soil is so thin that total quantities of water and nutrients are severely limited.

It would seem that the extreme heat, low water storage capacity of the soil, and high wind velocities of the goat prairie would create an environment that is too dry to support the luxuriant stands of prairie plants that occur there. The likely explanation for the lush growth is that significant quantities of water are obtained from dew that collects on the slopes when they chill down at night, having quickly radiated their warmth, unhindered by insulating trees, into a clear sky. Even during the hottest and driest seasons, one's shoes usually get soaked with dew when walking the goat prairies early in the morning (Curtis 1959).

The vegetation of the goat prairies includes species typical of the western Great Plains. Grasses include little bluestem, sideoats grama, and hairy grama grass. Other plant species include downy paintbrush, ground plum, dotted blazing star, and pasque flowers.

These windswept slopes tend to remain as prairie not only because of their dryness and sun-baked character, but also because they have been grazed (initially by bison and elk, but later by cattle and sheep) and rejuvenated periodically by burning that set back encroaching trees and brush. Goat prairies are both natural and cultural artifacts of the region. The fires that maintain them have been sparked naturally by occasional lightning bolts, but most often, carelessly or intentionally, by humans.

In the past fifty years or so fires have been suppressed, and steepland grazing has proven unproductive. Now, in their absence, sumac, red cedar, and exotic buckthorn are invading and overwhelming the goat prairies. The dense stands of red cedars are actually virgin forests because most of the goat prairies had been treeless for thousands of years.

In times of heavy snow cover, deer move around to the goat prairies, congregate among the cedars, and bask on the sunny, snow-free slopes. They browse the cedars but especially go after the invading sumac that thrives on the periphery of the goat prairie. When the deer walk onto the flat land above the prairie, they are usually in a cornfield. Corn is the principal component of their winter diet. Winter starvation of deer is unheard of in the Unglaciated Area.

Freed from the regimen of fire and grazing, the youthful forests of the Unglaciated Area have obscured many of the once-familiar rock formations so prominently indicated on maps of the 1800s. Casual observers no longer see the landmarks, but they are rediscovered yearly by the legions of deer, grouse, ginseng, and mushroom hunters that prowl the bluffs.

Trout Streams

The Unglaciated Area is rich in trout streams that are highly productive due to the fertility of the watershed and the abundance of cold, mineral-rich spring water that gushes out of aquifers exposed by stream downcutting. In a classic stream, brook trout (and watercress) inhabit the cold water of the headwaters with brown trout and

Native Brook Trout

Companies B, H, and I of the First United States Dragoons visited Winona in 1835. Lieutenant Albert M. Lea, the expedition's topographer, wrote: "Desiring to visit Wabasha's band, the officers directed our course toward Lake Pepin and about the first of July we encamped on a small rivulet which empties into a river [the Zumbro] that enters the Mississippi four miles away, just below Lake Pepin. We camped on the bank of this stream three days, and during that time our whole force of 164 men had as much speckled trout [native brook trout] as we desired, all taken from a single brook only a step wide. One of my men took 130 in four hours with an improvised hook and line" (letter to the editor, *Freeborn County Standard*, June 21, 1877).

rainbow trout successively occupying the warmer waters downstream. The brookie is the area's only native stream trout.

Microhabitats

Climate is moderated in the deepest, narrowest valleys of the Unglaciated Area. Even balsam fir, a northern species, can be found around springs arising in cool, secluded valleys in southeast Minnesota. In the narrowest, uninhabited, steep-sided valleys, mosquitoes are seldom a problem because there is insufficient standing water for breeding. Timber rattlesnakes, once abundant, still occur in the blufflands, and diminutive massasauga rattlesnakes can be found in the Chippewa River delta, Trempealeau National Wildlife Refuge, and along the lower reaches of the Trempealeau River. Stands of white cedar occur as vertical forests on some north-facing cliffs where they are sheltered from fire and deer browsing and where snows persist longer in the spring. Gardeners who study plant hardiness zone maps know that zone 4 extends northward up the Mississippi River Valley for about a hundred miles to the Twin Cities.

The Unglaciated Area itself and the rugged blufflands that border it exhibit great ecological diversity. Extreme southeast Minnesota, for example, is the warmest, wettest, most geologically diverse part of the state. Historically, it was a place where the "big woods" of the east met and intermingled with the great prairies that stretched westward toward the Rockies. Minnesota's blufflands comprise only 3 percent of the state's land area, yet it is home to 43 percent of the plant and animal communities that are listed as state endangered, state threatened, or of state special concern.

Within the Unglaciated Area and in the surrounding blufflands exist north-facing, *algific talus slopes* that retain ice just beneath their surfaces all summer long. These rare landforms create cool, wet microclimates that harbor plants like balsam fir, Canada yew, and yellow birch that are normally found farther north.

A River Tour along the West Edge of the Unglaciated Area

The most dramatic examples of the Unglaciated Area's unique geological, biological, and anthropological features can be seen by taking a trip down the Mississippi by boat or via the highways that flank the river. There are many scenic overlooks where the river panorama can be viewed from the bluff tops. In the following account, pertinent modern features like dams and power plants are also discussed.

Red Wing, Barn Bluff, and Abandoned River Channels

About sixty miles downstream from St. Paul, the business district of the city of Red Wing lies at the upstream end of an abandoned glacial channel of the Mississippi River. Barn Bluff and the isolated bluffs southeast of the city were islands in the glacial river. Highway 61 and the Milwaukee and Soo Railroad tracks follow the low-gradient, abandoned river channel as it runs southeastward from Red Wing and opens onto Lake Pepin near the town of Frontenac (which also lies within the old river channel). Segments of the old channel presently exist as the shallow lakes and marshes that flank Highway 61. Downstream from Red Wing, a branch of the old channel passes through the village of Wacouta where most houses are built upon the level, elevated surface of the Mankato Terrace.

Famous Lake Pepin and the Chippewa River

Bounded by cliffs, steep bluffs, and the mouths of tributary rivers, Lake Pepin was often described by European explorers as being surrounded by mountains. The most famous lake on the Upper Mississippi River, Lake Pepin is twenty-two miles long and three miles wide. With no appreciable current, it is a true lake within the river.

Lake Pepin was born about 9,500 years ago. As the Wisconsin ice sheet wasted away, the outlets of Lake Agassiz and Lake Duluth shifted to the north and east. Cut off from their glacial headwaters, the Glacial River Warren and the Glacial St. Croix River lost their great flows, shrinking in size to become the modern Minnesota, St. Croix, and Mississippi Rivers (Zumberge 1952; Wright and Ruhe 1965).

Deprived of the rushing floods of glacial meltwater and, with a riverbed that slopes only three inches per mile, the shrunken Mississippi could not transport the heavy sand and gravel load contributed by the Chippewa River. Cutting through extensive glacial terraces with a slope of nearly three feet per mile, the Chippewa enters the Mississippi from the east, creating a huge sand delta that partially dams the Mississippi, impounding it, and creating Lake Pepin.

One would think that Lake Pepin would have been filled with sediment by now, but its life has been spared (temporarily) because upstream tributaries carry relatively little coarse sediment. The Mississippi Headwaters has been the main contributor of coarse sediments to Lake Pepin, and it has formed a delta at the lake's upper end, slowly decreasing the length of the lake. The modern Minnesota River has a large drainage area of silty drift but a low gradient and thus carries a relatively low sediment load consisting mainly of clay and silt. The St. Croix River, the other major tributary above Lake Pepin, contributes little sediment to Lake Pepin because it has a small forested drainage area composed of sandy drift and flows through hard igneous and metamorphic bedrock for some distance. Sand bars are scarce on the St. Croix.

About 9,200 years ago, Lake Pepin extended upstream to near St. Paul where fine lake sediments are buried by sixty-five feet of coarser river sediments contributed by the Headwaters. The lake sediments are buried fifty feet deep at Hastings and fifteen feet deep at Red

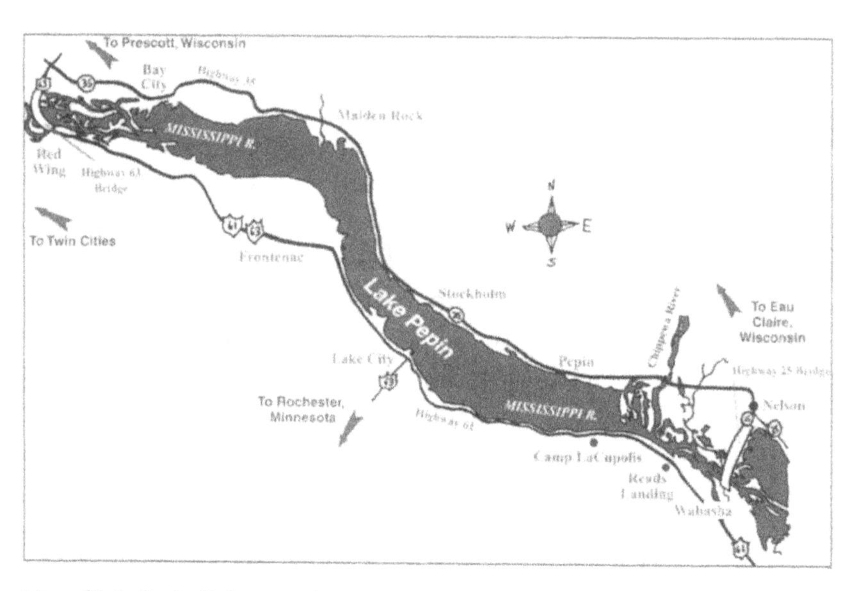

Map of Lake Pepin (Reimer 2001).

Wing. The delta at the head of the lake moves downstream at a rate of five to seven yards per year (Wright 1972a), so when Native Americans occupied the area around Red Wing between A.D. 1050 and 1300, the head of the lake would have been five to seven miles farther upstream than it is today (Dobbs and Mooers 1991).

Lake Pepin is destined to be filled as the delta at the head of the lake continues to extend farther into the lake. An interesting side effect of this delta extension is the sandbar formed across the mouth of the St. Croix River, creating Lake St. Croix in its lower thirty-five miles (Wright 1985).

On September 21, 1805, after making it upstream through Lake Pepin, Zebulon Pike reached the confluence of the Mississippi, Minnesota, and St. Croix Rivers. He made an interesting commentary on their transparency and color: "The Mississippi became so narrow this day, that I once crossed in my batteaux with forty strokes of my oars. The water of the Mississippi, since we passed Lake Pepin, has been remarkably red; and where it is deep, appears as black as ink. The waters of the St. Croix and St. Peters [Minnesota] appear blue and clear, for a considerable distance below their confluence" (Coues 1965, 77–78).

How times and water quality have changed! As explained in chapter 5, we now know that the Mississippi's red color was due to "bog tea" contributed by the Headwaters.

Because it has no current to erode away late winter's two-foot-thick layer of ice, Lake Pepin has historically been the last area between Cairo and St. Paul to become ice free in the spring—as much as two weeks later than the river upstream or downstream.

Most of the river upstream of the lake is now open year round due to thermal pollution from the Twin Cities and from a nuclear power plant at Red Wing. We shall see in subsequent chapters that the Lake Pepin ice barrier has been a factor in commercial river transportation for 175 years, giving Winona, which lies downstream from Lake Pepin, a competitive early shipping advantage over Red Wing and Twin Cities ports.

Lake Pepin has an imposing wind fetch because it is straight, twenty-four miles long, and has no islands. Boaters quickly learn that it can be dangerous during strong winds and sudden storms. On the evening of September 16, 1805, on his way upstream, Zebulon Pike had decided to sail up the lake in spite of his interpreter's advice that it was safer to run the lake at night after the wind had subsided:

> However, the wind serving, I was induced to go on; and accordingly we sailed, my boat bringing up the rear, for I had put the sail of my big boat on my bateau, and a mast of 22 feet. Mr. Frazer embarked on my boat. At first the breeze was very gentle, and we sailed with our violins and other music playing; but the sky became cloudy and quite a gale arose. My boat plowed the swells, sometimes almost bow under. When we came to the Traverse, which is opposite Point De Sable, we thought it most advisable, the lake being very much disturbed and the gale increasing, to take harbor in a bay on the east. One of the canoes and my boat came in very well together; but having made a fire on the point to give notice to our boats in the rear, they both ran on the bar before they doubled it, and were near foundering; but by jumping into the lake we brought them into a safe harbor. Distance 40 miles.

Never one to give up, Pike decided to try again the next day: "Although there was every appearance of a very severe storm, we embarked at half-past six o'clock, the wind fair; but before we had hoisted all sail, those in front had struck theirs. The wind came on hard ahead. The sky became inflamed, and the lightning seemed to roll down the sides of the hills that bordered the shore of the lake. The storm in all its grandeur, majesty, and horror burst upon us in the Traverse, while making for Point De Sable; and it required no moderate exertion to weather the point and get to the windward side of it. Distance three miles" (Coues 1965, 61–65).

Because it is so large, deep, and free of rock channelization structures like wing dams, Lake Pepin is a mecca for modern sailors. It is the most intensively sailed portion of the entire Upper Mississippi River. On a good day, there may be as many as a hundred sailboats on the lake.

The massive Chippewa delta at the foot of the lake is a mosaic of sloughs, marshes, islands, and channels. The Chippewa delta pushed the Mississippi against the Minnesota shore, eroding the soft sandstones and quarrying the overlying dolomite, thus creating the scenic cliffs and steep bluffs that loom over Highway 61 and Lake

Precipitous sand terraces loom over the outsides of bends on the lower reach of the Chippewa River, one of the Mississippi's most important tributaries. Deposited as outwash during Wisconsin glaciation, the terraces supply unlimited quantities of sand to the Mississippi River where it must be periodically removed by dredging to keep the commercial navigation channel open for barge traffic. The surface of the terrace is the height to which the Mississippi floodplain was elevated prior to scouring by the Glacial River Warren and Glacial St. Croix River. Terraces on the outsides of river bends are most vulnerable to erosion and slumping because the current is fastest there, especially during floods.

Pepin's outlet. The delta constricts the river at the outlet of Lake Pepin, increasing the velocity of the current. The rushing water has scoured a hole about fifty-five feet deep that has maintained itself for over one hundred years. The river's unusual depth and current at this spot combine with the unusual physical properties of water to produce a remarkable phenomenon.

Water is densest at 39°F. In winter, this heavier, relatively warm water from the lake's bottom upwells in the constricted lake outlet and, in concert with the swift current, keeps the river ice-free throughout the winter at Reads Landing, Minnesota. In some years the open water extends downstream beyond Wabasha, Minnesota, a distance of more than five miles.

During winter, hundreds of bald eagles gather in the area to fish the open water. They are easily seen from Highway 61, resting in roadside trees and on the ice, creating a growing attraction for tourists and casual bird-watchers. To accommodate this interest, there is a popular eagle-watch facility on the river in downtown Wabasha.

Reads Landing, a famous steamboat and logging port during the late 1800s, overlooks the confluence of the Mississippi and Chippewa Rivers. Here, a great mound of dredged sand can usually be observed on the Wisconsin side of the river. The dredging, done by the U.S. Army Corps of Engineers, is needed to keep the Mississippi open for barge traffic.

The steep-gradient Chippewa is the main sand-generating tributary in the area, and its coarse, glacial outwash sands comprise most of the dredged sediment to be seen along the Mississippi as far south as Lock and Dam 8. Chippewa sands usually can be identified by characteristic shiny, black, flat pebbles about the size of a fingernail. Below Lock and Dam 8, river sands are finer and are derived mainly from weathered sandstone.

Aerial photo of Lake Pepin and the Chippewa River delta in late winter (courtesy of Miranda Photography).

Viewing the sand-clogged Chippewa River of today, it is hard to believe that shallow-draft steamboats ran up the Chippewa as far as Durand, Wisconsin, until 1914. At this site in 1805, Zebulon Pike described the unspoiled Chippewa: "At the entrance of Lake Pepin, on the E. shore, joins the Bateaux (Chippewa) River, which is at

Lake Pepin and Chippewa Delta in Late Winter

In the photo on page 84 we look upstream toward the confluence of the Mississippi and Chippewa Rivers. Lake Pepin lies at top center, still frozen and snow-covered. The Chippewa River enters the Mississippi from the right, flowing through its massive, wooded delta on the Wisconsin side of the Mississippi. The delta was formed because the steep-sloped Chippewa delivers more sediment, mainly sand, than the Mississippi can carry away. It constricts the master stream, damming it to create Lake Pepin. The river immediately below Lake Pepin is kept open in winter by the upwelling of warm water (39°F) from the lake bottom and by the swift current due to the constriction. Hundreds of bald eagles scavenge dead and dying fish in the open water all winter long. The greatest numbers can be seen can be seen from late winter until the river becomes ice free in early April.

Over geologic time, the Chippewa delta pushed the Mississippi against the Minnesota shore where it cut into soft sandstones, undermined the overlying dolostone, causing pieces to break off, forming high, picturesque cliffs.

The snow-covered uplands seen in the upper left are the top of the ancient sedimentary plateau dissected by the Mississippi and its tributaries, creating an unusual topography where there are no hills, just valleys.

The city of Wabasha, Minnesota, lies in the center of the photo, atop a sandy terrace formed in glacial times by outwash from the Chippewa River. The terrace was once part of the glacial floodplain. The site has a rich history of Native American occupation.

The irregular islands in the foreground are composed primarily of sand dredged from the Mississippi by the Corps of Engineers in their effort to keep this "problem area" open for navigation. The crescent-shaped marks on the islands are stands of young willows and cottonwoods that sprang up from seeds on the periphery of the spoil piles where the sand was moist.

least a half a mile wide, and appears to be a deep and majestic stream. This river is in size and course, for some distance up, scarcely to be distinguished from the Ouiscousing [Wisconsin]" (Coues 1965, 305–6).

Below Lake Pepin during postglacial time, the Chippewa River's heavy burden of coarse sediments caused a transporting problem for the Mississippi. To transport the sand and gravel, the Mississippi was forced to braid. Instead of one main channel, the Mississippi formed many smaller channels that individually had a higher current velocity than a single main channel. Thus, the coarse load of the Chippewa River could be carried by the Mississippi. Many of the small braided channels can still be seen on topographic maps and aerial photographs. Being braided is characteristic of overloaded glacial rivers worldwide.

Buena Vista Bluff

Buena Vista Bluff at Alma, Wisconsin, is easily accessible by car. Rising 540 feet above the river, it provides the best view of a lock and dam system (L&D 4) on the entire Upper Mississippi River. Visible are the dam's lock, roller gates, tainter gates, tailwaters, the pool above the dam, and the dam's earthen dike that extends 1.1 miles across the river bottoms to high ground in Minnesota. Pool 4 (the impoundment above the dam) is extraordinarily long. It includes Lake Pepin and extends forty-four miles upriver to L&D 3 south of Red Wing. The bluff also provides a bird's-eye view of a large coal-burning power plant that is supplied by rail and by barge. Peregrine falcons were reintroduced to the area by using a nest box on one of the tall stacks.

Pool 5

Pool 5, which extends downstream from L&D 4 at Alma, Wisconsin, to L&D 5 near Minneiska, Minnesota, deserves special mention because it is arguably the most beautiful and least abused of all Upper Mississippi River pools. The main channel is flanked by beautiful sand beaches (courtesy of the Chippewa River and Corps of Engineers dredgers) that are popular with swimmers, boaters, and campers. Because Pool 5 is included in the Upper Mississippi River National Wildlife and Fish Refuge, virtually all of the sand beaches are public. There are no large cities on the pool, leaving it relatively uncrowded. Water quality is good because the Metropolitan Twin Cities has vastly improved its sewage treatment in recent years and because Lake Pepin and upstream navigation pools act as settling basins for sediment and other pollutants. Fishing is excellent for a wide variety of game fish.

Trempealeau Mountain and Trempealeau Bluffs

Trempealeau Mountain, one of the most celebrated landmarks on the entire Mississippi River, lies within the Unglaciated Area about

---

The Rock in the House

Maxine Anderson walked through her home in Fountain City, Wisconsin, on the morning of April 25, 1995, photographing the rooms that she and her husband Dwight had just remodeled. Then all hell broke loose.

At 11:38 A.M. a boxcar-sized hunk of the dolomite cliff immediately behind their house broke off and thundered down the tree-covered, 45-degree slope at the base of the cliff. The original piece broke into a spray of smaller pieces that bounded down the slope, shattering trees and furrowing the ground. One forty-five-ton fragment, measuring about fourteen feet in diameter and five feet thick, cartwheeled down the slope in great bounds, crashed through the roof of the Anderson bedroom, and lodged on edge in the basement floor. The rock was so massive that much of it protruded through the roof of the house. Luckily, Maxine had just stepped out of the bedroom into the dining room a minute before and escaped unscathed. A similar wheel of rock lost momentum and came to rest one hundred feet short of the house. Another of equal size stopped thirty yards short of the neighbors' kitchen.

Remarkably, shortly after midnight on April 5, 1901, a rock probably underestimated at five tons crashed through the roof of the Henry Tobler bedroom, killing Elizabeth Tobler, who was asleep beside her blind husband. That killer rock and the one that landed in the Anderson house came to rest about twenty-five feet from each other. It is noteworthy that both events took place in April (*Winona Daily News*, Thursday, April 25, 1995; *Winona Republican Herald*, April 5, 1901).

The two events described above were separated by a span of

twelve miles downriver from Winona. The "mountain" and the adjacent Trempealeau Bluffs are mentioned in the journals of most early river travelers; they are islands of bluffland that lie within the Mississippi's broad glacial floodplain, isolated from other high land.

The Winnebago Indians called the whole rocky eminence "Hay-nee-ah-chah" or "soaking mountain" (Martin 1965). Accordingly, the French continued the name in the form "La Montagne qui trempe a l'eau," or "the mountain which soaks in the water." The modern name Trempealeau is made up of the last four words of the French phrase. Trempealeau Mountain and much of the Trempealeau Bluffs are contained in Perrot State Park, a place of great geological and historic interest. It was named in honor of Nicholas Perrot, who wintered there in 1685–86.

Trempealeau Mountain and the Trempealeau Bluffs were originally part of the Minnesota upland, but are separated from the Minnesota Bluffs by the Mississippi River and from the Wisconsin Bluffs by a broad expanse of sandy bottomland that stretches eastward for three and one-half miles. The Mississippi River used to flow in the broad abandoned trench on the Wisconsin (east) side of Trempealeau Bluffs and received tributaries from the Minnesota side.

(cont. from p. 87)
ninety-four years, but in geological time the two events were virtually simultaneous. As we drive through the Unglaciated Area we commonly see large rocks, some of monstrous proportions, scattered at the bases of the bluffs. They tell us that there was once a cliff at the site, but that the cliff has retreated, leaving its telltale rocky debris behind. It is important to remember that cliffs are temporary geological features. Pulled by gravity, and often cracked loose by successive freeze–thaw episodes in the spring, pieces of the cliff tumble down onto the steep talus slope at the base of the cliff. This causes the cliff to retreat and to become shorter. Eventually, the talus slope will extend upward to the bluff top, and the cliff will have disappeared. This process is especially fast if the dolomite cap rock is underlain by soft sandstone that is easily eroded by wind, rain, and freezing as was the case at Fountain City.

Unbelievably perhaps, the erosion of the sandstone is accelerated by animals as insignificant as mice, bank swallows, and digger wasps.

One tributary, Cedar Creek, severed Trempealeau Mountain from the Trempealeau Bluffs. At the time, Trout Creek flowed southward carving a channel where the Mississippi runs now. In time, glacial floods deposited so much sand and gravel in the old Mississippi channel east of Trempealeau Bluffs that the rising riverbed forced the Mississippi to invade and erode the valleys of Cedar and Trout Creeks, where it has flowed ever since (Martin 1965).

Trempealeau Bluffs and Trempealeau Mountain are capped by Oneota Dolomite that is underlain by thick, orange layers of Jordan Sandstone. Because the sandstone cliffs face the setting sun, they can be brilliant when seen from the river or Highway 61 in late afternoon, especially on a brisk winter day when the air is clear and the sandstone's redness contrasts sharply with green cedars, blue ice, and white snow.

Trempealeau Mountain is no longer isolated by water. Trempealeau Bay, which separated it from the Trempealeau Bluffs, was transected in 1865 by the fill associated with the Chicago, Burlington, and Quincy Railroad that ran up the middle of the gorge.

In 1912 the Trempealeau River was diverted into Trempealeau Bay, then about twelve feet deep, as part of an ill-fated scheme to "reclaim" the adjacent bottomland for agriculture by diking and draining. Sediment carried by the river rapidly filled in the bay. Although water levels in the bay were raised about six feet with the completion of L&D 6 in 1936, the bay is now a wetland, filled with sediment that was deposited mainly in the 1912 to 1936 period.

Trempealeau Mountain has a huge scar on its west side, created when fill was excavated to repair railroad right-of-way washed away during the great 1965 Upper Mississippi Flood.

The Trempealeau formations are not entirely unique on the Upper Mississippi. There is a similar group of three large, detached rock bluffs in Minnesota at the northwestern end of Lake Pepin, but there the main river reentered its original valley, abandoning the lateral gorges (Martin 1965).

Queen's Bluff and Great River Bluffs State Park

Queen's Bluff lies about twelve miles downstream from Winona, Minnesota, within Great River Bluffs State Park. Honoring Zebulon

George Catlin's 1823 drawing of Queen's Bluff (Catlin 1841, plate 249).

Pike, George Catlin descriptively called this spectacular geological formation "Pike's Tent." Early maps of the area label it "Guinn's Bluff," which, when sufficiently mispronounced over the years, became "Queen's Bluff." The adjacent bluff just upstream was therefore named King's Bluff, and both are now often called "Twin Bluffs."

Queen's Bluff is a product of stream erosion. It is the end of a truncated spur rising between the Mississippi River and a deep valley that parallels it. The steep north-facing cliff near the crest of the bluff is composed of the same dolomite beds that cap all high river bluffs from Red Wing southward to the Minnesota–Iowa state line. These beds once extended in level, unbroken strata across the valleys and gullies, and from one side of the river to the other.

An exceptionally large stand of white cedar, long considered a relic from the climate of the last ice age, crowns the bluff. It exists far south of the white cedars that occur in many northern bogs and along Lake Superior. It forms a vertical forest in which most of the cedars are rooted on the cooler, moister north-facing slope, but which peek over the crest and overlook the goat prairie. The stand is not indicated on Catlin's drawing, perhaps because the tops of the cedars had been burned off repeatedly by fires sweeping up the goat prairie. The white cedars have grown luxuriantly in the past half-century because wild fires have been suppressed. Some of the cedars have even invaded the northern edge of the prairie. Remarkably, pasque flowers (prairie indicators) bloom in spring among the cedars at the crest of the Queen's Bluff. Several nearby north-facing bluffs and rock quarries now also have lush growths of white cedar, and they are rapidly spreading to other cool, moist, north-facing slopes.

The oldest white cedar snags on the cliff face were cored for tree-ring analysis by rappelling down from above. None was older than one hundred years. Hence the mystery—if the stand is not a glacial relic, where was the seed source? Present studies indicate that the source may have been white cedars brought into the area by pioneers in the 1850s (Carol Jefferson, personal communiqué, 2001).

To the delight of bird-watchers, Queen's Bluff had its first pair of nesting peregrine falcons in the spring of 2000. Peregrine falcons were successfully restored to the Upper Mississippi River as part of a raptor recovery program that was initiated after their near extermination by chlorinated hydrocarbon pesticides like DDT and polychlorinated biphenyls (PCBs).

Queen's Bluff lies within Great River Bluffs State Park, and the bluff is so fragile (and dangerous) that access to it is limited. In years past, climbers made deep ruts as they ascended the goat prairie, destroying fragile plant communities and causing extensive erosion. Once on top, they built fires, and cut fifty-year-old, stunted oaks for wiener sticks. It is now unlawful to climb the prairie and overland access is limited.

La Crosse: City on a Terrace

La Crosse, Wisconsin, occupies a Pleistocene terrace on the east bank of the Mississippi River. Along this section of the river, the valley walls are steep because Oneota Dolomite protects underlying sandstones from erosion. The river valley is wide here because erodable

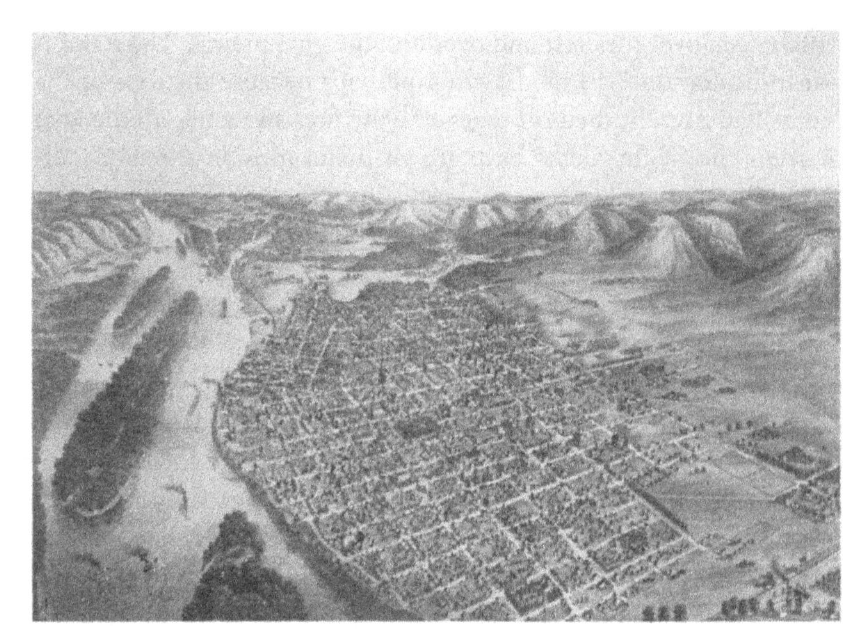

In this remarkable bird's-eye lithograph by C. J. Pauli, we peer down upon La Crosse, Wisconsin, a typical river city in 1876. Incorporated in 1856, La Crosse occupies a broad, flat terrace bordered on its river side by a steep west facing scarp rising about thirty feet above the flood plain. The terrace is bisected by the modest valley of the La Crosse River, separating the communities of North La Crosse and La Crosse. The larger Black River, whose headwaters extend northeastward into the rich Wisconsin pineries, enters the Mississippi just below the Chicago, Milwaukee and St. Paul Railroad bridge that cuts across the lower end of French Island. Two timber rafts are headed downstream, one pushed by a steamboat and the other steered by men using long sweeps. The artist has simplified the labyrinth of islands and sloughs that lie between La Crosse and Minnesota, but its present complexity is due mainly to flooding of bottomland by the 9-Foot Channel Project of the 1930s. In this view, the Mississippi's main channel sweeps around the lower end of French Island, joins with the Black and runs along the La Crosse waterfront. Perceptively, the artist has drawn the largest island (Barron Island), as a "crab claw island," characteristic of rivers overloaded with sediment. The head of La Plume Island is seen at the bottom of the drawing.

The La Crosse terrace had been virtually treeless as late as the early 1800s, its prairie vegetation maintained by climate, fire, and the grazing of elk and bison. A field of vegetated sand dunes, typical of many terraces, lies between the city and the base of the Wisconsin bluffs. Most islands are forested because the rivers protect them from fire. The city's young trees are mainly American elms, formally planted for shade and beautification. Strategically located at the confluence of the Mississippi, two Wisconsin rivers, and Minnesota's Root River, the La Crosse terrace had been a major site of Native American encampments for centuries (Wisconsin Historical Society, WHi-5716).

sandstones occur at river level, allowing the river to cut into the bluffs as it meandered laterally. Atop the bluffs, the Oneota Dolomite extends eastward to form a flat, partially dissected upland. Locally, valleys are called *coulees* (French for deep streambeds with steep sides).

The La Crosse terrace provided an expansive agricultural setting for Oneota Native Americans. The terrace is rich in artifacts and evidences of their culture. The city takes its name from the French term *(la crosse)* for the Indian ball-and stick game played on the prairie.

Effigy Mounds National Monument

Nearly two hundred prehistoric ceremonial and burial mounds are preserved in this 1,481-acre park located about six miles upstream from the confluence of the Mississippi and Wisconsin Rivers near MacGregor, Iowa. Most of the park is located atop the bluffs that overlook the river. On these bluff tops Indians arduously built most of the park's 191 ceremonial and burial mounds.

The earliest mounds were circular and date back to about 450 B.C. They gave way to cigar-shaped linear mounds between A.D. 300 and 700. Later mounds (effigies) took the form of animals, including birds (probably peregrine falcons) and bears. For unknown reasons, mound building ceased around A.D. 1300.

Pikes Peak State Park

The observation platform that juts out from the precipice at Pikes Peak State Park provides one of the most spectacular scenes along the entire Mississippi. Located on the Iowa side of the river, just downstream from Effigy Mounds National Monument, it is not separated from the river by a highway or city, and it overlooks the confluence of the Wisconsin and Mississippi Rivers. The Mississippi River's commercial channel travels along the base of the bluff immediately below the overlook, allowing spectators to look down upon passing towboats and their strings of barges.

Zebulon Pike recognized the park site as a strategic point, and an excellent location for a fort. The government agreed on the vicinity but selected the prairie terrace around Prairie du Chien for the fort.

From Pikes Peak, on the Iowa side of the Mississippi, George Catlin sketched the Prairie du Chien, Wisconsin, area in 1823. The settlement, fur-trading post, and Fort Crawford sprawl across the prairie in the center of the drawing. Note that the south- and west-facing bluffs are virtually devoid of trees, as are the largest islands (Catlin 1841, plate 253).

There are effigy mounds within the park, but they are not as extensive or large as the ones within Effigy National Monument.

Wyalusing State Park, Prairie du Chien, and the Wisconsin River

Atop a 530-foot bluff above the confluence of the Wisconsin and Mississippi Rivers, visitors to Wisconsin's Wyalusing State Park overlook the site where, in 1673, the French explorers Marquette and Joliet first entered the Mississippi River valley.

From the park's overlooks and its extraordinary campsites, visitors can view the two rivers and the broad prairie terrace that historic Prairie du Chien was built upon, beginning in 1673.

The city's name, French for "prairie of the dog," refers to the Native American chief Alim whose name meant "dog." The settlement flourished as a fur-trading center, with Native Americans bringing furs from great distances to barter with the French. From 1685 to 1831, forts were built and occupied successively by French, British, and American militia.

The Mineral District

The mines of the mineral district in southern Wisconsin, northern Illinois, and eastern Iowa predate the coming of the white man. Mesquakie Indians on the Iowa side of the Mississippi around Dubuque began mining lead in the 1500s. The Indians had no use for lead but used it for barter.

In 1690 Native Americans of the Potosi, Wisconsin, area showed French explorer Nicholas Perrot a cavern whose walls were rich with veins of lead. The Indians and French worked Perrot's "Snake Cave" for more than a century, but it was not until the early 1820s that the "lead rush" began. Later developed as the St. John mine, it produced 450,000 pounds of ore from 1827 to 1829. Lead, and later zinc, was mined in a small area bounded roughly by Potosi, Galena, Illinois, and Dubuque, Iowa. Potosi grew into a leading Mississippi River port that flourished until 1848 when a cholera scare and the California gold rush enticed many early residents to leave.

In 1788 Julien Dubuque, a French Canadian fur trader, settled in the area and made a treaty with the Fox Indians to work their mines. In 1796 he received a land grant from the governor of New Spain, who resided in New Orleans, to work the 189-square-mile area named the Mines of Spain. (Louisiana, all land drained by the western tributaries of the Mississippi, had been ceded by France to Spain in 1762). The present 1,380-acre Mines of Spain State Recreation area was designated as a National Historic Landmark in 1993.

Named for the lead-sulfide ore galena, the city of Galena, Illinois, was officially chartered in 1827 when it was more populous and busier than Chicago. Dubuque, Iowa's oldest city, was established in 1833.

This prosperity of the mineral district was the direct result of the lead and zinc deposits concentrated in a zone about fifteen miles long and ten miles wide. Until about the end of the Civil War, Galena and the surrounding area were the nation's lead-production center. In 1845, this mining district supplied 90 percent of America's needs. The transport of lead accounted for much of the Mississippi's steamboat trade from about 1820 until the early 1860s.

With the discovery of lead and silver in the western United States in the 1860s, the importance of lead and zinc mining in the mineral district diminished. The last of the large lead mines in Iowa closed in 1910 (Wiggers 1997; Anderson 1983).

7

After the Glaciers

Retreat of the Wisconsin Ice Sheet

About 18,000 years B.P., the Wisconsin ice sheet began to shrink. The retreat was swift. From its southern terminus at Des Moines, Iowa, it only took about two thousand years—a geological blink—to melt northward across the state. By 10,000 B.P., the "official" end of the Pleistocene epoch, glacial ice had disappeared from the Lake Superior basin, leaving the Great Lakes region ice-free. By about 6,500 B.P. the great Laurentide ice sheet had disappeared except for ice caps that still persist in Arctic islands (Pielou 1991).

The ice did not retreat in an orderly fashion. Indeed, the speed of ice disappearance varied from place to place, and from time to time. There were many temporary advances. "If a time-lapse moving picture could have been taken from a stationary satellite, the great ice sheets would have looked like active amoebas, with undulating outlines, wobbling unsteadily as they contracted to nothing" (Pielou 1991, 107).

The modern Upper Mississippi River came into being about 9,000 B.P.—a scant five thousand years or so before human "civilizations" emerged in the "Old World." Unlike the Middle and Lower Mississippi Rivers, the Upper Mississippi from St. Anthony Falls to St. Louis has not meandered since its supply of glacial meltwater ran out.

Most natural features of the river floodplain are now obscured by man-made features including cities, navigation dams, and their broad impoundments, piles of dredge spoil created by maintenance of the Nine-Foot Navigation Channel, flood levees, railroads, highways, bridges, and agricultural activities.

Postglacial Climate

The climate of immediately postglacial Midwestern America has no modern analog (Pielou 1991). The present interglacial period, called the *Holocene* or *Recent,* was triggered by an increase of only about 10°F or 11°F in the earth's mean annual temperature. However, warming was not a gradual, continuous process.

The warmest time interval in our interglacial, called the *Hypsithermal* (high heat*) Interval,* was warmer than now. It began about 10,000 B.P., lasted until about 6,000 B.P., and was followed by cooler temperatures that favored several episodes of advance and retreat of mountain glaciers (Pielou 1991).

Cold returned about A.D. 1350, causing the *Little Ice Age* that lasted until about A.D. 1870. It caused the temporary expansions of glaciers and ice caps, and southward shifts of vegetative zones—and it severely impacted Native Americans and invading Caucasians.

Postglacial Vegetation and Wildlife

As with previous glaciations, the Wisconsin glacier had destroyed everything in its path, and the cooling climate ahead of the advancing ice caused the southward migration of vegetative zones and the animals that inhabited them (LaBerge 1994).

As glaciers melted, plants and animals moved northward again, recolonizing the areas devastated by glaciers. Postglacial warming and drying began first in the westernmost parts of the Great Plains and progressed eastward in the wake of receding ice (Pielou 1991).

The land south of the ice sheet was probably not an unbroken band of tundra, as once was surmised, but a complex mosaic of tundra and coniferous forest. This mixture of environments provided habitat for many mammals—both in numbers and species (Wiggers 1997).

The fossil remains of giant beavers, mammoths, and mastodons are often found in glacial till throughout the Upper Mississippi River Basin. Species like caribou, musk oxen, and lemmings that are now limited to tundra areas were probably here too. Many of today's mammals and other species were already here at the time of the last glaciation, but probably moved south ahead of the ice to more moderate climates and returned as the ice receded (Tester 1995).

As glacial ice retreated northward, climatic zones and vegetation also shifted to the north. Deciduous forests replaced Iowa's coniferous forests, and they, in turn, gave way to prairie grasslands. Fossil wood found in Iowa that dates from 24,000 to 10,500 B.P. is dominated by conifers including larch, hemlock, fir, and yew, but especially by spruce. Beginning about 10,500 B.P., deciduous forests dominated, but prairie began replacing them by 9,000 B.P. Iowa was dominated by prairie during the interval from 7,700 to 3,200 B.P. From 3,200 B.P. until the state was settled by whites it was a mixed prairie environment with deciduous forests in favorable areas, especially in valleys and on steep north facing slopes (Anderson 1983).

As the ice retreated through southern Minnesota, plant succession began on the barren outwash plains, moraines, and sand dunes. Gale-force winds blew along the ice front, carrying seeds of hardy pioneer plants. Aspen, willow, juniper, and shrubs appeared first, but they were succeeded mainly by forests unlike any modern forests. White spruce was the dominant species, but aspen, birch, ash, elm, and oak were also common (Haapoja 1997).

Red (Norway) pine and jack pine subsequently invaded the area. White pine came later, having survived the most recent glacial onslaught by retreating to glacier-free refugia in the Appalachian Mountains and on continental shelves that had been exposed by falling ocean levels during glacial maxima. Evidence from fossil pollen shows that white pine reentered Minnesota, the western edge of its range, about seven thousand years ago (Haapoja 1997).

An Incomprehensible Interval of Time

We have covered an incomprehensible span of geologic time in these chapters. Perhaps the magnitude of geologic time can be better appreciated if we compare the earth's geologic time scale (4.6 billion years) to a calendar year of 365.25 days.

Precambrian time would take up most of the year (January 1 to late November). The Upper Mississippi's oldest and most abundant sedimentary rocks (Paleozoic) would have accumulated in the three-week period from late November until mid-December. The river's younger strata (Mesozoic) would have formed from mid-December to the day after Christmas.

The Pleistocene Ice Age would have begun about 9:00 P.M. on New Year's Eve and ended sixty-nine seconds before midnight. The entire postglacial period (Recent Epoch), which includes the invasion of North America by humans and the advent of human civilizations, would take place in a little over one minute of revelry prior to the singing of "Auld Lang Syne."

8

The Mississippi's Prehistoric Native Americans

Hunters, Gatherers, Farmers, and City Planners

The First Americans

In just three thousand years, between 12,000 B.P. and 9,000 B.P., about forty species of large North American mammals became extinct. Never before in geologic history had so many diverse species of mammals disappeared in such a short period of time.

The list includes the dire wolf, mammoth, mastodon, giant bison, five species of horse, western camel, giant beaver, ground sloth, stag-moose, giant short-faced bear, and others. The cause of the extinctions is still debated. They may have been caused by habitat changes, some unknown catastrophe, or overkill by spear-wielding hunters who had evolved their skills in Asia and invaded North America about fourteen thousand years ago when Ice Age glaciers were just beginning to melt (Pielou 1991).

During the most recent glacial period when sea levels were low, a land bridge connected Alaska and Siberia along the Aleutian Island chain for sixty thousand to seventy thousand years. It created an avenue for the exchange of many species of large mammals, as well as humans, between North America and Asia. Traditionally, it has been thought that the first Americans crossed the land bridge and made their way southward between retreating sheets of ice or along an ice-free corridor along the Pacific coast. Rising ocean levels about 15,500 B.P. rejoined the Pacific and Arctic Oceans, creating the present Bering Strait (Pielou 1991).

Recent archeological discoveries indicate that the first Americans may have arrived from Asia in fits and starts even earlier than 14,000 B.P. Moreover, they may have come by sea rather than by land, paddling boats along the Ice Age coast to Peru and beyond. The newcomers may have colonized South and Central America first, then expanded north on the heels of retreating glaciers. Unfortunately, most evidence of their maritime culture would have been submerged when sea levels rose at least three hundred feet at the end of the Ice Age. Also, they had low archeological visibility because of their reliance on wood and skins (Wright 1999; Parfit 2000).

Evidence from the Monte Verde site in southern Chile seems to prove that humans inhabited that area about 14,400 to 13,600 B.P.—after a presumed long period of southward migration (Gore 1997; Fiedel 1999). Some anthropologists suggest that the earliest groups may have been in the New World more than thirty thousand years ago (Johnson 1988).

It is certain, however, that after 14,000 B.P. Paleoindians rapidly expanded their range throughout North America. Because they were nomadic hunter-gatherers and made no permanent camps, they left very few artifacts as evidence of their presence.

The earliest human inhabitants of the Upper Mississippi basin probably hunted large, extinct mammals like mammoths and mastodons as well as other animals. Paleoindian inhabitants (*Clovis* people) have been documented toward the end of Wisconsin glaciation, about 10,000 B.P., in direct association with mastodon remains at Kimmswick, about one hundred miles south of St. Louis (Graham et al. 1981). The evidence for Clovis presence is widely scattered distinctive lanceolate spear points that have one or more longitudinal flakes or *flutes* removed from the points. The bow and arrow had not yet been invented, but the early hunters used *atlatls* to increase the throwing distance and speed of their spears (Dobbs and Mooers 1991).

By 9,000 B.P., the postglacial, warm, dry period began. It lasted for several thousand years and peaked about 7,000 B.P., causing the expansion of prairies and prairie animals, including bison, well east of modern ranges. As the warm period waned, conditions gradually changed until they were similar to the present. By 4,000 B.P., the Upper Mississippi River had become a resource-rich corridor. During the Archaic period that lasted until about 1000 B.C., *exchange networks* developed whereby people traded goods over great distances, usually by water.

Organized, Riverine Transportation Routes

The period after 1000 B.C., until about A.D. 700, is known as the Woodland period. The climax of development of horticulture and the building of earthworks was between 500 B.C. and A.D. 500 and is known as the Middle Woodland or, more commonly, the Hopewell period (Young and Fowler 2000).

During the Hopewell period, the Mississippi and its tributaries became organized transportation routes, facilitating the trading of "status goods" like copper from Lake Superior's Isle Royale, lead ore (galena) from Illinois and Iowa, catlinite (pipestone) from southwestern Minnesota, quartzite from Wisconsin, obsidian from the Yellowstone, sheet mica from North Carolina, grizzly bear teeth from the western prairies and foothills, alligator teeth from the Mississippi Delta, and welk and conch shells from the Gulf of Mexico. There were extensive trade networks in place on the Mississippi River long before the European-American invasion. The rivers were also avenues for the diffusion of cultural influences long distances from their points of origin. The great trade network reached its peak around A.D. 300, but broke down after A.D 500, not to reappear until resurrected by the leaders of the great cultural center of Cahokia almost five hundred years later (Young and Fowler 2000).

By about A.D. 1050, a sweeping series of economic, technological, and social changes had transformed nomadic aboriginal societies throughout much of eastern North America from hunting and gathering groups to populations relying on corn horticulture as well as hunting and gathering. Referred to as the Late Woodland period, it marks the first appearance of the bow and arrow in native North

American cultures. Other technical advances included the use of ground clamshells for tempering pottery, adoption of intensive horticulture based on corn, squash, beans, sunflowers, tobacco, and other crops. Hoes and other specialized tools, new methods of food preparation and storage, and a settlement pattern based on semisedentary villages were developed and adopted (Dobbs and Mooers 1991).

The glacial outwash terraces that flank the Mississippi River floodplain were ideal sites for habitation. They provided easy access to the uplands as well as the river and its resources, excellent views of the areas for defense and observation, and dry, level building sites that tended to be breezy and mosquito-free (Dobbs and Mooers 1991). Terraces were extensively used, for example, in the vicinity of Red Wing and Lake Pepin in Minnesota and La Crosse and Prairie du Chien in Wisconsin.

Lake Pepin served as a regional transportation hub. The Red Wing locality was strategically convenient to several distinctive natural environments that provided a wide variety of food and resources, while the Upper Mississippi and its tributaries provided natural transportation corridors into them.

The Cannon River provided rapid access to the bison hunting grounds on the prairies to the west, while the Mississippi Headwaters were a route to the extensive wild rice marshes of north and central Minnesota. The native copper deposits of the western Lake Superior region were available via the St. Croix River, and the Chippewa River provided access into the pineries of western Wisconsin. Especially important, the Mississippi itself was a conduit for the cultural influences that emanated from the Mississippi site of Cahokia, five hundred miles to the south (Dobbs and Mooers 1991).

Cahokia

A thousand years ago Native Americans built a great city that flourished for three hundred years on the Illinois side of the Mississippi River within sight of the soaring Gateway Arch at St. Louis, Missouri. Known today as Cahokia, it was the center of the most sophisticated prehistoric Indian civilization north of the Rio Grande, and it acted as an intense cultural reactor that profoundly touched and influenced aboriginal groups throughout the Mississippi Basin (Young and Fowler 2000).

Cahokia lies in the American Bottom, a ten-mile-wide portion of the Mississippi floodplain extending eighty miles from the mouth of the Illinois River on the north to the mouth of the Kaskaskia River on the south. Like the valley of the Nile, the American bottom was a gift of a river and its fertilizing floodwaters. It provided an almost inexhaustible supply of plant and animal resources for the Cahokia population.

It was no accident that Cahokia, the largest city in prehistoric North America, and modern-day St. Louis developed on the same site. St. Louis had its beginnings as much on the east side of the Mississippi as on the west side where downtown St. Louis now stands. Strategically located near the confluence of the Mississippi, Missouri, and Illinois Rivers, both cities depended on water transportation. They were gateways to the vast, rich interior of North America—almost equidistant between the Great Lakes and the Gulf of Mexico.

Prehistoric Indians of the Late Woodland culture first inhabited Cahokia about A.D 700. Between A.D 800 and 900 the Mississippian culture emerged, developing an extensive agricultural system with corn, squash, and several other seed-bearing plants as principal crops. This stable food base, supplemented by hunting, fishing, and gathering wild food plants, enabled Cahokians to develop highly specialized social, political, and religious organizations. Cahokia probably reached its highest point of development about A.D. 1100–1200. By A.D. 1150 it was a sprawling, planned city of residential areas, public plazas, sacred areas, and solar observatories (Young and Fowler 2000).

Cahokia occupied about six square miles. The main agricultural fields lay outside the city. Cahokians moved an estimated fifty million cubic feet of earth to construct huge mounds, mainly as platforms for ceremonial buildings and residences for the elite.

Large dugout wooden canoes made it possible to traverse the whole length of the Mississippi Valley and to transport goods in quantity. Streams of people from the hinterlands brought items to barter in the great mound center at St. Louis (Chapman and Chapman 1983). Most of the goods imported into Cahokia were raw

Corn: The Wonder Plant Developed by Indian Horticulturists

No other plant has had an impact on the Upper Mississippi and its basin like that of corn. Corn (maize) is native to North America and was first domesticated in central Mexico. Paleoindians and American Indians cultivated and improved it for over five thousand years.

The original cornlike plant was a wild grass with kernels that could fall to the ground separately, enabling the plant to reseed itself without human intervention. It is unknown what the Indians did to turn a self-seeding grass, with small self-contained grains, into a plant with huge podless kernels growing in tight rows on cobs covered with a single sheath. Christopher Columbus was hospitably offered corn and tobacco when he landed on the island now called Cuba in 1492, and on a single day the two fateful crops were introduced to Europe and the rest of the world (Visser 1988).

In contrast to the usual broadcasting of seed that was standard procedure in Europe, American Indians carefully planted the precious kernels of corn individually in mounds, equally spaced, in well-tended gardens. Many Indians discovered that corn was much more productive with massive doses of fertilizer, in the form of fish buried in the mounds, or hills.

Since there were no draft animals in North America, all land clearing and mound building were done by hand. Common practice was to plant corn, beans, and squash in the same hill. The beans and squash had also come from Mexico. The corn grew straight and tall, the beans climbed the corn, and the squash plants trailed down the mounds covering the flat land between the mounds, suppressing weeds. Using slash-and-burn techniques, this trinity of crops is still farmed in the same way in many areas of Central America, South America, and throughout the islands of the Caribbean. With intensive cultivation of corn and beans, American Indians were able to attain combined yields of thirty to forty bushels per acre (Chapman and Chapman 1983).

Because corn originated in the subtropics where there were about equal amounts of daylight and dark, it took centuries for it to adapt to the longer hours of daylight in north temperate areas. It was not until about A.D. 750–800 that it became an important food-producing plant in the Upper Mississippi Valley. During the Mississippian period, Indian farmers developed new strains that could survive where there were only 120 frost-free days. By the time of European contact, corn had materials—clay, hematite, copper, mica, shells, chert, diabase, and salt—used in Cahokia's specialized workshops to manufacture statuettes, pigments, pottery, tools, projectile points, beads, and ritual paraphernalia (Young and Fowler 2000).

Its size and influence put Cahokia in the upper ranks of the world's preindustrial communities of the time. From A.D. 800 to 1300, Cahokia was the central community of the American heartland. At its peak, the city proper had a population of about twenty thousand, and thousands more lived in satellite communities within the American Bottom under the influence of the central community. Cahokia and its satellite communities controlled the traders and warriors who navigated the region's three great river systems (Young and Fowler 2000).

By A.D. 1200, even though the people were living well, the urban behemoth was beginning to fail. The decline of Cahokia was not a single catastrophic event; rather it was probably due to a general degradation of the environment. Wood was not plentiful in the American Bottom, which was mainly a mixture of prairies, savannas, lakes, creeks, and sloughs. But an urban civilization demanded large numbers of trees for fuel, houses, public buildings, palisades, furniture, and watercraft. It has been estimated that deforestation occurred in a two-hundred-mile radius around Cahokia.

---

been further modified to accommodate the climate, temperature, and day length as far north as Minneapolis. Because it was high in caloric value and could be stored for the winter, it supported public works such as mound building, clearing of fields, and the building of fortifications—but also the support of religious or social elites (Young and Fowler 2000).

By selective breeding, Indians developed pure corn varieties that suited their climate and growing conditions. Throughout the New World there were six main colors of corn—red, blue, black, yellow, white, and variegated. Having lost their wildness, and having their seeds tightly enclosed in a strong, enveloping husk that prevented seed dispersal, all of the varieties depended upon humans for their survival. If humans ceased to unwrap the husk, strip kernels from the cob and plant them, corn would cease to exist (Visser 1988).

When supplies of local trees were exhausted, oak and hickory were brought down from uplands, and red cedar and bald cypress came from as far south as the mouth of the Ohio River. Clear-cutting the uplands probably led to erosion, rapid runoff, and the flooding of cornfields and residential areas.

Political failures and rivalries, exacerbated by ecological problems, probably contributed to Cahokia's demise. The crisis reached a climax between A.D. 1200 and 1275. By A.D. 1350, Cahokia was virtually abandoned, leaving the American Bottom largely empty of human habitation until the arrival of Europeans in the seventeenth century (Young and Fowler 2000). Interestingly, the latter date marks the beginning of the Little Ice Age with widespread ecosystem crashes in Eastern North America and the end of Norse habitation of Greenland.

By the time intensive archaeological studies were begun at Cahokia in the 1960s, "The area around Cahokia was a flat, biologically monotonous, agro-urban environment littered with the detritus of American economic life. Gone was the network of wetlands, swamps, and sloughs that had once supported that great diversity of plant and animal life. Drive-in-theaters, carpet sales outlets, trailer parks, asphalt parking lots, weed-choked empty lots, and oil storage tanks lined the highways and interstates. In the midst of this wasteland, the mounds rose suddenly from the flat land like islands of calm, their mysterious presence a radical contrast to the urban frenzy surrounding them" (Young and Fowler 2000, 113).

Red Wing

At about the same time Cahokia reached its peak, Minnesota's first native farmers, the Oneota people of Red Wing, established their village complexes on high terraces. Located just upstream from Red Wing, Minnesota, this exceptionally dense concentration of Oneota habitation sites was noted in 1823 during the explorations of Stephen Long.

From about A.D. 1050 to 1300 the Oneota occupied some fifty-eight square miles where the Cannon River meets the Mississippi. They hunted deer, elk, and bison, and relied heavily on the Mississippi River for fish, turtles, clams, and plant foods. The Red

Wing area was environmentally rich, with the Cannon River and a number of smaller steams emptying into the Mississippi, replenishing the soil with nutrients, and creating habitat for wild rice and waterfowl.

Like Cahokia, the Oneota villages were located near a major delta of the Mississippi River, created by sedimentation of the upper end of Lake Pepin. The rich soils were ideal for growing large amounts of corn. Directly or indirectly, Cahokian technology made its way to Red Wing, La Crosse, and Trempealeau. Cahokians may even have journeyed there as colonists (Young and Fowler 2000).

When the Oneota culture was at its peak, the head of Lake Pepin would have been five to seven miles farther upstream than it is today, providing inhabitants access to both lake and delta. The pottery at the Oneota sites had distinct Cahokia-like design motifs. Nonlocal artifacts included galena (lead ore) and a Cahokia tri-notched arrowhead. The area contained more than two thousand mounds and earthworks, and some mounds originally had the flat-topped pyramidal shape typical of Cahokia (Dobbs and Mooers 1991).

From about A.D. 1250 to 1350 the Oneota culture shifted downriver

Effigy Mounds

By A.D. 600, the Late Woodland, or Effigy Mounds, culture had emerged throughout the Unglaciated Area of the Upper Mississippi Valley. Prehistoric peoples built mounds at many times and places throughout the Americas, but only in northeast Iowa, southern Wisconsin, northwestern Illinois, and southeastern Minnesota was there a culture that specialized in ceremonial mounds called effigies. They left more than ten thousand effigy mounds scattered over the landscape (Young and Fowler 2000).

Usually built on terraces and atop majestic bluffs, the earthen effigies represented a variety of wildlife including birds, turtles, snakes, and bison, but especially bears. Mounds were usually constructed in an open, prairie setting, but because of over a century of fire suppression, they are now mainly seen within mature hardwood forests.

Mound building ended along the Upper Mississippi River around A.D. 1300. No one knows if the Effigy Mounds people departed the region, or if they stayed and underwent cultural change. Effigy mounds are best seen in Effigy Mounds National Monument near McGregor, Iowa.

to new settlements at La Crosse. Their agricultural communities sprawled across the La Crosse landscape for about two centuries.

The most economically important animal by all pre-European groups in the La Crosse, Wisconsin, area was the white-tailed deer harvested near villages within the valley and its tributaries. Elk and bison were also utilized but were mainly hunted on the prairies, deboned, and transported back to the settlements.

Other consumed animals included muskrats, beaver, domestic dogs, ducks, geese, and turtles. Turkeys were harvested southward from Prairie du Chien, and their stout primary wing feathers, usually still attached to the entire dried wing, were traded to northern groups as fletching for arrows. Captured fish included freshwater drum, catfishes (some over fifty pounds), black bullheads, northern pike, suckers, gar, and bowfin. Many of these species were taken during spawning or by seining or trapping in shallow backwaters (Theler 1999). Turtle eggs were harvested in large numbers (Kane et al. 1978).

Freshwater mussels (clams) were an important seasonal food source, and their shells were sometimes made into tools (Arzigian et al. 1994) or crushed to make a tempering agent for pottery. One shell midden near Prairie du Chien contained more than a million shells, the result of many seasons of use (Theler 1987).

The Oneota people abandoned the area just prior to French exploration and moved westward into Iowa and southern Minnesota.

Freshwater Mussels: Edible if You're Hungry Enough

Freshwater mussels (clams) are edible but rubbery, and their taste is often reminiscent of the smell of the sediment they were collected from. Mississippi mussels probably tasted better in the days before civilized folks defiled the river with their sewage. They were a staple, seasonal food of American Indians. Some of Zebulon Pike's crew subsisted on mussels when they were separated from the expedition for several days. Small mussels are quite tasty if collected from clean northern lakes, placed in fresh water for a few days to cleanse themselves of grit, gutted, and then fried or made into chowder. However, eaters are cautioned that mussels are great concentrators of environmental contaminants like PCBs and mercury.

The French first made contact with them during the late 1600s when they were recorded as the Ioway Tribe. They may have abandoned their ancestral lands for a variety of reasons: disease, population pressures from the east, the lure of bison hunting on the prairies to the west, or because of agricultural failure resulting from the Little Ice Age. Agriculturists were probably more sensitive to minor climatic changes than were hunters.

During the past one thousand years the climate has changed several times, alternating from warm/moist to warm/dry, to warm/moist followed by the much cooler/moist conditions of the Little Ice Age, which lasted from about A.D. 1350 to 1870.

2

Exploration and Early Exploitation

9

The Quest for Furs and a Northwest Passage

Changing Ownership of the Mississippi River Basin, 1673–1803

The Spaniards established the existence of the Mississippi River, approaching it from the south and west in a quest for gold. The French approached the Mississippi Basin from the north and east, explored its length, and developed the fur trade. But it was the English and their colonists who came across the Allegheny Mountains, settled it, and incorporated it into the rest of American society.

No Gold, Just Furs

Arriving rich and arrogant from bloody conquests of Mexico and Peru, the Spanish were the first white men to see the Mississippi River.

Hernando de Soto, with a force of 620 conquistadors, blundered through the wilderness in search of cities of gold, killing and enslaving the Indians as they went. On May 8, 1541, after a seven-day march from his winter camp in northern Mississippi, de Soto and his men stood atop the Chicasaw Bluffs below present-day Memphis, Tennessee, to view the Lower Mississippi. It is likely that the river came as no surprise to him, as it had appeared on Spanish maps in 1513, probably based upon intelligence gained from the Indians. A Portuguese member of the expedition penned what is perhaps the earliest recorded description of the Mississippi: "The river was almost halfe a league broad. If a man stood still on the other side, it could not be discerned whether he were a man or no. The River was of great depth, and of strong current; the water was alwaies muddie; there came downe the River continually many trees and timber" (Petersen 1968, 29).

De Soto's men built barges, crossed the Mississippi, and pushed westward burdened with armor and beset with fever, hunger, heat, biting insects, and hostile Indians.

They made winter camp about two hundred miles west of the Mississippi, having lost 250 men and 150 of their 223 horses since they left their ships. De Soto died on May 21, 1541, and two months later the gaunt, half-naked remnant of his army reached the New Spain settlement of Tampico, about halfway down the coast of the Gulf of Mexico. The defeat of the Spaniards had been so crushing that few explorers dared venture beyond the coast, and for 132 years the wilderness was relatively quiet—peopled only by Indians (Childs 1982; Petersen 1968).

By the seventeenth century, Europe's three "superpowers"—England, France, and Spain—were competing to establish colonies and control the New World. It was apparent that North America lacked treasures like those that Spanish conquistadors had plundered from the Aztecs and Incas in Central and South America, but intrepid Frenchmen learned that there were riches to be had, not in metals but in animal skins that could be exported to Europe to make luxurious clothing.

The fur trade changed the map of North America, and it irreversibly left its imprint on the Mississippi River. Profits from the North American fur trade financed historic, world-shaping wars involving Great Britain, France, and Spain. It played a major role in the development of the United States and Canada for over three hundred years.

Each country hoped to discover a river that flowed into the Pacific Ocean so it could have a monopoly on a short, lucrative trade route to the Orient and its treasures of silk, spices, and jewels. In addition, they needed forts to protect their colonies and trading posts where they could barter with the Indians. They all needed inflows of cash, mainly to finance wars with each other. Their explorers craved adventure, glory, and riches. French Roman Catholic missionaries were inspired to convert Native American heathens to Christianity.

The St. Lawrence River: Gateway to the Interior of North America and the Mississippi Basin

Unknown to most Americans, the story of the exploration and exploitation of the Mississippi River, its valley, its resources, and its peoples begins not at the Mississippi's mouth in the Gulf of Mexico, but on the coast of the North Atlantic Ocean around the mouth of the St. Lawrence River. There, in the early 1500s, the French began trading manufactured goods to Indians in exchange for furs. Fur trading quickly became profitable, and posts were established at Quebec (City) in 1608 and at Montreal in 1642.

The St. Lawrence River led to the Great Lakes and to the rivers that drained into them. Soon it became the main portal through which European manufactured goods were distributed to the interior of North America. The French paddled up the tributaries of the St. Lawrence and the Great Lakes, portaged over low continental divides, and entered the drainages of the Mississippi River (which flows southward into the Gulf of Mexico) and the Red River of the North (which drains northward, ultimately into Hudson Bay via the Nelson River).

Penetrating ever more remote areas, the traders bartered their stocks of manufactured goods for the dried pelts of beaver, otter, mink, fox, marten, fisher, lynx, and other animals captured by the Indians. The furs were transported back by water to Montreal and Quebec for shipment to Europe.

The Indians were well acquainted with the Mississippi and its tributaries, having lived on them for countless generations. They depended on the rivers for life itself and traveled them in canoes of bark or skins, or in pirogues hollowed out of cottonwood logs pointed at both ends. Paddled or poled, their craft could handle heavy loads with a small draft, yet remain maneuverable. Frenchmen were quick to adapt these boats to their own exploration and fur trading.

Down to the Mississippi River

The French were the first Europeans to penetrate the Upper Mississippi when the fur trader Louis Joliet and his party, which included Father James Marquette, began an epic journey. With high hopes that their expedition would yield a new route to the Orient, the explorers cast off from St. Ignace, at the junction of Lake Michigan and Lake Huron, on May 17, 1673. They hugged the north shoreline of perilous Lake Michigan with heavily laden canoes, entered Green Bay, and paddled upstream on the Fox River until they were able to portage about a mile and half over a low, marshy continental divide into the Wisconsin River that unites with the Mississippi just south of Prairie du Chien, Wisconsin. (The city of Portage, Wisconsin, is named for the historic portage between the Mississippi and Great Lakes watersheds.)

After they had paddled the length of the Wisconsin River, Father Jacques Marquette wrote, "We safely entered Missisipi on 17th of June, with a joy I cannot express" (Ogg 1904, 71).

Supposedly, two Miami Indian guides who spoke an Algonquin dialect gave the name "Misisipi" to Joliet and Marquette. The explorers kept that name for the river as they went southward. Meaning "big river," it has been called thus ever since, with extra *s*'s and *p*'s added for emphasis (Madson 1985).

Floating southward to the mouth of the Arkansas River, Joliet concluded that the Mississippi flowed into the Gulf of Mexico and not the Pacific Ocean. Disappointed that the Mississippi was not a direct link to the Orient, the party returned by going up the Illinois River, through the low Continental Divide, and down the Chicago River to Lake Michigan, noting that a canal could be constructed to connect the two waterways. Although they had not found a short cut to the Orient, the exploration of Joliet and Marquette helped establish France's claim to the interior of the continent.

Centuries before Joliet and Marquette "discovered" the Upper Mississippi, Indians from the prehistoric city of Cahokia had traded goods and may have sent settlers as far north as Wisconsin along this waterway. In 1673 Indian villages dotted the shores of the Mississippi

River from northern Minnesota to the Gulf of Mexico. Collectively, Indians knew the course of the Mississippi from its source to its mouth. They also knew routes that connected continental drainages. The Fox-Wisconsin waterway, for example, that links Lake Michigan with the Mississippi River had been used for thousands of years by Native Americans who canoed the corridor's length, fished its waters, and lived along its shores. Recognizing this wealth of knowledge and experience, all of the explorers depended on Native Americans for information and as guides.

The French presence on the Mississippi River was firmed in 1678 when René Robert Cavelier, sieur de La Salle, was authorized by the French government to explore the Mississippi and establish forts where necessary. Based in Montreal, La Salle was a prosperous fur trader who had traveled extensively throughout the Great Lakes area.

In 1680, La Salle sent Michel Accault and Antoine Auguelle, with Father Louis Hennepin, a Franciscan priest, on a trip to the Mississippi Headwaters, which they reached by canoeing and portaging from Lake Superior. After being held captive for two months by Sioux Indians at Lake Mille Lacs in east-central Minnesota, Auguelle and Hennepin canoed down the Mississippi and viewed the Falls of St. Anthony, the only significant waterfalls on the entire Mississippi River (Kane 1987).

La Salle and his party were the first Europeans to travel the length of the Mississippi from the mouth of the Illinois River to the Gulf of Mexico. Near the Mississippi's mouth on April 8, 1682, he claimed the entire Mississippi drainage area for France and named it *Louisiana* in honor of King Louis XIV. His claim was more comprehensive than even he could have imagined; he did not know of the immensity of the Missouri River drainage or of the existence of the Rocky Mountains. The drainage of the Mississippi included the entire wilderness from the Appalachian Mountains on the east to the Rocky Mountains on the west, and from the Great Lakes on the north to the Gulf of Mexico on the south. He also claimed all of Louisiana's seas, harbors, ports, bays, adjacent straits, all nations, people, provinces, cities, towns, villages, mines, minerals, fisheries, streams, and rivers (Childs 1982). He returned to Montreal via the Illinois River–Great Lakes route, establishing Fort St. Louis on the Illinois River on his way.

By today's standards La Salle's odyssey was incredible, having been made without outboard motors, motels, or restaurants. But he wasn't through. With the dream of establishing a colony at the mouth of the Mississippi River he sailed from Montreal to France in 1684, returning with four ships and more than three hundred colonists in 1685. The expedition sailed past the mouth of the Mississippi due to a navigation error. (Longitude measurements were imprecise at that time, due mainly to inaccurate methods of telling time.) In desperation, La Salle established a colony called Fort St. Louis east of the present site of Corpus Christi, Texas, but disaster haunted the enterprise. Indians threatened the colony while colonists sickened and died. In an attempt to return to Montreal to seek aid, La Salle and a small contingent began an overland march to the Mississippi, planning to return to Montreal, but couldn't find the river. Frustrated, some of his men rebelled and killed him.

French Domination of the Wilderness

The French were more suited to life in the wilderness than the Spanish who had come as conquistadors demanding gold from primitive peoples. The French respected the Indians' knowledge of their environment, their ability to utilize its bounty, and their fortitude. Frenchmen lived like Indians, quickly adopted Indian ways of travel, and took Indian wives. Voyageurs, priests, trappers, and traders made their way along Indian pathways into the interior of the continent.

By the 1730s the French had established military and trading posts on the Mississippi northward from the Ohio River as well as on major tributaries. The French military presence was not meant to subdue the Indians or pave the way for future settlement. The French were interested in furs, and the forts were there to keep trade routes open and to keep the British out (Phillips 1961).

In the first part of the eighteenth century, vague or nonexistent boundaries encouraged eastern Louisiana to become a field of competition between rival groups of traders. The northern boundary was indefinite but generally thought to reach to New France. On the west it extended to the Mississippi River and the unknown region comprising the territory of New Spain. Its eastern border was the unknown line marking the limit of the English colonies.

By the mid-eighteenth century, French trading posts were well established throughout the midcontinent, supporting French territorial claims and demonstrating that the Mississippi was navigable along its entire course. The names of cities along the Upper Mississippi are testament to the far-reaching influence of the French: Cape Girardeau, St. Louis, Prairie du Chien, La Crescent, La Crosse, Trempealeau, and Lamoille, to name a few.

France Loses Its North American Empire

The long-simmering rivalry between the French and British in the territory between the Allegheny Mountains and the Mississippi River boiled over into open warfare in 1754. The war spread from the Americas to Europe, where it is known as the Seven Years War. Known to Americans as the French and Indian War, this conflict would shape the future of the North American continent.

Defeated by the British, and with its treasury depleted, France ceded all of Louisiana west of the Mississippi to Spain in 1762. In 1763 the Treaty of Paris brought France's imperial ambitions in North America to an end.

The Treaty of Paris put nearly all of North America's known fur country in the hands of the British Empire. The Union Jack now flew over most of Canada and the territory south of the Great Lakes, and between the Appalachian Mountains and the Mississippi River. France ceded to Great Britain all of Louisiana east of the Mississippi River, except for a small Spanish-owned area south of Lake Maurepas and Lake Ponchartrain that included New Orleans. Then called the "Isle of Orleans," it was bounded by lake, river, and gulf (Carter 1942). The vast fur country east of the Mississippi River that for nearly a century had been a battleground of French, Spanish, and British traders was at last in British hands, and they had free navigation of the Mississippi from its source to the sea.

Even as the Mississippi Valley changed hands, French fur traders continued to dominate mid-America. In 1764 Frenchmen simply crossed the Mississippi into Spanish territory to found the city of St. Louis.

St. Louis was founded by Pierre Laclede, a partner in a French fur-trading company, on the first elevated spot south of the junctions of the Illinois and Missouri Rivers with the Mississippi. Protected from floods, it provided access to all northern tributaries from which furs came. The fact that this was also the point where the channel of the Upper Mississippi deepened resulted in the city becoming the terminus and cargo transfer point for small boats operating on the Upper River and larger boats operating on the Lower River (Tweet 1983).

The British triumph after a century and a half of conflict left France with no fur lands of her own, but the fur country comprised vast regions occupied by Indians and lawless fur traders who had long relied upon the French for their supplies and markets, and who held little affection for the British.

Acquisition of the Northwest Territory by the United States

The British North American monopoly would prove short lived. Only thirteen years after the Treaty of Paris was signed, signatures were affixed to another document that would irrevocably reshape the continent.

In 1776, the British North American colonies declared their independence. In 1783, subsequent to her defeat in the War of Independence, Britain ceded its empire south of the Great Lakes and east of the Mississippi to the infant United States. Replacing the Union Jack with the Stars and Stripes was of little consequence to established British commercial interests in the region. British traders continued to dominate the collection of furs in the area, shipping them to London via the St. Lawrence or down the Mississippi and through Spanish New Orleans via Lake Maurepas and Lake Pontchartrain to the Gulf of Mexico (Carter 1942). The immense Spanish territory west of the Mississippi held vast wealth in furs, but Spain had neither the power nor the interest to exploit its economic potential (Phillips 1961).

The Louisiana Purchase

Warily eying the westward expansion of the United States as the eighteenth century drew to a close, Spain deemed its Louisiana holding most useful as a buffer between the expansionist Americans and her empire in Texas, California, and the Far West.

In 1800 the Spanish secretly ceded the territories they had nominally held since 1762 to Napoleon Bonaparte in hopes that a renewed French presence would create the barrier they sought. With his attentions focused on imposing French hegemony on Europe, Napoleon was fairly indifferent to his North American holdings. In March 1803, the French made their three-year-old agreement with Spain public. A month later, seizing an opportunity to bolster the French treasury, Napoleon sold Louisiana to the United States.

With the Louisiana Purchase, Americans now controlled all of the land drained by the Mississippi River and its tributaries—except for a very small portion of what is now southern Alberta and Saskatchewan—the entire Mississippi Basin! The Louisiana Purchase not only doubled the size of the young United States, it fueled for many Americans the feeling that the United States was destined to eventually own all the land to the Pacific Ocean.

Curious to discover what lay in the largely unexplored expanses his country had bought, President Thomas Jefferson dispatched the

Fort Snelling: Wilderness Outpost

In reporting on his 1805 expedition up the Mississippi, Zebulon Pike recommended a fort be built atop the bluff commanding the confluence of the Mississippi and Minnesota Rivers.

In 1819 Fort St. Anthony was begun with temporary buildings on the site Pike identified. In the following year, work was begun by American soldiers to convert Fort St. Anthony's temporary works to a permanent fort. Renamed Fort Snelling (after Josiah Snelling, who supervised the reconstruction of the fort), it was completed in 1825 and quickly became a center of industry and culture, as well as military duty. Explorers often used the fort as a base of operations for expeditions up the Mississippi and Minnesota Rivers.

Lewis and Clark expedition. Between 1804 and 1805 they worked their way up the Missouri River, across the Rocky Mountains, on to the Pacific Ocean, and back. The expedition put an end to any lingering hope of an easy water route to Asia, but at the same time paved the way for the fur trade in the Far West.

Fur traders needed to adapt to different business conditions in the Far West. Accustomed to hunting bison on the prairies, Indians of the West had little interest in trapping or trading in furs. Lacking native trappers, the fur companies hired adventurous white "mountain men" to do the trapping in the mountainous headwaters

100,000 Acres for a Song

Lieutenant Zebulon Pike made an eloquent speech to several Sioux chiefs and about 150 warriors at the site of Fort Snelling on September 23, 1805, imploring them to sell nine square miles of land at the junction of the Minnesota and Mississippi Rivers to the United States for a military post and trading post. He also asked them to make peace with their blood enemies, the "Chipeways," who were allies of the British. He concluded his speech saying, "I now present you with some of your father's tobacco and other trifling things, as a memorandum of good will; and before my departure I will give you some liquor to clear your throats" (Coues 1965).

In his report to his commanding officer, General James Wilkinson in St. Louis, Pike stated, "You will please to observe, General, that we have obtained about 100,000 acres for a song. You will please to observe, General, that the second article, relative to consideration, is blank. The reasons for it were as follows: I had to fee privately two of the chiefs, and beside that to make them presents at the council of articles which would in this country be valued at $200, and the others about $50."

In 1819, the Indian Agent at St. Louis received instructions from the War Department to supply "a certain quantity of goods, say $2,000 worth . . . in payment of lands ceded by the Sioux Indians to the late Gen. Pike for the United States." The payment was finally made in 1821, more than twenty-five years after ratification of the treaty by the U.S. Senate, thus finalizing the great land swindle. In his wildest dreams, Pike could never have imagined the now-bustling nine square miles replete with multilane highways, interchanges, bridges, Twin Cities International Airport, towboat traffic, and thousands of homes.

of Missouri River tributaries, hastening the American westward migration.

The War of 1812 marked a watershed for American influence in the Mississippi Valley. Although the United States had legal claim to the lands ceded to it after the War of Independence, British influence remained strong in the Northwest Territories until resolution of the second armed clash between the ex-colonials and the former mother country.

Following the war, American settlers poured over the Allegheny Mountains into the eastern Mississippi Basin. Farther west, St. Louis traders began a cautious advance up the Missouri. Their efforts were richly rewarded, and by 1832 St. Louis was the unchallenged capital of the American fur trade of the Far West (Phillips 1961).

Three centuries passed between De Soto's discovery of the mouth of the Mississippi in the Gulf of Mexico and the location of its source in the wilds of northern Minnesota. Many explorers, including Zebulon Pike in 1805, searched for the river's source without success. It wasn't until 1832 that Henry Rowe Schoolcraft, guided by an Ojibwa Indian, finally "discovered" the true source of the Mississippi at Lake Itasca in northern Minnesota.

New Orleans Under Four Flags

A Frenchman, Jean-Baptiste Le Moyne, sieur de Bienville, founded New Orleans in 1718. In 1722 it was made the capital of the French colony of Louisiana that stretched across the central third of what would become the United States.

In subsequent years, the flags of France, Spain, the Confederate States, and the United States have flown over New Orleans. A fifth flag, the British Union Jack approached the city in the waning days of the War of 1812 but was decisively turned back by the forces of General Andrew Jackson and the pirate Jean Laffite on January 8, 1815, eight days after the signing of the Treaty of Ghent, which had formally ended the war.

Less than two years earlier, Brigadier General Zebulon Montgomery Pike, explorer of the Upper Mississippi, the Mississippi Headwaters, the Louisiana Territory, and New Spain, was killed at the head of his troops during the assault on York, Upper Canada, on April 27, 1813. Pike was thirty-four years old.

In this regard, I am reminded of a cartoon of a native African and a pith-helmeted English explorer viewing a great waterfall. Turning toward the explorer, the African says, "They are lovely aren't they? I had always hoped that someone would discover them."

Early Surveys by the U.S. Army Corps of Engineers

The U.S. Army Corps of Engineers has played a pivotal role in the Mississippi Valley virtually from its acquisition by the United States.

The Corps of Engineers began extensive surveys of the Upper Mississippi with Lieutenant Zebulon Pike's exploratory voyage upriver from St. Louis in 1805. From 1817 to 1823, Major Stephen H. Long explored the river, looking for ways to improve it for settlement and commerce. He mapped the Upper Mississippi northward to Prairie du Chien as well as the Wisconsin–Fox River waterway to Lake Michigan. Passing through the village of Chicago he noted, as had Joliet in 1673, the advantage of a canal connecting the Chicago and Illinois Rivers. In 1824 Congress assigned responsibility to the Corps for managing the Mississippi and improving it for navigation.

10

Prairie, Savanna, and the Big Woods

The Riverine Landscape at the Time of Exploration and Settlement by Americans and Europeans

In mid-September 1805, journeying upstream below Lake Pepin, Zebulon Pike penned a vivid, concise description of the blufflands, prairies, and savannas flanking the river:

> The shores are more than three-fourths prairie on both sides, or, more properly speaking, bald hills which, instead of running parallel with the river, form a continual succession of high perpendicular cliffs and low valleys; they appear to head on the river, and traverse the country in an angular direction.
>
> Those hills and valleys give rise to some of the most sublime and romantic views I ever saw. But this irregular scenery is sometimes interrupted by a wide and extended plain, which brings to mind the verdant lawn of civilized life, and would almost induce the traveler to imagine himself in the center of a highly cultivated plantation. The timber in this division is generally birch, elm, and cottonwood; all the cliffs being bordered by cedar.
>
> The navigation unto Iowa River [Upper Iowa River] is good, but thence to the Sauteaux River [Chippewa River] is very much obstructed by islands; in some places the Mississippi is uncommonly wide, and divided into many small channels which from the cliffs appear like so many distinct rivers, winding in a parallel

> course through the same immense valley. But there are few sandbars in those narrow channels; the soil being rich, the water cuts through it with facility. (Coues 1965, 306)

Nineteen years later, similar landscapes were eloquently described by George Catlin: "The whole face of the country is covered with a luxuriant growth of grass, whether there is timber or not; and the magnificent bluffs, studding the sides of the river, and rising in the forms of immense cones, domes, and ramparts, give peculiar pleasure, from the deep and soft green in which they are clad up their broad sides, and to their extreme tops, with a carpet of grass, with spots and clusters of timber of a deeper green; and apparently in many places, arranged in orchards and pleasure-grounds by the hands of art" (1841, 129).

The Myth of the Ecologically Invisible Native American

Although the popular mind-set holds that North America existed as a virtually untouched wilderness before European settlement, much of the landscape was already the product of human culture.

A popular misconception is that American Indians were ecologically invisible, living in perfect harmony with the environment as hunter-gatherers. On the contrary, many Indians were farmers. By A.D. 1500 they had cleared large areas to produce corn, beans, squash, tobacco, and other crops. The impact of Indian agriculture was masked by the devastating effects of Old World diseases on native populations that had little resistance to them.

Devastated by smallpox, introduced by the Spanish in the early 1500s, and other exotic diseases like diphtheria, typhoid fever, measles, and cholera, the Indian population of North America collapsed from an estimated eighteen million in 1500 to less than one million by the late 1700s. What had been intensely cultivated fields and gardens reverted to forest, prairie, and savanna. Many Indian agricultural lands had two to three centuries to reforest before the first permanent European American settlers arrived. When the first waves of European American settlers poured westward over the Appalachian Mountains the landscape looked more "pristine" than it had in more than one thousand years (MacCleery 1996).

Fire: A Prime Ecological Force and Horticultural Tool

Vast areas of the Mississippi Basin were very much cultural landscapes. Although the prairie was mainly a product of climate, much of it owed its existence to grazing and prairie fires that kept invading forests in check. Trees standing in prairies were prime targets for lightning, which often ignited them and/or the dry grasslands. Fires were also set by Native Americans, either accidentally or purposely for a variety of reasons including making the grasslands more attractive to grazing animals like elk and bison (buffalo).

Roving bands of bison worked in concert with fire to maintain a "patchwork" landscape, maintaining a diversity of prairie plants and animals. Bison tended to graze areas recently burned; and fires tended to burn areas not recently grazed. In nature's rotational grazing plan, bison grazed and moved, grazed and moved, their hooves churning and cultivating the soil. Nature's multitudes of diggers—mainly rodents and insects—facilitated the process.

Flooding has long been considered the principal factor influencing plant community types on the floodplain. However, it is now known that fire, either natural or human-caused, played an important role in maintaining floodplain prairies, savannas, and open woodlands (Nelson et al. 1998).

Everywhere in North America, Indians regularly set fire to millions of acres to improve game habitat, facilitate travel, reduce insect pests, remove cover for potential enemies, enhance conditions for berries, and drive game. Frequent, low-intensity fires shaped the famous oak savannas of the Midwest. They existed as components of the landscape prior to Indian intervention, but Indians' actions greatly expanded the extent of such habitats (MacCleery 1996).

For Native Americans, fire was a prime horticultural tool. It was easily and quickly employed, and it could be used to work large areas. Applied periodically for centuries, fire was a force as profound as weather in its ecological impact. Most Indian fires were set in spring and fall when soil moisture was high and conditions were favorable for low-intensity burning of forests. This tended to create plant communities adapted to low-intensity fires and to reduce the number of high-intensity fires caused by lightning. Burning increased the range of pines, oaks, and other forest types that flourish under a frequent fire regime (MacCleery 1996).

Indian use of fire as a game management tool in the Winona, Minnesota, area was documented by Lafayette Bunnel: "After a very cold spell until late in the fall, that had closed Lake Pepin, there came several days of mild, dry weather, and then a sudden change and a strong westerly wind. In a few hours time it was almost as dark as night. All of the men folks were away but myself, and I had just returned, when Matilda told me that she did not know what to do with Mrs. Kennedy, for the coming darkness and smoke had led her to believe that the world was coming to an end sure enough. Just then an old squaw with some of her people came up to the house, and asked what was the matter, and Mrs. Kennedy told her. Indians do not swear, but they have strong expressions of contempt, and the Sioux woman withheld none of her language, and ended her harangue by saying: 'Thou foolish white woman, canst thou not smell the burning grass of our buffalo prairies? Thinkest thou that our people are fools not to prepare early food for them?'" (1897, 225).

In 1833, C. J. Latrobe described a prairie fire near Maiden Rock Bluff on the Wisconsin shore of Lake Pepin: "We calculated at this time that the fire spread over a tract nearly twelve miles in length, while the distant glare upon the clouded horizon showed that it was raging far inland. The whole evening the lake, the Maiden's Rock, the clouds, and the recesses of the glen were illuminated by the flames, while, gaining the rank growth on the borders of the lake and the brow of the distant mountains, the country opposite blazed like tinder in the wind: and from the summit of Maiden's Rock, which we had again ascended before we retired to rest, the scene was fearfully grand" (Curtis 1959, 297).

"The importance of fire became evident when immigrants moved out onto the prairies and cut off prairie fires: Millions of acres of open oak savannas and even treeless land to the east of these farms became dense woodlands or forests within two decades. Across North America ecosystems changed rapidly as Indian burning stopped. Prairies became woodlands, savannas became dense forests, and dense undergrowth invaded open forests" (MacCleery 1996, 46–47).

Indians had employed fire to control plant communities. White settlers eliminated most fires, but controlled plant communities with the ax, the plow, and the cow. The changes wrought by the Indians were at least as great as those produced by the white man, but the actions of the Indians left the soil covered with a dense growth of plants so that soil erosion and soil depletion were not problems. The white man, on the other hand, bared the soils of his agricultural lands, opening them to the evils of erosion and nutrient depletion. Additionally, the white man "short-circuited" the water cycle, reducing percolation, reducing water storage, increasing runoff rates, and causing floods (Curtis 1959).

Vegetation in the Early 1800s

The descriptions of plant communities recorded by explorers like Zebulon Pike, George Catlin, and Stephen Long have been verified by an interesting mix of technologies.

In 1785, the U.S. General Land Office (GLO) initiated the Rectangular Survey System to dispense land to settlers in western territories. It divided the landscape into townships containing thirty-six sections, each section being one square mile. At each section corner and midway between section corners, GLO surveyors pounded a steel post into the ground. In timbered areas they referenced the post's location by selecting two nearby trees, and recording the direction and distance to them, the trees' common names, and their diameter breast high. If no trees were present, the post was set into an earthen mound, and field notes recorded the area as prairie. After each surveyed mile, the surveyors recorded type of terrain, soil, plants composing the undergrowth, and tree species.

The GLO surveys from 1830 to 1860 reflect the changes that had occurred in the previous two hundred years under the influence of unstable and varied Indian populations, but they do not properly indicate prehistoric conditions. Settlement of the land by Europeans was largely a postsurvey operation, since title to the lands could not be granted until they had been surveyed and recorded. Therefore, the vegetation as deduced from the survey can be thought of as an intermediate stage in the transition from the prehistoric equilibrium Indians had with the land and the modern balance between white men and the land (Curtis 1959).

More recently, as part of the U.S. Geological Survey's Upper Mississippi River Long Term Resource Monitoring Program, GLO survey records have been digitized to construct computer-generated maps of the former forests, savannas, prairies, marshes, and areas of open water that existed along the Upper Mississippi River over 150 years ago.

In 1805, at the time of Zebulon Pike's exploration, the Headwaters pineries extended southward to about the city of Brainerd, in central Minnesota. There, the Mississippi River entered an area characterized by a mosaic of prairie, savanna (grassland interspersed with fire-resistant trees), and extensive stands of "big woods" (mature deciduous forests that included maple, oak, basswood, walnut, hickory, and many other species).

South of St. Paul, the Upper Mississippi River flowed through an eastern extension of the Great Plains known as the *Prairie Peninsula,* which extended eastward all the way to Pennsylvania. Along the

Fire-Resistant Oaks

The bur oak is a fire tree *par excellence;* it not only has stump-sprouting ability, but older trees are tolerant of ground fire, due to their thick, fire-resistant bark. The swamp white oak is also fire resistant, often forming part of the savanna in poorly drained soils along the river (Curtis 1959). The acorns of the swamp white oak are important food for wood ducks that gorge on them, swallowing them whole, filling their crops, and subsequently grinding them in their gizzards. Wild turkeys, recently introduced to the area, also feast on the acorns as do squirrels and deer.

Old bur oaks may have incredibly large root systems. Standing alone in a field, a 350-year-old bur oak with a five-foot-diameter trunk may have lateral roots that radiate outward for a quarter of a mile. When it rains, those roots take only about fifteen minutes to capture the moisture (Weflen 2001).

river corridor, easily burned areas like bluff tops, broad terraces, broad valley floors, and large islands tended to be prairie or savanna.

Named by the French, the savannas were "parklands" of prairie punctuated with groves of fire-resistant trees like bur oak. English-speaking explorers and early settlers usually referred to them as "oak openings." Most steep, dry, southwest-facing slopes existed as

Savanna

Savanna has become the general name for parklike plant communities where trees are a component but where their density is so low that it allows grasses and other herbaceous vegetation to become the actual dominants in the community. Throughout the Unglaciated Area, savanna was one of the most widespread communities in presettlement times. The most familiar was the *oak opening*. In the uplands, the usual combination was bur oak and moist prairie, although black oak and white oak were sometimes present. Swamp white oak and wet prairie formed the community on low grounds. A rough estimate of the area occupied by the oak openings in the early 1800s in Wisconsin is 5.5 million acres. The oak openings were derived from preexisting forests and do not represent a migration of oaks across an open prairie (Curtis 1959).

When settlers stopped the prairie fires, a rapid change took place in the oak openings. In Wisconsin, this process began mainly in the 1845 to 1855 period.

Grubs, whose tops may have been burned back for centuries, sprouted into clumps of small, same-age trees, and within a decade the openings were filled by saplings and brush so dense that it was difficult to walk through them. Every trace of the sunny openings vanished. Within twenty-five to thirty years, dense oak forests were present. The new trees overtopped the original oak-opening trees, shade-killing their lower branches, and exposing them to wood-rot fungi. The old veterans, derisively called *wolf trees* by foresters, were usually destroyed by forest owners or snapped off by windstorms. With the deaths of the huge, open-grown trees went a graphic story of plant ecology (Curtis 1959). Today, occasional wolf trees can still be found, usually enveloped by a young forest. They can be identified by their thick trunks, long lateral branches, and knobs formed where lateral limbs have broken off. The trunks are usually hollow, providing homes for squirrels and raccoons as well as nesting for wood ducks.

"goat prairies." Hardwood forests were most prevalent in areas protected from fire, mainly deep valleys, north-facing slopes, and smaller islands.

Uplands were predominantly covered with savanna communities of fire-tolerant white oak, bur oak, and black oak. Some of the savannas had a parklike distribution of trees with a grassy understory. In others, oak groves were interspersed with open prairies and dense thickets of fire-stunted oak and hazel brush. Fire-sensitive sugar maple–basswood forests were restricted to steep, moist ravines and north-facing slopes, the last places to lose their protective blanket of snow during the spring fire season.

Floodplains flanking the river had communities similar to both islands and the surrounding uplands. Bur oak, tolerant of both fires and floods, was the dominant tree species on both floodplains and uplands in 1848.

In the Pool 4 area above Alma, Wisconsin, for example, prairies once dominated the floodplain. Forests were generally restricted to smaller islands, the banks of the Mississippi and its tributaries, valley slopes, and ravines. Island forests were dominated by flood-tolerant species like elm, silver maple, willow, bur oak, birch, and ash. Because the GLO surveyors did most of their work along the Mississippi during the winter when trees were leafless, they may not have always distinguished bur oak from swamp white oak. The barks of the two species are similar. Presently, silver maple is the dominant floodplain species in Pool 4.

Farther south, at the confluence of the Illinois and Mississippi Rivers, records from 1815 to 1817 show that about 56 percent of the presettlement floodplain consisted of forest and savanna dominated by hackberry, pecan, elm, willow, and cottonwood. Prairies covered about 41 percent of the floodplain. Between 1817 and 1903, all of the higher-elevation moist prairies were converted to agriculture (Nelson et al. 1994). Species diversity decreased in floodplain forest communities after impoundment in the 1930s, and silver maple is now the dominant species.

The floodplain forests of today represent only a small portion of presettlement forests that decreased rapidly (and were simplified) in the 1800s because of the conversion to agricultural land and the harvesting of trees for fuel and lumber (Wiener et al. 1998).

After 1830, steamboat traffic increased rapidly, creating an enormous demand for the fuelwood that lined the riverbanks. Woodyards became "big business," and many farmers supplemented their incomes by harvesting and selling cordwood from bottomland forests. Hardwoods with the highest fuel value were selectively harvested. These included oak, ash, maple, elm, pecan, and hackberry. Willow and cottonwood, having low fuel value, were less desirable, just as they are now for woodstoves and fireplaces (Havighurst 1964).

The European background of the first explorers of the Upper Mississippi River Valley had conditioned them to the viewpoint that trees were the natural covering of any nonarctic lands left free from disturbance. The treeless prairies they encountered in the midcontinent were sources of wonder that called for some clear-cut explanation. The earliest writers were almost unanimous in their conclusion that prairie fires were responsible for the lack of trees. After the Civil War other explanations were offered, including soils, climate, and grazing by bison. "The earliest settlers preferred the forest to the prairie because they were suspicious of the treeless lands, were unable to plow them properly with the tools then available, and were generally short of the capital necessary to build the essential buildings with purchased lumber. This situation changed fairly rapidly when the true richness of the prairie began to be realized and the trend then shifted to settlement of the prairies and savannas in preference to the heavy forests. There was a nearly complete occupation of the prairies by 1880" (Curtis 1959, 465–66).

Tragically, many settlers mistakenly believed that the most fertile soils lay under forests, thinking that prairie soils were too poor to grow trees (Curtis 1959). Scandinavians, in particular, settled in the Mississippi Headwaters not only because the area looked so much like their homelands, but also because they thought they could successfully farm the area once it had been cleared of forest. They were unaware that the soils of the recently glaciated area were poor to begin with, and were further impoverished by podsolization (the leaching of the soil by acidified rainwater under pine forests).

My Finnish immigrant grandparents made the mistake of homesteading and trying to farm the meager soils near Jenkins, Minnesota, in the Mississippi Headwaters. They eked out a living and managed to raise twelve children before they abandoned the farm and moved to Brainerd, where my grandfather found work in the Northern Pacific Railroad repair shop.

In Wisconsin's savanna country, which includes the Unglaciated Area, the steepest, rockiest, and wettest sections of farms were allowed to develop into forest as a source of fuel and building materials, resulting in widespread occurrence of ten- to sixty-acre woodlots on lands that formerly had been brushland or savanna. By 1950 there were more closed forests than there had been at the time of settlement 120 years earlier (Curtis 1959).

Today, the prairie heritage of the Upper Mississippi Basin is reflected in the names of its cities and towns—Mound Prairie, Long Prairie, North Prairie, Belle Prairie, Belle Plain, Plainview, Eden Prairie, Prairie de la Crosse (La Crosse), Prairie du Chien, and Blooming Prairie. If not named for the prairies, settlements were

A Visit with the Chief

Zebulon Pike's encounter with an Indian chief on March 18, 1806, in a dense sugar maple forest of the Mississippi Headwaters near Little Falls, Minnesota, says much about wildlife of the time, and also of the character of Native Americans who hadn't yet been abused.

> Here there was one of the finest sugar-camps I almost ever saw, the whole of the timber being sugar tree. We were conducted to the chief's lodge, who received us in patriarchal style. He pulled off my leggings and moccasins, put me in the best place in his lodge, and offered me dry clothes. He then presented us with syrup of the maple to drink, and asked whether I preferred eating beaver, swan, elk, or deer; upon my giving the preference to the first, a large kettle was filled by his wife, in which soup was made; this being thickened with flour, we had what I thought a delicious repast. After we had refreshed ourselves, he asked whether we would visit his people at other lodges, which we did, and in each were presented with something to eat; by some, with a bowl of sugar; by others, a beaver's tail, etc. After making this tour we returned to the chief's lodge, and found a berth provided for each of us, of good soft bearskins nicely spread, and on mine there was a large feather pillow. (Coues 1965, 184–85)

often named for groves of trees that provided shelter, fuel, and building material for pioneers—Walnut Grove, Soldier's Grove, Maple Grove, Cherry Grove, Inver Grove, and Spring Grove.

Wildlife in the Early 1800s

At the onset of the nineteenth century in the primeval Headwaters area, woodland caribou and elk roamed the mature stands of red and white pine. Farther south, in the mosaic of prairie, savanna, big woods, and wetlands there existed a wealth of wildlife including elk, bison, white-tailed deer, antelope, black bear, lynx, and gray (timber) wolves. There were even a few mountain lions (cougars) and grizzly bears. White-tailed jackrabbits, badgers, and prairie chickens (referred to as pheasants by Zebulon Pike) were abundant in the savannas and prairies. The area teemed with ducks, geese, coots, swans, and shorebirds during spring and fall migrations. Prairie songbirds like meadowlarks, bobolinks, and loggerhead shrikes added beauty and variety to the landscape. Peregrine falcons and Mississippi kites nested in cliffs bordering the river. Timber rattlesnakes up to four feet long were abundant in the bluffs, and diminutive massasauga rattlesnakes inhabited the Chippewa River delta. Wetlands teemed with muskrats, but beaver had become less abundant, having already suffered more than two hundred years of exploitation.

Lafayette Bunnel stated that antelope were once common on the prairies of Wisconsin and were still quite often killed by the Sioux of the Wapasha band a few miles west of Winona, until driven out

Roasted Beaver Tails

Native Americans and Zebulon Pike considered roasted beaver tails a delicacy. Hungry explorers, voyagers, and keelboat crews ate prodigious amounts of fatty meat (up to ten pounds per day per man) because the fat "stuck to their ribs," providing high-calorie fuel for long days of extreme physical labors. The beaver's broad tail contains a pencil-thin, muscular, ratlike tail that runs down the middle, but the wide lateral areas are mainly fat that melts at low temperature (as in the mouth). The author has eaten roasted beaver tails, and concluded that they are certainly edible, flavorful, and rich in calories, but that most modern fat-hungry Americans would prefer a pizza, a greasy hamburger, or French fries.

farther by the approach of white men in the 1850s. He also commented on bison, elk, and black bear: "Buffalo were very numerous in the early days, ranging from the prairies of Illinois over those of Wisconsin, across the river of that name, and up the east side of the Mississippi as far as the prairie lands extended. The last seen by white men east of the river was a small band on the Trempealeau prairie, sighted from the deck of a steamer going to Fort Snelling in 1832. . . . Ten years after the disappearance of the buffalo east of the Mississippi, the writer was able to trace their trails along the Buffalo Channel of the Chippewa River, where countless herds had roamed. Elk were also abundant there, in the Mississippi bottoms, on the prairie, and in the oak thickets below and east of Eau Clair, extending their range over the headwaters of all the streams south of the pine-belt as far as the Black River. . . . Bears at that date were quite numerous" (1897, 329–30).

Passenger Pigeons by the Millions

Passenger pigeons, a species now extinct, were inconceivably abundant in the Mississippi River Basin. They traveled in flocks of millions and formed huge nesting colonies in the trees of mature hardwood forests. They depended upon deciduous forests for much of their food, especially acorns and chestnuts, but also foraged on the seeds and berries of plants of the prairies and savannas.

On April 26, 1806, on his return trip from the Mississippi Headwaters, Zebulon Pike observed incredible numbers of passenger pigeons near Cincinnati Landing, Illinois (RM 296), documenting that the food of the young birds included foods typical of the savanna: "Stopped at some islands about ten miles above Salt River, where there were pigeon-roosts, and in about 15 minutes my men had knocked on the head and brought on board 298. I had frequently heard of the fecundity of this bird, and never gave credit to what I then thought inclined to the marvelous; but really the most fervid imagination cannot conceive their numbers. Their noise in the woods was like the continued roaring of the wind, and the ground may be said to have been absolutely covered with their excrement. The young ones which we killed were nearly as large as the old; they could fly about ten steps, and were one mass of fat; their craws were filled with acorns and the wild pea. They were still reposing on their nests, which were merely small bunches of sticks joined, with which all the small trees were covered" (Coues 1965, 212).

In the spring of 1842, Lafayette Bunnell observed a passenger pigeon rookery of almost incomprehensible proportions near Winona, Minnesota.

> I was returning in a canoe from a trip up the river, and as I came in sight of the oak timber then growing on the Wisconsin side below the site of the lower bridge, I saw clouds of pigeons settling to roost, when crash, would fall an oak limb, and then the noise would follow like the letting off of steam. It did not occur to me at first, what it was that made the latter noise, but as I approached nearer, and saw limb after limb fall, some of them of very large size, and then heard the increased noise, I saw, and heard, that it was numberless pigeons breaking down the limbs and chattering in glee at having overloaded and broken them down. Some of the young Sioux were watching the "roost" to see if any had commenced laying, for some were already building nests, and when I told Mr. Reed of the Indians being there and not a shot being fired at the pigeons, he told me that the Indians never disturbed pigeons or ducks by shooting at them when nesting, and that the life of a man doing so would not be safe among the Sioux, as the whole tribe would feast upon the squabs as soon as big enough. The pigeon-roost extended for twenty five miles below La Crosse, as reported to us by up-coming steam boats, and where there was heavy timber, the same scenes were repeated that I had witnessed—the whole length of the roost being about forty five miles. (1897, 186–87)

11

Early Exploitation of Natural Resources

Explorers and settlers found the Mississippi River Basin incredibly rich in wildlife, forests, and fertile soils. But by the late 1800s, beaver and bison were threatened with extinction. The great pine forests of Minnesota and Wisconsin had been raped and left smoking by the early 1900s. The world's last passenger pigeon died in 1914, and even the river's clams were disappearing. Poor logging and farming practices caused severe soil erosion and flooding.

The Fur Trade: Near Extinction of the Beaver

The fur trade in North America had begun in the early 1500s when early explorers in the area of Nova Scotia, New England, and the lower St. Lawrence River traded manufactured goods to Indians for furs, but the great fur trade of New France began in about 1600 along the St. Lawrence River, forming the basis of the French empire in North America. France's empire included the Upper Mississippi Basin.

In the 1600s and 1700s, trade routes were established from Lake Superior up the St. Louis River (at the present site of Duluth,

Minnesota), and then overland to the Mississippi Headwaters. Access from the Mississippi to the far north was gained by ascending the Minnesota River to the headwaters of the Red River of the North that flows northward toward Hudson Bay. The Red River was also accessed by portaging from Lake Superior up the Pigeon River and into the Rainy River drainage that flows westward (along the present Minnesota–Canada border) into the Red.

Those who think of the three-hundred-year fur-trade era as a simpler time are advised to read Paul Phillips's two-volume *The Fur Trade* to be apprised of the international intrigue and political maneuvering, wheeling and dealing, monopolies, skullduggery, exploitation of Native Americans, and wars that were incidental to the virtual extermination of some fur bearers, especially the beaver.

The gold and silver that the first adventurers hoped to pick up like leaves on the forest floor could not be found, but fur-bearing animals were abundant. Moreover they could be purchased at low

To Make a Beaver Hat

Beaver pelts were first processed by trappers—stretched, scraped free of muscle and fat, and dried. Sent to Europe, mainly England, some were tanned to make beautiful garments that were warm and durable, but most were processed to make felt for expensive, stylish, top hats. The long, relatively stiff guard hairs were removed by pulling or clipping, leaving the soft, luxurious underfur attached to the skin. The fur was dampened repeatedly with a mixture of thirty-two parts quicksilver (mercury) and five hundred parts aqua fortis (nitric acid), which curled the outer two-thirds of the hair length and raised the microscopic scales that covered the hair surface, enhancing the furs' felting properties. Thus treated and dried, the fur was sheared from the skins, which were often rendered to make glue (Anonymous 1829; Thomson 1868).

Mercury solutions also served as biocides to control the molds and bacteria that caused problems in the wet felt-making process. The mercury also gave the felt a lustrous sheen, but caused mercury poisoning in hat makers, causing them to become "mad as a hatter," as described in the "Mad Tea Party" in Lewis Carol's *Alice in Wonderland*. The beaver trade prospered until the mid-1800s when beaver became scarce, and silk "stove pipe" hats became more fashionable than beaver hats.

prices from Indians for manufactured trade goods that included cheap fabrics, blankets, hatchets, knives, mirrors, combs, bells, awls, needles, pipes, metal pots, beads, buttons, fish hooks, fire steels, guns, ammunition, Jew's harps, and trinkets. Indians who had previously killed principally for food began to kill for furs alone. Undoubtedly, the trade item most destructive to the Indians was "firewater"—rum, whiskey, brandy, and high wines. In 1782, an Indian brave could trade four beaver skins for a yard of broadcloth and four dozen buttons to surprise his wife—or treat himself to a gallon of brandy (Phillips 1961).

Nicollet's Great Map

In June 1836, Joseph N. Nicollet left St. Louis on the steamboat *St. Peter's* traveling upriver to Fort Snelling accompanied by a number of tourists (Fertey 1970). The Mississippi would never again be the same.

From Fort Snelling, Nicollet ventured upstream, exploring the Mississippi almost to its source at Lake Itasca in northern Minnesota. On subsequent explorations he charted the lakes and tributaries within the Upper Mississippi River watershed. Unlike his unschooled predecessors, Nicollet was a gifted mathematician and astronomer. He filled notebooks with thousands of latitude and longitude measurements made by celestial navigation and chronometer. Altitude was measured by barometer, a technique new to America in the 1830s. Nicollet's expedition resulted in the first accurate map of the Mississippi Basin. It facilitated exploitation.

Transportation was a leading topic of conversation among men of affairs in the United States in the 1830s, and railroads were beginning to carry passengers, but Nicollet continued to think of a country laced with canals like his native France. He thought that waterways would always control the movements of goods and people. Thus, his interest in the Mississippi and its watershed reflected practical concerns as well as the romantic view of the great river, which was shared by many Frenchmen. After all, they had first discovered, explored, and claimed it!

Nicollet's reputation as a scientist and geographer has rested for over 150 years on his remarkable "Map of the Hydrological Basin of the Upper Mississippi River." The original plates for making the map are still in the possession of the U.S. Army Corps of Engineers. Using them, the Minnesota Historical Society has published a modern reprint of this classic of American cartography.

Beaver was in the greatest demand to make felt for the high hats of the rich and prominent. During the 1697 to 1713 period about three hundred thousand beaver skins were sent yearly from the New World to Europe, increasing the wealth of the nation that controlled the beaver trade. Bear, fox, marten, otter, fisher, mink, weasel, wolf, and raccoon were also exploited. Most furs from the Mississippi Basin found their way to England, France, Germany, Russia, and China (Phillips 1961).

The main countries involved in the fur trade were France, Britain, and later the United States. Neither Spain nor Spanish colonies

Demise of the Buffalo Herds

At their peak, in about 1600, North America's buffalo population numbered somewhere between thirty million and two hundred million head that occupied nearly half of North America, including virtually the entire basin of the Upper Mississippi River. Their range extended from the Rockies to the Alleghenies, south to Texas, and north into Manitoba.

By 1812 most buffalo bands east of the Mississippi River had been destroyed, mainly by Indian hunters but also by whites as the frontier moved westward. Until President Jefferson bought Louisiana in 1803, the United States was without a buffalo herd.

The Missouri River traffic in buffalo robes became more important when beaver hats went out of fashion in the 1830s. In Mackinaw boats about thirty-six feet long and eight feet wide, made of whipsawed lumber from the base of the Rockies, furs and buffalo products were sent down the Platte River, into the Missouri River, and on to St. Louis. In 1848 alone, 110,000 buffalo robes were sent to St. Louis along with hundreds of barrels of buffalo tallow and 25,000 salted buffalo tongues. Tongues were highly prized. Not only were they the easiest part of the buffalo to harvest and ship, they were all meat, tender and delicious, containing no bones, tendons, or ligaments.

By 1890 the buffalo herds that had once numbered at least thirty million had dwindled to a mere one thousand animals; by 1900 it seemed certain that buffalo were doomed to extinction like the passenger pigeon. Of the vast herds that once thundered across the plains, only a few hundred animals were left, held mainly in zoos or as captives on western cattle ranches (Haines 1975). The bedlam of thousands of hooves had been silenced forever.

afforded any market for furs. They had no means of protecting furs from their warm climates and no facilities for processing pelts, nor did Spain produce goods desirable for the Indian trade.

During this period, a steadily increasing demand for skins for shoe leather was already leading to the extermination of deer and elk along the frontiers of the whites. Most deerskins went to Europe, but some went to the tanneries of Philadelphia and New York.

Returning from their Voyage of Discovery in 1806, Lewis and Clark reported that the headwaters of the Missouri River and the streams of the Rockies were immeasurably rich in beaver; this set off a frenzied fur rush up the Missouri.

In 1826, buffalo robes became a great source of profit at St. Louis. Indians had been generally averse to the indiscriminant killing of buffalo because they depended on them for food, shelter, and clothing. But when their scruples were overcome, they entered into the hunt as sport (Phillips 1961). The fur trade continued to prosper at Prairie du Chien during the 1830s and 1840s under the leadership of men such as Joseph Rolette, Hercules Dousman, and Henry Rice (Hartsough 1934).

By 1850, elk, bison, and antelope had been eliminated from most of the eastern half of the Upper Mississippi River Basin. Beaver seemed doomed because of over two hundred years of relentless harvest by trappers. A closed season was declared in 1910 because it was feared that beaver would become extinct.

We Cut the Top Off Minnesota and Wisconsin and Sent It Down the River

As settlers poured into the treeless prairies of the midcontinent in the 1840s and 1850s, the need for lumber for houses, barns, and sheds became acute. Fortuitously, the upper watersheds of the Mississippi Headwaters, St. Croix, Chippewa, Black, and Wisconsin Rivers were forested with mature stands of white pine and Norway (red) pine. The great pine forests of Wisconsin and Minnesota were believed to be inexhaustible, created for the sole benefit of humans. What's more, the immense pineries were laced with big rivers that seemed to be designed for the sole purpose of floating logs. At the peak of the ensuing logging era in the 1880s, there were more than one hundred thousand lumberjacks, armed with double-bitted axes and two-man crosscut saws, in the winter forests of Minnesota and Wisconsin. Between 1835 and 1915, virtually all of the usable white pine and Norway pine logs in the states of Minnesota and Wisconsin were floated to sawmills.

The great pine forests had been born in cataclysmic times over a century earlier. Norway pines, which require full sun and bare mineral soil for seedling development, invaded areas after they had been devastated by horrendous forest fires, destructive winds, or a combination of both. The Norways formed dense even-aged stands, growing straight and tall. Deprived of sunlight, their lower branches died and rotted off. This self-pruning caused the tall tree trunks to be virtually knot-free.

White pine had entered the area from eastern sanctuaries about four thousand years ago. Being tolerant of shade, they seeded the forest floor under the stands of Norways. They, too, grew straight, tall, and knot-free, forming mixed stands with the Norways, and producing prime fodder for the lumbermen. Knot-free (clear) lumber is more desirable than knotty lumber because it is stronger and easier to work.

The magnificence of the forest can be gauged by the fact that at first loggers would take no trunk that measured less than eighteen inches in diameter at the smaller end. The standard log was sixteen feet long, and four such logs were sometimes cut from a single tree. White pine, the more valuable of the two, was exploited first. When the whites became scarce, the loggers turned to the Norways. It took only forty years to log off the world's finest stand of white pine (Madson 1985).

We may think that the loggers' tools were primitive because they didn't include chain saws, but their razor-sharp, two-man crosscut saws and double-bitted axes were carefully maintained, precision instruments—and the lumberjacks worked like teams of surgeons. One specialist was responsible for the exacting job of sharpening saws, but each logger was responsible for sharpening his own ax—and he could shave with it. I suggest that readers interested in precise two-man axmanship see the classic documentary movie *The River* listed in the bibliography. It was the source of the title for this section.

If they had a mind to, sawyers could fell and dismember a huge pine in less than ten minutes. With their crosscut saw, two expert lumberjacks could cut through a thirty-inch log in twelve seconds. They were experts with mechanical devices including inclined planes, wheels, pulleys, levers, ropes, and cables. They had steam donkeys and railroads. Their massive logging sleds transported immense loads of sawlogs on iced winter roads, pulled by oxen, mules, or horses.

Most of the early cutting was done in the winter when swamps were frozen and there were no mosquitoes or deer flies. Many farmers logged in winter but returned to their farms in spring.

Logs were hauled by sleigh over iced roads or by railroads to riverside landings where they were stacked to wait for the spring floods that would transport them to sawmills or downstream to the Mississippi. Logs and rough-sawn lumber were made into rafts for their float trip to Winona, La Crosse, Clinton, Rock Island, and other river cities. During the peak of the lumbering period in the late 1800s, there were more than eighty sawmills located on the Upper Mississippi River and about 120 located on tributary streams. Winona, with no pines of its own, claimed to be the world's largest pine sawmilling center in the 1890s.

Beef Slough: One of the World's Busiest River Ports

Shortly after the Civil War, Mississippi River lumbermen looked to Wisconsin's Chippewa River as a source of pine logs. Beef Slough was a crucial staging area in the transit of logs harvested from the vast pineries of the Chippewa basin. A distributary of the Chippewa, Beef Slough departs from the Chippewa's main channel at Round Hill, three miles below the village of Durand, Wisconsin, and flows about twenty-five miles before joining the Mississippi River just above the city of Alma, Wisconsin (Kohlmeyer 1972).

On the morning of September 12, 1863, the first raft to be towed by boat below Lake Pepin left the mouth of the Chippewa River (Russel 1928). In 1879, at the peak of the logging boom on the Chippewa, twelve hundred to fifteen hundred men of many nationalities were employed at Beef Slough, daily sorting and assembling four million feet of logs into rafts. By 1893, seventy-five steamboats were exclusively engaged in log and lumber rafting on the Upper Mississippi. With as many as 837 raftboat departures in one season, Beef Slough was one of the busiest river ports in the world.

In 1888, accumulated deposits of sand and sawdust formed a shoal that blocked the entrance to Beef Slough. Forbidden by the War Department from constructing rafts on the mainstem Chippewa because they would impede steamboat traffic, the only alternative for lumbermen was to drive loose logs down the Chippewa to West Newton Slough on the Minnesota side of the Mississippi, about eight miles below the mouth of Beef Slough. The Beef Slough installations that had cost one million dollars were abandoned, and the complex booming operations were transferred to West Newton in 1889. In 1890, a record 660,764,000 board feet of logs were rafted at West Newton and 149,934,000 feet at Beef Slough where high water had allowed a final year of operations (Kohlmeyer 1972).

Log rafts departing West Newton for Rock Island were as long as 1,625 feet and as wide as 275 feet. Lumber rafts occupied as much as six acres of space and contained as much as nine million board feet of lumber (a board foot is one foot square and one inch thick), the

Reads Landing: Metropolis of The Upper River

Reads Landing, Minnesota, once one of the most famous cities on the Upper River, was strategically located at the lower end of Lake Pepin, at the confluence of the Chippewa with the Mississippi. During its heyday during the late 1850s and 1860s, the city was a metropolis on the Upper River, with over two thousand inhabitants and more than twenty hotels. In the spring of 1858, thirty-two steamboats lined up along its waterfront waiting for the ice to go out on Lake Pepin; they hailed from Cincinnati, St. Louis, Galena, and Dubuque. Lines of stagecoaches carried impatient passengers to St. Paul. The city also catered to the loggers who brought their great log rafts down the Chippewa from the pineries of Wisconsin. Reads Landing issued bonds for public improvements and deemed itself a rival of St. Paul. With the coming of the railroad and the demise of the logging industry, Reads Landing virtually died (Russel 1928; Hartsough 1934). Today there are less than a hundred inhabitants. Motorists on Highway 61 scarcely slow down for it, oblivious of the "hot times" of yesteryear.

equivalent of nine hundred railroad cars (Russel 1928). Big rafts had to be "unzipped" down the middle so the two halves could pass through bridges with narrow spans.

Beef Slough has long since reverted to its original wilderness, with no vestige remaining to indicate the bustling place it once was or the significant role it played in the history of the lumber industry. Similarly, all evidences of the lumber boom are gone at West Newton.

The Art of Log and Lumber Rafting

Log rafts were usually only one course of logs deep; the largest ones were 1,550 feet long, covered almost ten acres, and contained over two million board feet of wood (Russel 1928). To make a raft, logs were placed side by side lengthwise in the stream. A large hole was bored at each end of each log. A birch limb was laid across and a split bur oak was bent over it as a staple, and pegs were driven into the holes to make the staple fast. Each section of logs thus bolted together extended the length of the raft and was called a string. At the end of each string was a great sweep oar made of a plank bolted to the end of a young tree. A raft with ten oars was a ten-string raft. Lumber rafts contained greater volumes of wood because they were made of carefully stacked boards, and were deeper. The largest ones held as much nine million board feet of rough sawn lumber (Madson 1985).

The earliest log and lumber rafts floated with the current, guided around bends and obstacles by fifteen or so strong men at the bow, each with a sweep. If there was no current, as in Lake Pepin, a skiff hauled a large anchor and a longline out ahead of the raft, and dropped it. The men laboriously pulled the raft to the anchor, raised the anchor, put it in the skiff to repeat the process—from daylight until dark. Each raft man had a small, crude shelter to protect him from the elements. There was an open-air "kitchen" for preparing the beans, salt pork, pancakes, and coffee (Russel 1928).

Improved methods of propelling and maneuvering the huge, unwieldy rafts became urgent because free-floating rafts steered by oarsmen were in danger of being smashed against the piers of the new railroad bridges that spanned the river, beginning in the 1850s. At first, side-wheel steamboats pushed the rafts, but they were replaced by stern-wheelers that used steam donkeys and cables to twist

The introduction of bow boats made it possible to maneuver massive rafts of lumber or logs through the twists, turns, and bridges along the river. Responding to whistle signals from the pilot in the raft boat at the stern, the bow boat moved forward or backward, swinging the bow of the raft left or right. Here, the pontoon bridge built between Prairie du Chien, Wisconsin, and Marquette, Iowa, in 1874 has been opened to allow passage downstream. During the 1875 to 1915 period, 145 raft boats and thirty-three bow boats operated on the Upper Mississippi (Fugina 1945; photo, circa 1910; Murphy Library, University of Wisconsin–La Crosse, negative 3810).

the raft around the bends. In 1890 a new towing technique was developed in which two steamboats were used to steer the ponderous rafts. A large steamboat pushed the raft while a smaller bowboat was lashed crosswise at the head of the raft. The bowboat steered the raft by going forward or backward. The bowboat technique made it possible to transport enormous rafts, some as large as ten acres. For example, in 1901 the Knapp, Stout and Company sent a raft with 9,152,000 board feet of lumber, and one million board feet of shingles and lath piled on top, to St. Louis (Kohlmeyer 1972).

Bridges were always a peril for rafters, especially in currents of five miles per hour that were alive with crosscurrents. At the upper and lower rapids (Rock Island and Keokuk) and at railroad bridges, the channel was too narrow and dangerous to let the whole raft through. This made it necessary to double-trip the raft, splitting it, and taking one half through at a time. Often, in spite of Herculean efforts of crew and pilots, the raft broke up, requiring days of back-breaking labor to round up the logs and reassemble the raft (Hartsough 1934).

The Logging Industry Withers and Dies

The peak of the logging boom came in 1892, and then it began to wither. It became apparent that the supply of white pine was not inexhaustible after all. One by one the sawmills were abandoned. Rafts became fewer and were always made of smaller logs, usually of material that would have been scorned twenty years earlier. The rafting steamers were sold into the lower river or the Ohio trade or were broken up for sidewalks (Russel 1928).

In August 1915, on its way to the J. C. Atlee mill in Fort Madison, Iowa, the steamboat *Ottumwa Belle* snaked the last lumber raft down the Mississippi River, gathering remnants left at mills that had gone belly-up. Like the first raft of the white pine boom, it was a lumber raft. When the raft passed Albany, Illinois, it picked up Stephen Hanks who had ridden one of the first lumber rafts on the Mississippi and piloted the first log raft. Hanks rode in the funeral procession as far as Davenport, saw the railroad bridge swung for the last time on a raft boat, had a reception from old river men, and was escorted back to Albany. Hanks had seen the whole thing—the beginning, the culmination, and the end—all in one man's lifetime. He died on October 17, 1917, two days before his ninety-sixth birthday (Russel 1928).

The logging boom in the Mississippi watershed had sputtered and died. For about fifty years lumber had moved westward from the river cities across the prairie, as far as Denver, building a hundred cities, a thousand towns, and innumerable farms. But by the 1920s the lumber flow had reversed, with the river cities getting their lumber from the Pacific Northwest. Ironically, some of the same logging companies that had raped the virgin white and Norway pine forests of Minnesota and Wisconsin had leapfrogged the great plains and set up shop in Washington, Oregon, and Idaho.

Removal of pine forests by clear-cutting in the St. Croix, Chippewa, Black, and Wisconsin watersheds disrupted the stability of soils that had been held in place only by dense vegetative cover. The resulting erosion was exacerbated by early farming practices. Rains and snowmelt washed eroded soils into Mississippi tributaries and into the Mississippi where much of the sediment was trapped between wing dams. Because the watershed had lost much of its ability to store water, fluctuations in runoff rate were magnified, causing increased numbers of floods and higher flood crests.

The Funeral of the Blue Meteor

Dubbed the "blue meteor," the passenger pigeon was the most impressive species of bird that man has ever known. It moved about and nested in such enormous numbers as to confound the senses. Its doom was sealed by this tendency to group in large numbers. It never traveled alone.

The extinction of the passenger pigeon was especially shocking to Americans, many of whom could remember flocks that darkened the sky for hours. In 1813 John James Audubon had mathematically calculated that a single flock in the Ohio River Valley contained more than 1,115,000,000 birds. The last wild passenger pigeon in Wisconsin was shot at Babcock in September 1899, and on September 1, 1914, the world's last passenger pigeon died in the Cincinnati Zoo. Named "Martha," she died at age twenty-nine (Office of Communications 1976).

On May 11, 1947, a monument commemorating the passenger pigeon was dedicated in Wyalusing State Park, Wisconsin, just downstream from Prairie du Chien. Inscribed upon it are the words of the great naturalist Aldo Leopold, who reflected upon the horrendous ecological tragedy:

> We have erected a monument to commemorate the funeral of a species. It symbolizes our sorrow. We grieve because no living man will see again the onrushing phalanx of victorious birds, sweeping a path for spring across the March skies, chasing the defeated winter from all the woods and prairies of Wisconsin. The pigeon was a biological storm. He was lightning that played between two

> opposing potentials of intolerable intensity: the fat of the land and the oxygen of the air. Yearly the feathered tempest roared up, down, and across the continent, sucking up the laden fruits of forest and prairie, burning them in a traveling blast of life. Like any other chain reaction, the pigeon could survive no diminution of his own furious intensity. When the pigeoners subtracted from his numbers, and the pioneers chopped gaps in the continuity of his fuel, his flame guttered out with hardly a sputter or even a wisp of smoke.

The passenger pigeon's demise was an ominous indicator of the wildlife disasters that were occurring due to the Caucasian invasion. It is generally believed that their extinction was caused by a combination of market hunting (netting, trapping, shooting, and removal of nestlings) and habitat destruction. The telegraph provided a means for market hunters to track migrating flocks. The new railroads provided fast shipment of barrels of iced pigeons to Chicago, New York City, and other eastern markets. In 1882, squabs were a glut on the Milwaukee market where the retail price ranged from thirty-five cents to one dollar per dozen. Dried gizzards were sent to the Orient as aphrodisiacs. Pigeons were rendered to make lard. Adult birds were used for target practice. Importantly, farmers and woodcutters fragmented the expansive, mature hardwood rookeries that were indispensable for accommodating millions of nesters at a time.

Like a Great Climatic Change: Consequences of Land Mismanagement

The exploitation of the Upper Mississippi River drainage basin, beginning in the 1840s, by immigrants and their descendants changed the face of the land. Their farming and forestry practices bared the land, creating the equivalent of a great climatic change. This was especially true in the rugged Unglaciated Area where pioneer farmers cultivated the flattest land they could find—the Mississippi Valley floor, the floors of tributary valleys, terraces, and the top of the plateau.

The steel plows of European settlers sliced into the thick black sod that had been slowly developing for ten thousand years on the fertile loess soils that were the legacy of glacial retreat. Neat patches of wheat, oats, barley, buckwheat, hay, and corn soon replaced the sea of native prairie grasses and forbes (broadleaved flowering plants).

In preparation for sowing their crops, farmers intensively groomed their land, first plowing it, then harrowing it and pulverizing it until it was as "smooth as the kitchen floor." At harvest time, shocks of small grains were hauled to the threshing machine, leaving little straw on the land. Straw was piled at the threshing site, and often burned, ensuring that there would be no addition of humus to the soil or plant debris to break the force of raindrops or the wind. All of this rendered the soil particles vulnerable to dislodgement and transport by water and wind.

Cattle, the "white man's buffalo," grazed the "wasteland" that was too steep or forested to cultivate. The four- to-five-foot-tall big bluestem and Indian grasses of the prairies and savannas were forced onto smaller and smaller plots of land. Erosion rates increased so much that the floor of the modern Upper Mississippi Valley was blanketed by soil that washed into the river and its valley as a result of steep-land agriculture.

Early farming practices were extremely destructive to the land because little care was given to the precipitous "sidehills." With horse-drawn machinery and the moldboard plow that had been introduced in 1837, farmers opened land that should never have been cultivated. In the Unglaciated Area, bluff tops (ridges) were plowed to the extreme edges of the bluffs—and often over the edges. Sure-footed horses traversed areas that today's tractors cannot.

Conservation to the early farmers usually meant letting no portion of their land lie unproductive. Land too steep to be plowed was referred to as wasteland to be logged and grazed. White settlers often carried on the Indian tradition of burning to discourage tree growth and to stimulate the production of additional grasses for grazing. Bluffland fires were so common that they sometimes lighted the routes of steamboats, enabling them to travel at night. Dairy cattle and horses were the principal grazers, but sheep and goats also helped denude the hillsides. Soil conservation measures such as contour tillage, strip cropping, and terracing were unknown.

Adding insult to injury, the gullies that inevitably developed on steep land were often filled with topsoil scraped from the eroded fields so that cultivation could continue. Soil conservation practices improved after the 1930s, but wetland drainage accelerated, as did the practice of straightening streams and digging them deeper to hasten runoff.

Early farmers often moved to new lands after they had depleted the soils of their homestead. It was often said that a hard-working farmer could go through two farms and three wives in his lifetime.

Soil and rocky debris washing down from the uplands caused most tributary valley floors to raise, often burying the original fertile soils of the valley floor. This, in turn, increased the gradients of streams that ran through the valleys, causing them to run straighter and faster, cutting their channels deeper. The lower reaches of tributaries and their deltas still store much of the eroded soil.

In Wisconsin, upland erosion and tributary sediment yields to the Mississippi were highest during the 1850s through the 1920s, with rates declining since then because of improved land management practices (Knox et al. 1975; Trimble and Lund 1982; Trimble 1983).

In addition to clearing natural vegetation that had slowed runoff, settlers began draining wetlands that had filtered nutrients and also regulated runoff rates. Wetland drainage accelerated in the 1900s, and by the end of the century an estimated 34 to 85 percent of wetlands in Wisconsin and Minnesota had been lost, and 85 to 95 percent in Iowa, Illinois, and Missouri (Dahl 1990).

The wetlands of the Minnesota River Basin were especially easy to drain because the Glacial River Warren had carved the Minnesota River valley so deep. Virtually all of that basin's prime wetlands disappeared along with the muskrats, mink, ducks, shorebirds, and pheasants that depended on them.

Because of wetland destruction, stream channelization, agricultural drainage, and urbanization, floodwaters reached the Mississippi River faster, causing flood stages that were higher and lasted longer, but also low water stages that were lower.

With horse-drawn machinery, the settlers transformed the native tall-grass prairie into one of the most productive agricultural regions in the world. Second-, third-, and fourth-generation farmers have since used fossil-fuel powered machinery and an ever-expanding arsenal of fertilizers, herbicides, and insecticides, turning the Great

Plains into an inland ocean of corn and soybeans. Increasingly, conservation-minded farmers are being replaced by absentee corporate farmers that are usually poor stewards of the land.

Today, most agricultural land in the basin has been tiled, wherein perforated plastic pipes are laid within the soil, below plowing depth, to drain "excess" water from the land into ditches. If they could be seen, these tile fields and ditches would resemble immense beds of capillaries and veins that collect the life blood of the land and send it to the river—and ultimately to the ocean.

Fish Rescue

By 1870, after years of uncontrolled exploitation, it was obvious that the Mississippi's fishery resources were rapidly declining, and in 1871 Congress established the Office of the United States Commissioner of Fish and Fisheries. Within four years, the states of Iowa, Minnesota, Wisconsin, and Missouri had established their own Fish Commissions. The fisheries activities of these groups basically fell into three categories: fish rescue, fish stocking, and artificial propagation.

Prior to the formation of the Nine-Foot Channel impoundments in the 1930s, water levels in the Mississippi River fluctuated greatly throughout the year. Spring floods submerged lowland areas, and as the floodwaters receded, pools and lakes were formed, cut off from the main channel of the river. Conditions were favorable for the growth of newly hatched fish in such floodplain lakes, but the stranded fish usually died as water levels receded and the lakes dried up. Those landlocked fish that survived the summer were usually killed by freeze-outs during the winter.

Fish rescue programs began in Iowa in 1876. Their purpose was to salvage fish stranded in backwaters by seasonally receding water levels. Here was a practically inexhaustible supply of ready-hatched fish, which could be harvested at small cost compared to that of hatchery fish. Similar programs, begun in 1889 by the U.S. Fish Commission peaked between 1917 and 1923 when there were thirty-five rescue stations. Most fish were transferred to the main river channel, but also to inland waters of states bordering the river. Special aquarium railroad cars, hauled by fast passenger trains, transported "black bass" and "pike" to New York, Pennsylvania, Michigan, and Colorado. In

1921, at the program's peak, 176 million fish were rescued (Carlander 1954). Inadvertently, fish rescue apparently facilitated the widespread distribution of exotic carp. Impoundment of the Upper Mississippi in the 1930s stabilized water levels and ended fish rescue operations.

Early Introductions of Exotic Fish Species

American shad and Atlantic salmon are popular fish in various rivers along the Atlantic seaboard. Fish stocking began on the Upper Mississippi in 1872 with introduction of 25,000 American shad a few miles above St. Paul. In the ten years from 1874 to 1884, 1,340,000 more American shad and thousands of Atlantic salmon were unsuccessfully planted in the Mississippi (Carlander 1954). The rationale for these introductions was that the highly desirable fish would spawn in fresh water, return to the sea to mature, and a few years later swim back up the river where they could be harvested. The absence of dams on the Mississippi made all of this seem feasible.

Mississippi River habitats proved unsuitable for American shad and Atlantic salmon, but they were ideal for juvenile carp stocked during the same period. Because carp have had more ecological impacts on the waters of the Mississippi Basin than any other exotic species of fish, they are given special treatment in a subsequent chapter.

The Pearl Button Industry Devastates Clam Populations

A pearl button industry, which originated in Muscatine, Iowa, in 1889, flourished concurrently with fish rescue operations. The clam fishery rapidly expanded northward and southward to supply pearl button factories at several cities along the river, including Wabasha, Minnesota; Muscatine and Keokuk, Iowa; and Canton, Illinois. By 1899, clamming was the most important fishery in Wisconsin with over sixteen million pounds of shells harvested. Lake Pepin produced thousands of tons of shells annually during the first decade of the twentieth century.

The clam beds were depleted in a few years in most areas. Within a decade after the first factory was built, there were signs of exhaustion of the Muscatine clam beds. Two or three years were all that some beds could stand under the intense fishing pressure. Beds near New Boston, Illinois, produced more than ten thousand tons of

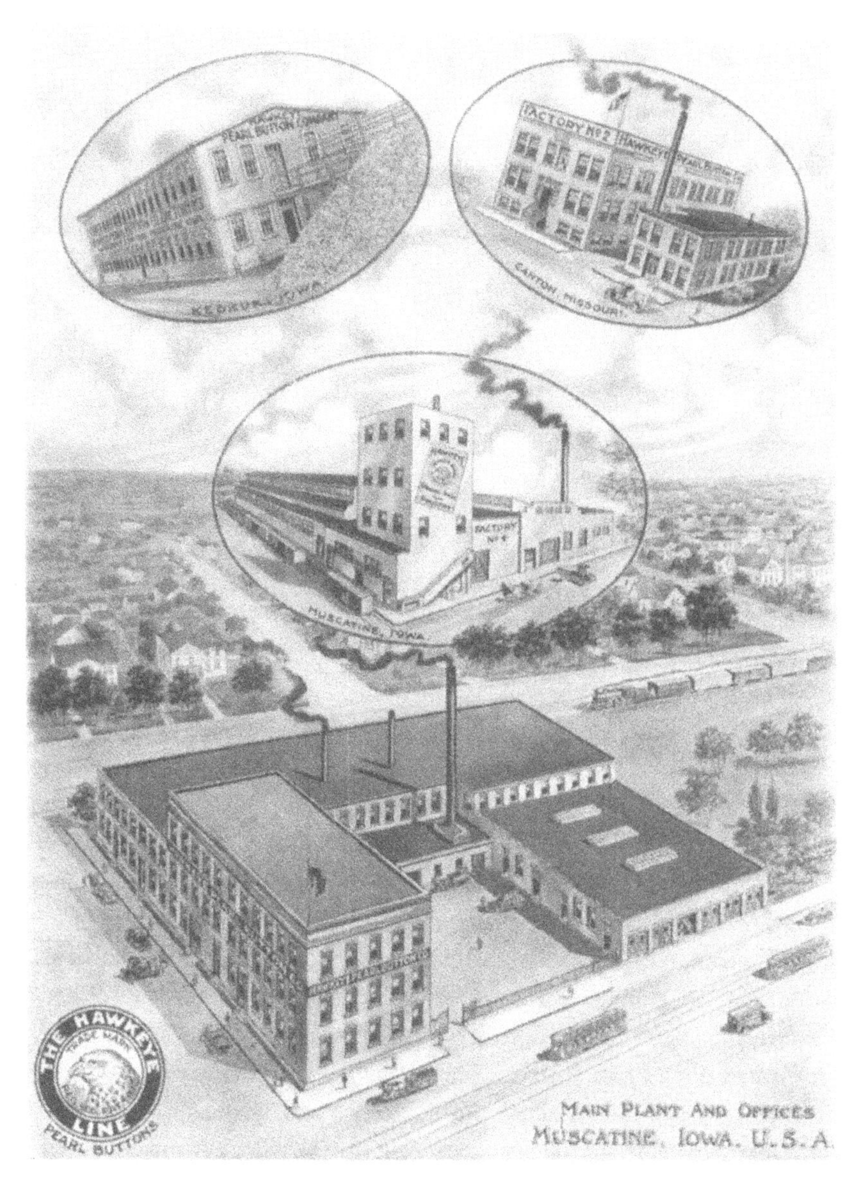

Black coal smoke was in vogue in the late 1800s when plumes were dubbed into factory scenes of the Hawkeye Pearl Button Company to portray progress and industriousness. Note the pile of clamshells at the upper left (undated Hawkeye Pearl Button Company trade publication).

Button blanks were sawn from clamshells with hollow toothed drills, then tumbled, polished, and often dyed. Most production work was done by nimble-fingered women, their long hair gathered under bandanas so it wouldn't be caught in machinery that lacked safety guards (undated Hawkeye Pearl Button Company trade publication).

shells between 1894 and 1897, but were then abandoned. It became apparent that the mussel fishery was doomed unless methods of artificial propagation could be developed. Accordingly, the Fairport Biological Station (Iowa) was established by the U.S. Bureau of Fisheries in 1908. Although artificial propagation was successfully employed on a large scale, depletion of the beds continued. Water pollution and siltation accelerated the process, and by 1950 the few remaining button factories were making buttons cut from shells collected from streams in Tennessee and Arkansas. The development of synthetic buttons in the early 1950s was responsible for the demise of the industry (Carlander 1954; Nord, 1967).

The magnitude of the clamming industry can be judged by the activities of one of the largest pearl button manufacturers, the Hawkeye Pearl Button Company, which was established in 1903 at Muscatine, Iowa. By 1918 it had two factories in Muscatine, one in

Keokuk, another in Canton, Missouri, and a sales office in New York City. The plants used four thousand tons of clamshells per year, produced seventy-five thousand buttons per week, and employed 971 factory workers. Processed shells were crushed and sold as poultry feed. Clamshell dust was sold to farmers to lime their soil.

3

Caging the Giant

12

Steamboats 'Round the Bend

From Canoes to Keelboats

Prior to the 1850s, North America's rivers were its superhighways, the prime carriers of commerce, news, and knowledge to Native American settlements and early white immigrants. As trade and traffic increased, cities grew up along the nation's rivers, especially at the junctions of navigable rivers or where rivers met the sea.

Traffic moved slowly, propelled by wind, muscle, and the power of the river itself. A variety of ingenious craft worked their way up and down the rivers, each suited to a particular task and to river conditions.

Native Americans, early explorers, and fur traders relied largely on birch-bark or dugout canoes. They were light, moved well even against the current, and could be carried across portages, but they had little space for freight, were open to the weather, and often capsized in rough water.

The pirogue was little more than an enlarged dugout canoe. Made of a cottonwood, sycamore, or cypress log, hollowed out by fire or adze, it could be made in any size. The largest ones were fifty feet or more long and five feet wide.

Bateaux were flatboats of whipsawed lumber. Often called "Mackinac boats," they were wider through the middle than the pirogue, with tapered ends, up to seventy feet long. They were propelled by oars and usually rigged with a mast and sail. They could handle cargoes of forty to fifty tons and were widely used in the fur trade.

The keelboat evolved later. It was widely used on the Ohio River well before 1800, subsequently becoming common on the Upper Mississippi River. During its heyday the keelboat was the clipper of inland navigation.

The boat was built on a keel, with side planks and ribs, sharp at both ends, with a flat bottom. Keelboats ranged from forty to eighty feet long, were eight to ten feet wide, and had a mere twenty to thirty inches of draft. There was a roofed cabin amidships with narrow, cleated walkways on both sides, on which the crewmen labored when poling the boat upstream. The crew usually numbered ten men, and the cargo was about three thousand pounds per man. Even when fully loaded it could navigate narrow, shallow channels.

Keelboating was a toilsome, dreary, slow means of navigation. Most of the time a keelboat was poled against the current, the crew endlessly trudging bow to stern braced against the cleats in the walkway. In water too deep for poles, crewmen went ashore with long lines and the boat was cordelled or warped by a line of heaving, grunting, sweating men. Boats were often rigged with a mast and sail that could be used when there was a favorable wind. When not under sail, fifteen to twenty miles per day was the norm going upstream; twenty-eight miles was extraordinary. Under sail, going upstream, Zebulon Pike's keelboat once made forty-three miles in a single day, a cause for rejoicing.

The monotony of a keelboatman's work was matched by the monotony of his diet. High in calories to sustain day-in, day-out physical effort, a crewman's daily fare was usually fried corn hominy for breakfast, salt pork and biscuits for dinner, and a pot of mush containing a pound of tallow for supper (Time-Life Books 1975).

In 1751 French boats on their way to Fort Chartres (below the mouth of the Missouri) were the first keelboats to ascend the

Mississippi River above the mouth of the Ohio (Hartsough 1934). In 1828 a line of keelboats ran between St. Louis and the mouth of the Rock River, taking furs, lead, and grain downstream and bringing gingham, sugar, and gun powder upstream (Russel 1928). Zebulon Pike and his party used a seventy-foot keelboat when they began their epic journey upstream from St. Louis in 1805, but traded for two bateaux and several canoes at Prairie du Chien (Coues 1965).

On larger deeper rivers, barges operated like overgrown keelboats. Up to a hundred feet long and twenty feet wide they were capable of carrying sixty to a hundred tons of freight in a cargo box located in the center of the boat. Each barge carried a crew of about fifty and was propelled by poling, warping, towing, rowing, or sailing. Later, they were pushed by steamboats (Hartsough 1934).

By the time of the War of 1812, the entire length of the Mississippi was in American possession. Traffic was heavy, mainly on the Lower Mississippi, but essentially one-way. Rafts, flatboats, and keelboats floated to the Mississippi via its tributaries, then drifted to New Orleans where their cargoes were unloaded. Rafts were disassembled and sold, and the backwoodsmen went home via some overland route (Clay 1986). The most famous of these was the Natchez Trace, following an old Indian trail from Natchez, Mississippi, to Nashville, Tennessee.

Steamboats Invade the River

The steamboat era on the Mississippi began in the 1820s and reached its zenith in the 1850s and 1860s. During those years the river and its tributaries were the main arteries of transportation and communication within the interior of the continent. Although the period 1860 to 1880 has been called the "Golden Age" of river transport, it was during the late 1860s and 1870s that the aura of steamboat invincibility was lost. In the years following the Civil War, canals, Great Lakes routes, east–west roads, and especially railroads took their toll on river shipping. Trade routes shifted from north–south to east–west, which brought the agricultural West into closer relationship with the industrial and financial East. The shift in trade routes was accentuated not only by the Civil War but also by a strong prejudice that developed against using the Lower Mississippi for shipments of grain to New Orleans, as many shippers came to believe that the heat and dampness of the southern region damaged their grain.

By the late 1890s steamboat trade on the Upper River was in steady decline, and by 1900 only about 6 percent of grain shipments from St. Louis went by river. After 1900 the decline was rapid, and by 1920 most steamboats were excursion boats operating between cities. Mark Twain returned to the Mississippi in 1882 for a nostalgic trip from New Orleans to St. Paul. When he passed St. Louis after an absence of twenty-one years he lamented, "half a dozen sound-asleep steamboats where I used to see a solid mile of wide-awake ones! This was melancholy, this was woeful!"

Advent of the Steamboat

Keelboats had served as the main freight haulers on the Upper Mississippi since the late 1700s, but their days were numbered. On April 16, 1790, John Fitch and his longboat without sails had performed the impossible on the Philadelphia waterfront. Propelled only by the force of a steam engine, the *Experiment* ran out into the Delaware River, outdistancing fast sailboats and strongly rowed oar boats. Fitch's steamboat made regular runs along the Delaware River throughout the summer and early fall of 1790 at speeds up to eight miles per hour. Yet, stagecoaches were still faster and cheaper (Harris 1989).

Seventeen years later, Robert Fulton launched the steamboat *Clermont.* With the financial backing of Robert Livingston, one of the negotiators of the Louisiana purchase, Fulton quickly secured a monopoly on steam navigation of the Hudson River. Flush with success, the partnership looked to duplicate their success on the rivers of the expanding West.

In 1809 Nicholas Roosevelt traveled by flatboat from Pittsburgh to New Orleans, familiarizing himself with the Ohio and Mississippi Rivers in preparation for a repeat journey by steamboat. Fulton and Livingstone were hopeful of extending their monopoly to the western rivers after the state of Louisiana granted them monopoly rights to New Orleans and the mouth of the Mississippi—though other states along the Ohio and Lower Mississippi would prove less compliant—assuring free trade and navigation throughout the heart of the continent.

Backed by Fulton and Livingston, Roosevelt set to work building the *New Orleans,* the first steamboat to operate in western waters. The *New Orleans* left Pittsburgh in September 1811 and arrived in New Orleans in January 1812, surviving the great New Madrid earthquake along the way (Hartsough 1934).

Although steamboats penetrated the Lower Mississippi River in 1811, it wasn't until the *Zebulon M. Pike* made the trip from Louisville to St. Louis in August 1817 that a steamboat traveled the Mississippi above the mouth of the Ohio. Little more than a barge with an engine, the *Pike* was the second smallest steamboat documented on the Mississippi. So underpowered that she had to be poled, keelboat-style, in swift currents, it took the *Pike* six weeks to come up from Louisville (Tweet 1983).

Three years later, in 1820, the *Western Engineer* made its way upstream as far as the Des Moines Rapids at Keokuk, Iowa. It was the first steam-powered vessel to operate that far north, and it inaugurated navigation on the Upper Mississippi proper.

The days of the muscle-powered vessel were over.

Steamboat Construction

The steamboat had to be built and operated cheaply enough to float a cargo and make money, be powerful enough to master swift currents, and still be simple enough to be operated by uneducated engine-room crews.

The first steamboats on the western rivers were patterned after keelboats or lake steamers, but by 1820 every steamboat on western waters followed a design innovated by Henry Shreve, a pioneer steamboat captain.

Shreve pioneered his design on the *Washington,* built in 1816 at Wheeling, West Virginia. The *Washington* was designed from the keel up to operate in the shallow, debris-choked waters of the western rivers. The boat's hull was a shallow-draft platform with a sharply curved bow. In a radical departure from the design of steamboats built to operate in deep water, Shreve mounted the Washington's engine high in the hull, above the waterline, rather than setting it deep in the hull. The practicality of the design was demonstrated in 1817, when the *Washington* made a round trip from Louisville, Kentucky, to New Orleans in forty-one days (Hartsough 1934).

The river steamboats evolved into the lightest draft steam vessels ever floated. Flat bottomed, multitiered, designed to slide over the water rather than plow through it, burning prodigious amounts of cordwood to fuel the most powerful, most dangerous engines then known, they were different from any vessel heretofore.

River steamboats were built with a care that trimmed every ounce of excess weight. All superstructure and decks were made of pine rather than sturdier but heavier oak. Although the superstructure was of the lightest, flimsiest construction, many boats were beautifully designed and adorned into floating palaces (Russel 1928).

For the owner, less weight translated into greater profitability. The less a boat weighed empty, the greater load it could carry and still navigate the shallow, winding waterways of mid-America. Steamboats regularly operated on waterways that could easily be forded by a short man on foot. Many packet boats drew only twenty inches of water, some only twelve when light, allowing them to maintain regular schedules even in late summer when rivers ran very low.

For small tributaries, the boat builder's ingenuity was most taxed to provide something that would carry engine, freight, and passengers—and still get over sandbars. In the 1800s, steamboats regularly navigated streams that today would challenge a small fishing boat.

The hackneyed old claim about the western steamboat that could run over a field after a heavy dew was not too wild an exaggeration. One captain claimed that when his boat became grounded on one side, all he had to do was walk over to the other side of the boat to free it. Sometimes he had only to shift his quid of tobacco to the other cheek.

Some tributary boats were very small. At one landing far up the Des Moines River, a farmer joshed the captain of the diminutive side-wheeler *Michigan* saying, "My wife is sick in bed and has never seen a steamboat. Can't you let me put your boat on my wagon and

Decked out with bunting for a holiday, the jaunty packet side-wheeler *Alton* belches coal smoke as she cruises upriver between St. Louis, Missouri, and Alton, Illinois (courtesy of the Winona County Historical Society).

take it up to my house and show it to her? I promise to take good care of it and be back with it in two hours." The captain was mortified, and the jest remained with the *Michigan* for as long as she survived (Russel 1928, 27).

The vast majority of boats built were wooden-hulled craft. The first iron-hulled steamboat on the Mississippi was built near New Orleans in the late 1830s, but there were few others before the 1870s. As late as 1906 only about 5 percent of western riverboats had hulls of iron or steel.

A steamboat's hull was only a platform on which to place machinery and freight. The hull was usually empty of cargo except for maintenance supplies like oil, paint, hardware, rope, and cables. It also contained a two-man hand pump and a steam pump to dewater the bilge. There were bundles of wooden shingles that could be pounded into leaking joints in the hull. As they soaked and swelled, they plugged the leaks. The largest of the 1800s steamboats had no more than five feet of hold, most only had three and a half or four feet.

The hull of a steamboat was more than a simple flat-bottomed, watertight box. A carpenter's square was of little use in building a steamboat, since almost every major fitting involved a curve or an angle other than ninety degrees.

The hull of the boat curved from bow to stern, creating a reverse arch, called a "rocker." A rockered hull made the boat more maneuverable by reducing the boat's draft fore and aft, and the arched

---

The *Sultana* Disaster

On April 27, 1865, the United States suffered the worst maritime disaster in its history with over 1,600 deaths, more than were lost on the *Titanic*. Yet, the tragedy got few headlines. The Civil War was over, and about 2,300 war-weary Union soldiers who had suffered atrocities, disease, and starvation in the infamous prisoner of war camps at Andersonville, Georgia, and Cahaba, Alabama, were shipped by rail to the repatriation center at Vicksburg, Mississippi. From there they were to be transported upriver by steamboats toward their homes throughout the Midwest. Battle-scarred and emaciated, they were thrilled to be going home to be with loved ones, some after four years of war and twenty-three months of imprisonment. Many, having been wounded, were semi-invalids.

As many as two thousand soldiers (the exact number is unknown) were crammed onboard the 360-foot side-wheeler *Sultana* along with the crew, about one hundred civilians, at least sixty horses and mules, a hundred hogs, 120 tons of sugar, and ninety cases of wine. The incredibly overloaded boat was rated to carry only 376 passengers, leaving "standing room only," but the troops were in high spirits. The Army was paying their fares—five dollars for enlisted men and ten dollars for officers. They were going home!

The icy-cold Mississippi was in flood stage, swollen by snowmelt from the north. The *Sultana*, with decks sagging, a patched boiler, and perhaps the cooling tubes in its tubular boilers partially plugged with silt, churned upstream. Because telegraph lines were still in postwar disrepair, Captain J. C. Cass pulled in at river cities along the way to spread the word that President Abraham Lincoln had been assassinated.

Without warning, at about 2 A.M. on April 27, about seven miles above Memphis, Tennessee, a boiler in the vessel's midsection erupted in a violent explosion, killing, maiming, and scalding the crowded passengers

design dramatically increased the strength and cargo-carrying capacity of the hull. Steamboats had two big timbers set as uprights between the chimneys and the bow. These timbers were connected to bow and stern with steel tie-rods for additional support in maintaining the hull's rocker against the weight of the cargo. With the weight of the engines and paddle wheels aft, the bulk of the cargo was carried forward.

The rockered design could cause problems of its own. Today's *American Queen* was built with a rocker that caused rain from the decks to flood the dance floor located midship, causing the builders to be sued.

The engines were the most durable part of a steamboat, often outlasting the craft they were built to propel. Engines were often salvaged from sunken, wrecked, or dismantled boats, and they were commonly used on a succession of craft and even hauled ashore to power sawmills.

The high-pressure engine developed by Oliver Evans was used by almost all Mississippi River steamboats from the 1820s on. Steam was generated by batteries of two or more long, cylindrical boilers, perforated lengthwise by pipelike flues that conducted hot gases from the fireboxes through the water-filled boilers to generate high-pressure steam. The fireboxes opened forward to capture the draft created by the forward motion of the boat. Two iron chimneys towered as much as one hundred feet above the upper deck to carry smoke and sparks away from passengers and the flammable boat and cargo.

---

and crew. Others were burned in the ensuing conflagration, but the death toll was highest among those who were forced by the flames to leap into the dark Mississippi. Most could not swim, were incapacitated, or couldn't find debris for floatation. There were only 786 survivors, and over two hundred of them died in overcrowded hospitals during the next few weeks.

The disaster was scarcely covered by the Eastern press and was soon forgotten. After all, it only concerned folks that lived far on the other side of the mountains. Today, the charred remains of the *Sultana* lie buried in an Arkansas soybean field (Rabun 1997; A&E Television Networks 1998).

Even though Brownsville, Pennsylvania, on the Monongahela River was twelve hundred miles from St. Louis, it was the hub of riverboat construction in the 1870s because the heavy industry for engine building was in nearby Pittsburgh. Copies of the Boulton and Watts engine that Robert Fulton imported from England in 1807 powered early American vessels. These engines were heavy,

The *Sea Wing* Disaster

At 7:30 A.M. on Sunday, July 13, 1890, the *Sea Wing*, a 109-ton sternwheeler departed from Diamond Bluff, Wisconsin, with eleven passengers on a twenty-seven-mile, all-day excursion downstream to Lake City, Minnesota. Captain Wethern had an attached barge, the *Jim Grant*, alongside to help accommodate passengers on the pleasure cruise, which promised to be doubly enjoyable—passengers would visit the summer encampment of the Minnesota National Guard's First Regiment at Camp Lakeview, at Lake City, where thirteen companies of men were bivouacked. On her way downstream, the *Sea Wing* picked up twenty-two passengers at Trenton, Wisconsin, and 114 more at Red Wing, Minnesota, mainly people in their teens or early twenties, eager to see local boys encamped at Camp Lakeview where there would be military demonstrations, a band concert, and a dress parade. The day was oppressively hot and humid, so the breezy ride downriver was exhilarating. People lunched, laughed, and danced to the music of a four-piece band aboard the *Jim Grant*.

The steamer tied up at Lake City at 11:30 A.M. and stayed there until the festivities were completed at Camp Lakeview. Even heavy clouds and advancing rain failed to dampen the spirits of the crowd. At 8 P.M., with about 175 passengers, the *Sea Wing* ventured out onto Lake Pepin for the voyage home. Tornadic winds struck the Lake Pepin area soon after, capsizing the *Sea Wing* in mid-lake as she was trying to cut her barge loose and get into the lee of Maiden Rock Bluff. Ninety-eight people died in the disaster.

In those days it was commonly believed that explosions would cause drowned people to float to the surface. Despite detonation of over thirty dynamite charges, some as large as fifty to one hundred pounds, on Monday and Tuesday, most bodies did not surface until the warm lake waters began to give up their dead on Wednesday morning (Johnson 1990).

complicated, and low pressure, condensing steam to create a vacuum, thus employing atmospheric pressure to assist in moving the pistons within upright, stationary cylinders. High-pressure engines, whose cylinders were placed horizontally and had oscillating pitmans, replaced them. Steam was exhausted directly into the air. They weighed only five tons as compared to one hundred tons for the less powerful low-pressure engines and yet were far more powerful, especially if the engine crew "hard fired" the boiler to keep pressures rising.

On early boats, safety features were minimal. Water and steam pressure gauges did not come into general use until the middle 1850s, so crews had to use instinct to determine if boilers were running dry or if steam pressures were building to dangerous levels.

The boilers were interconnected so that they all had equal water levels, but if the boat developed a list, water from the highest boiler could drain into the others. Left dry and uncooled, the exposed flues would become red hot, soften, and collapse under the pressure generated inside the boiler, releasing a blast of superheated steam into the engine room.

Boiler explosions were the most feared of all disasters. On April 27, 1865, one of the five greatest marine disasters in history occurred on the Lower Mississippi seven miles above Memphis when the steamer *Sultana* blew up, killing 1,647 people on board, most of them exchanged Union prisoners of war, bound for home (Petersen 1968).

The engines drove the paddle wheel: either a single wheel churning at the stern, or a pair of wheels, one on each side of the boat.

The largest, most pretentious boats were the grand side-wheelers of the Lower Mississippi, which became the stereotype of the Mississippi River steamboat. However, these magnificent two- and three-story gingerbread palaces seldom ventured above the mouth of the Ohio and certainly not above St. Louis.

Above St. Louis, on the Upper Mississippi and its tributaries, the standard upper riverboat was a modest stern-wheeler. With paddle wheels removed from the side, the stern-wheeler could have a wider hull with less draft than a side-wheeler of the same tonnage. This enabled them to operate on narrow streams, in shallow water, and during the low-water season—traits desirable above St. Louis. The position of the paddle wheel at the stern also gave it a measure of protection from driftwood, logs, and ice.

On the other hand, side-wheelers had superior steering and handling qualities. On two-engine boats the wheels could be run in opposite directions, giving the boat the capability to virtually pivot in place. Early stern-wheelers were difficult to maneuver in strong head winds or crosswinds, conditions that often rendered them helpless. In the twenty years following the Civil War, technological improvements of the steering of stern-wheelers improved their handling qualities to the point that they became the design of choice for towing barges and in rafting logs and lumber (Hunter 1969).

As Pretty as a Wedding Cake

River men loved steamboats, admiring their elegance, grace, and beauty. Mark Twain stated, "The steamboat is as pretty as a wedding cake—without the complications."

Charles Russel gave this eloquent description:

> The old Mississippi River packet was among the comeliest of the works of men's hands. Compared to her harmonious lines and symmetrical sweep of designing, the average Hudson steamer is a lump, and boats on the Rhine, the Elbe, and other European waters seem without form and void. There was a certain grace about the sheer of the Mississippi boat, a kind of jaunty and insouciant grace, that has never been equaled elsewhere; her funnels were placed where they gave an effect of due proportion; her pilothouse rose above her texas [the small group of officers' cabins atop the third deck] with a cumulative touch of the impressive and the appropriate; her paddle-boxes were where they ought to be; her dainty upperworks were decked out with filigree, like lace. Between her chimneys she swung some bright device that added elegance and distinction, a gilded ball, a gilded star, or other heraldry; the top of her pilothouse might carry a pair of great deer's horns, painted red. The stern yawl was hung aft of the verge staff and swayed to the motion of the steamer, a kind of rhythmic finale to a beautiful conception. And then those paddle-boxes, if she was a side-wheeler! The painter's skill reserved its best for their luscious adornment—a great picture in colors of St. Anthony Falls, of Minnehaha rolling a slumberous sheet of foam below, of a buffalo hunt, of striking scenery like that of Trempealeau, of Phil Sheridan spurring to the battle of Winchester—works of art at which shore people were never tired of gazing, open-mouthed. (1928, 66)

Fuel

With abundant timber lining the banks of the river, fuel was abundant, inexpensive, and readily available, fortunate, since boats consumed wood at a prodigious rate. To save weight and reduce draft, steamboats rarely carried more than a twenty-four-hour supply, usually about twenty-five cords, generally "wooding up" twice daily.

In the very early years, crews went ashore daily to cut and gather wood, greatly prolonging voyages, but as the steamboat trade flourished, wood yards sprang up to satisfy the demand for seasoned cordwood. Farmers along the river often found cordwood to be their principal cash crop, going so far as to build chutes to convey hardwood from the bluff tops to the riverbank. To save time in "wooding" some boats going upstream took flatboats of wood in tow, transferred the wood "on the fly," then turned the empty flatboat loose to drift back to the wood yard.

With the opening of coalfields in Illinois and Missouri in the 1850s, the use of coal as fuel spread to the Upper Mississippi, but with ample wood available along the route, few steamboats used coal exclusively, most mixing it with equal quantities of wood.

After the Civil War, the return of luxury steamboats and steamboat races rekindled interest in the river. To put a fast, hot fire under their boilers, racing steamboats added pitch, pine knots, resin, lard, bacon, side pork, or any other fatty, greasy material to their usual fuel of dry oak and soft coal.

Steamboat Speed

The speed and reliability of the nineteenth century steamboat compared favorably with the capabilities of modern river transport. By 1845, fast steamboats were able to run from New Orleans to St. Louis, a distance of twelve hundred miles, in six to eight days.

In the early 1850s, virtually all boats ascending or descending the Upper Mississippi stopped at Galena, Illinois. The fastest boats made the trip from Galena to St. Paul and back in a little over three days, averaging two trips per week (Petersen 1968).

On August 17, 1858, burning soft coal and barrels of pitch,

Captain D. Smith Harris's *Gray Eagle*—the fastest boat on the Upper River—left Dunleith, Illinois, bound for St. Paul with momentous news.

He delivered mail at twenty-three landings and took on thirty-five cords of wood, yet attained an average speed of a little over eleven miles per hour upstream and probably thirteen miles per hour when underway (Merrick 1987; Petersen 1968).

Arriving in St. Paul, 265 miles upriver, Harris spread the news that the first telegraphic message, a greeting from Queen Victoria to President Buchanan, had flashed undersea by the Atlantic cable.

As late as 1858, a Mississippi riverboat was still the fastest means of communication or travel north of Prairie du Chien. Steamboats provided the first widespread, regularly scheduled transportation for mail, passengers, and cargo along the Mississippi and its major

Galena: The Metropolis of the Mineral Region

Between 1823 and 1848, no other single factor was so important in developing steamboating on the Upper Mississippi as the shipment of lead. During this quarter century, approximately 7,645 trips were made to the lead mines by steamboats, making the Upper Mississippi busier than either the Ohio or Lower Mississippi when measured by arrivals at St. Louis.

Called "Fevre (Fever) River" until 1827, Galena was the metropolis of the mineral region prior to 1850. In 1828 there were ninety-nine steamboat arrivals at the lead mines, shipping thirteen million pounds of lead out of Galena. In 1828, five years before Chicago became a town, Galena boasted a weekly newspaper, *The Miner's Journal*, and had forty-two stores and warehouses, twenty-two porter cellars and groceries, lawyers, doctors, mechanics, and seven hundred residents.

But Galena's days were numbered. Larger steamers virtually filled the narrow Fever River as they squirmed up it for seven miles to reach the city, especially as the river became shallower and narrower due to siltation. Because of their superior steering, side-wheelers were able to negotiate the river, but stern-wheelers could not. Some of the stern-wheelers were as long as 230 feet. The arrival of the railroad brought momentary prosperity to Galena in 1854, but its extension to Dunlieth on the east bank of the Mississippi across from Dubuque the following

Galena, Illinois, was at its peak of prosperity in 1856, but as the Fever (Galena) River silted in, it became increasingly difficult for stern-wheelers to turn around in Galena's harbor. Note that there is still enough room to accommodate side-wheel steamers that could "turn on a dime" (Whitefield lithograph courtesy of Galena/Jo Daviess County Historical Society & Museum).

year left Galena with only lead to depend on. Lead had been going overland since 1840, and the eastward flow increased with the coming of the railroad.

The "Port of Galena" ceased to exist in 1860 when the side-wheeler *W. L. Ewing* was unable to get to Galena on account of the mud. The city council of Dubuque, Galena's longtime rival, poked fun at Galena, suggesting a resolution that the bed of the Fever River be plowed up and potatoes planted (Petersen 1968).

The city that had once been acclaimed as the mighty commercial emporium of the northwest, the home of General Ulysses S. Grant and eight other Civil War generals, and the chief port of call for steamboats from as far away as Pittsburgh and New Orleans, was dying on the vine. Dubuque had exceeded her in size and population. At her peak of prosperity in 1850, Galena had a population of 9,241, but by 1870 her population had dwindled to less than four thousand. Today Galena thrives as a mecca for tourists and history buffs who relish her colorful past (Petersen 1968; Tweet 1983).

tributaries. The term *packet* was applied to vessels that made regular trips at stated intervals in contrast to transient or tramp boats. A *line* consisted of two or more steamboats offering packet service in a given trade.

By the early 1860s steamboats were running from St. Louis to St. Paul, a 683-mile voyage, in three days—just under ten miles an hour. (A contemporary diesel-driven towboat on the Upper Mississippi, pushing a fifteen-barge tow, will average twelve miles an hour downstream.) When pressed, steamboats could turn in truly impressive travel times.

In 1870, the *Robert E. Lee* set a speed record that was not bested until 1929—then by only three hours. In the most famous steamboat race on the Lower River, the *Robert E. Lee* and the *Natchez* raced upstream twelve hundred miles from New Orleans to St. Louis in a record time of three days, eighteen hours, and fourteen minutes—an average speed of over thirteen miles per hour (Clay 1986).

On July 6, 1876, the steamer *Far West* started downstream on the Missouri River from the mouth of the Big Horn River, carrying survivors of the Battle of the Little Big Horn. She made the seven-hundred-mile run to Bismarck, North Dakota, in fifty-four hours—averaging 13.14 miles per hour despite stopping for wood and making several other stops along the way (Time-Life Books 1975).

Hazards

The average life of a steamboat was only about five years, but owners knew that if they could operate their boat for four years they could buy or build two or three new ones from their profits, even without the aid of insurance (Merrick 1987). The boats' flimsy, flammable, wooden construction made them "accidents waiting to happen." Fires started from chimney sparks, grease fires in the galley, careless smokers, and from tipped or broken lanterns. Snags pierced their hulls, ice crushed them like eggshells, and boilers exploded. They ran aground on sand bars and smashed into rocks while running rapids. Where the channel ran close to shore, leaning trees menaced the chimneys and upper works of boats. After 1856, railroad bridges added to the river's natural hazards. By 1869 and the completion of the transcontinental railroad, collisions with the Rock Island bridge had already claimed sixty-four steamboats. Even before the bridge was completed, a steamboat crashed into one of the piers, burned, and sank. Of the 1,667 rafts and steamboats passing through the bridge in 1857, 55 collided with the structure (Tweet 1983).

To protect his boat, a steamboat pilot had to "read water"—to guess the speed of current in a bend, and to detect from surface swirls, slicks, and ripples the presence of concealed hazards. Snags were a perennial and persistent danger, although they were more of a hazard on the Lower Mississippi and Missouri Rivers. A tree trunk washed in during flood time could embed itself in the bottom with one stout branch sticking up under the surface where it could impale a steamboat. Especially deadly were "sawyers," waterlogged tree trunks that bobbed slowly and ponderously in the current with a sawlike motion. Anchored in the bottom and pointing downstream, they were like pickets that were submerged most of the time. When a steamer was pierced by a snag she was necessarily in shallow water and the passengers had a fair chance to escape, but the odds were against them in a boiler explosion or a fire.

Clearing the river of snags was the first order of business after the Army Corps of Engineers was given the task of maintaining the river for navigation in 1824. Powerful, twin-hulled snagboats, known as "Uncle Sam's tooth pullers," began removing snags on the Lower Mississippi River in 1829.

The first steam snagboat, *Heliopolis,* was completed in 1829. Its twin hulls and steam-powered winches were distinctive, but its most innovative feature was a heavy, wedge-shaped, iron-sheathed snag beam tipped with a massive hook. It could be used either as a battering ram or a dislodger. Such snagboats could remove snags weighing as much as seventy-five tons. In 1829 the *Heliopolis* raised a tree 160 feet long and 3.5 feet in diameter. By 1830 the snagboats' designer, Henry Shreve, had cleared the worst obstructions in the Mississippi from St. Louis to New Orleans. At least in part because of snagging, St. Louis became a competitive commercial city and the steamboat center of the United States before the Civil War (Dobney 1978). By 1867 the ancient snags of the Upper Mississippi River had been removed. After about 1870, the government provided regular service with snag-pulling steamboats.

Ice was a formidable killer of steamboats. Great chances were taken during the 1850s to get through Lake Pepin in the spring before it was free of ice. The river above and below was usually ice-free for about two weeks before the lake cleared. The navigation season was short at best, and before railroads reached St.

The Backwoods Lawyer and the Rock Island Railroad Bridge

On the morning of May 6, 1856, a crowd gathered at the waterfront of Davenport to greet the long, graceful, immaculate, side-wheeler *Effie Afton*. She charged upstream into the stiff current of the Rock Island Rapids, toward the new Davenport–Rock Island Railroad bridge that had been built oblique to the current. She signaled her approach by sounding one long, lovely chord of D, F-sharp, and A, and fought her way into the opening in the bridge. Suddenly her starboard wheel stopped, she swung sideways, hit a stone bridge pier on one side, rebounded and hit a stone pier on the other side, and careened badly until her great chimneys fell. Then, before the horrified spectators, she burst into flames. The blazing wreck floated down almost to Credit Island and sank. Luckily, rescuers were able to save all the passengers and crew. Less fortunate were the three hundred oxen tied to the lower deck for transport to Wisconsin logging camps. About a third of them survived; the rest suffocated or were so badly burned that they had to be killed to put them out of their pain.

In the court action that followed, the steamboat company sued the Chicago and Rock Island Railroad Company, owner of the bridge, alleging unlawful obstruction to navigation by reason of the bridge. After a thorough investigation of the accident, the railroad's articulate backwoods lawyer convinced the jury that if the steamer's starboard wheel had not malfunctioned, it would not have hit the bridge, and the railroad should not be liable. With astonishing frankness he admitted that pilots should look out for the bridge and be wary of it. The case had momentous repercussions because it established the principal that railroad bridges, serving a larger public interest, were of consideration *prior* to navigable streams. The decision spurred the development of railroads and assured the triumph of Chicago, railroad center, over its rival, St. Louis, river town. The backwoods lawyer was Abraham Lincoln (Russel 1928).

Paul an early start in spring gave great financial rewards to daring captains. The folks at the head of navigation had been cut off from the outside world since late November, and they eagerly anticipated mail, news, and shipments of manufactured goods. During the spring ice blockade, freight was unloaded at Read's Landing at the foot of the lake, hauled by team to Wacouta at the head of the lake, and reloaded on another steamboat for transport to St. Paul and other ports above the lake. Impatient passengers often walked the thirty miles around ice-bound Lake Pepin (Petersen 1968). To be the first through Pepin, some boats perilously sneaked along the shore where the ice rotted first. Others tried cutting a channel through the stronger ice in the center of the lake. In either case, boats could be crushed if a change in the wind moved the ice mass (Merrick 1987).

Surprisingly, the greatest ice disasters were not in the northern reaches of the Upper Mississippi, but in the southern extremities where ice gorges were more likely, often damming the river and piling ice as high as thirty feet. When the ice dam finally gave way it caused an immediate flood of water and ice downstream. St. Louis was a veritable killing place for steamboats because so many boats overwintered there. In just a few hours on February 27, 1856, about twenty steamboats and fifty small boats were destroyed and an equal number put out of commission when a sudden rise in river level caused a flood of ice two feet thick to devastate the levee area where the boats were moored. Similar disasters occurred at St. Louis in 1865 and 1876 (Merrick 1987).

With federal funds, an Ice Harbor was constructed at Dubuque, Iowa, midway between St. Louis and St. Paul. Completed in 1885, it could provide ice protection for fifty barges and twenty steamboats. The harbor soon became a premier shipyard, producing many steamboats; the most famous was the *Sprague,* the largest stern-wheel towboat ever to operate on the Mississippi.

Fire was an ever-present threat to boats made out of wood, and often disaster on one boat spread to others in the neighborhood. On the night of May 17, 1849, fire broke out on the steamer *White Cloud,* moored at the St. Louis levee. The fire spread from vessel to vessel along the levee and then ignited adjacent buildings.

Twenty-three steamboats and fifteen city blocks were consumed (Time-Life Books 1975).

Steamboats Open the Interior of the Continent

Overcoming the Rapids

The most formidable obstacle to navigation on the Upper Mississippi was part of the river itself. As the Wisconsin ice sheet melted away about ten thousand years ago, and flow from glacial lakes ceased, huge amounts of sand and gravel from tributaries filled the riverbed to about its present level as far south as St. Louis. Overloaded with sediment, the river from St. Paul to St. Louis became a maze of islands, shifting sandbars, twisted channels, and long shallow reaches. This glacial legacy, together with the Rock Island Rapids and the Des Moines Rapids hindered navigation on the Upper Mississippi for more than a hundred years.

The nimble steamboat with its flat bottom, light structure, and shallow draft was ingeniously designed to overcome most obstacles, but it had extreme difficulty with the Des Moines Rapids just upstream from Keokuk, Iowa, and the Rock Island Rapids running upstream to Le Claire, Iowa. Prior to 1820 most river men believed that the two rapids would forever prohibit steamboat travel from Keokuk north (Tweet 1980).

Both rapids impeded navigation, even for keelboats and rafts, but they were very different in age, structure, and the problems they created. The Des Moines Rapids were the older of the two, running through a channel at least 120,000 years old. Because there was no real channel through the rapids, the whole 11.25-mile stretch simply became impassable during low water (Tweet 1980).

Zebulon Pike and his crew made it through the Des Moines Rapids in their keelboat without mishap in 1805, but only after Sac Indians assisted them by removing fourteen of the heaviest barrels to lessen the keelboat's draft. Two Sacs piloted the boat through the rapids.

The faster, more perilous Rock Island Rapids proved more of a challenge. Pike lost his rudder on the first rapids and had to stop for repairs. Fortuitously, a favorable gale sprang up; the crew hoisted a sail and negotiated the rapids without a pilot and without touching a rock. Pike stated, "had we struck a rock, in all probability we would have bilged and sunk. . . . Those shoals are a continued chain of rocks, extending in some places from shore to shore, about 18 miles in length. They afford more water than those of the De Moyen, but are much more rapid" (Coues 1965, 25).

Geologically, the Rock Island Rapids were much younger than the Des Moines. Both of the last two glaciations, the Illinoisan and the Wisconsin, pushed the Mississippi westward south of Clinton, Iowa. It finally settled into its present channel about eighteen thousand to twenty thousand years ago, forming the Rock Island Rapids as it cut through glacial debris and hard limestone outcroppings. Because the river hadn't had time to scour a smooth, wide channel as at Keokuk, it formed more typical rapids with fingers, or "chains," of rock projecting outward from each shore. The Rock Island Rapids were much more extensive and dangerous than the Keokuk Rapids and ran from Le Claire, Iowa, down river to Rock Island (Tweet 1980).

The normal velocity of the Mississippi was about three miles per hour, but over the Rock Island Rapids it ran about five. The water ran in puzzling and unreasonable ways over the chains of rock, and the current patterns were different for every river stage, making the rapids among the most perplexing in the world (Russel 1928).

Boats and log rafts always ran the Rock Island Rapids by daylight, and it was customary to employ a special rapids pilot for this dangerous, fifteen-mile stretch. Most of the pilots lived in Le Claire, at the head of the rapids.

The first steamboat to encounter the rapids was the sternwheeler *Western Engineer* in 1820 carrying Major Stephen H. Long on an exploratory trip for the U.S. Army. At Keokuk he encountered the formidable Des Moines Rapids. The prevailing feature of the rapids was the flatness of the riverbed, formed by almost horizontal ledges of limestone that followed the slope of the river. At low water, the rapids were no more than twenty-four inches deep, a severe obstacle even for shallow-draft steamboats. The rapids had limited traffic on the Mississippi as far back as the eighteenth century when fur traders had to have their boats unloaded and their cargo carried across the rapids in smaller craft called "lighters." Indian villages had sprung up at Keokuk and Montrose in response to the need for labor.

The master of the *Western Engineer* considered it impossible to navigate the rapids and returned to St. Louis. But it would not be long before boats challenged the rocks and rushing waters at Keokuk (Hartsough 1934).

In May 1823 the diminutive steamboat *Virginia,* loaded with tourists, Indians, and military supplies for Fort Snelling, made its way through the Des Moines Rapids. It probed its way so slowly that passengers often walked along the shore and rejoined the boat upstream. Not only did her pilot have to thread his way through rocks, snags, and sandbars, the *Virginia*'s progress was interrupted frequently so those on board could cut or gather wood to replenish her fuel supply. She had to tie up at sundown each evening because navigation was too risky after dark—except for one night when a large wildfire lit the night sky for several hours Despite every caution, the boat still stuck fast in the Rock Island Rapids for two days before being freed by a sudden rise in water.

The *Virginia* successfully challenged the general belief that steamboats could not negotiate the rapids of the Upper River. Leaving St. Louis on April 21, 1823, she arrived at Fort Snelling in about twenty days (Hartsough 1934).

The *Virginia* completed the round trip to Fort Snelling, and then turned around to make a second round trip in June (Tweet 1980), solidly establishing the value and practicality of steam navigation on the Upper River.

Following the successful voyages of the *Virginia* and drawing on Stephen Long's reports of 1817 to 1823, which recommended, among other things, that canals be constructed around the rapids, Congress passed the General Survey Act in 1824. The act gave the president authority to use officers of the U.S. Army Corps of Engineers to make surveys of navigation routes. In addition, Congress assigned responsibility to the Corps of Engineers for managing the river and improving it for steamboats. The authority has rested there ever since (Madson 1985).

Wasting no time, the Corps of Engineers began to improve navigation on the river by removing snags and sandbars, by excavating rock to eliminate rapids, and by closing off sloughs to confine flows to the main channel. These alterations enabled shallow-draft steamboats to use the river and its tributaries as water highways. Concurrently, erosional processes in the watershed were accelerated as settlers began to log the forests, plow the prairies, and practice steepland agriculture. The tempo of these activities increased at a fever pitch in ensuing years.

The first reconnaissance surveys of the Des Moines and Rock Island Rapids were made in 1829, and recommendations were made to excavate the Des Moines Rapids channel to a depth of five feet, but no further work was done until 1837 when Lieutenant Robert E. Lee resurveyed the rapids and endorsed excavating the channel. Work was done sporadically until 1866 when it was apparent that excavation was not working. To improve the rapids the way suggested by Lee and others would have required an eleven-mile cut, two hundred feet wide through solid rock, resulting in a narrow sluice with an extremely strong current, making navigation difficult and dangerous.

Economic pressure to do something about the bottleneck on the river was increasing as the War between the States came to a close. The five states bordering the Mississippi north of the Des Moines Rapids were growing more than a third of the produce in the United States, and they wanted to ship it downriver. Most important, the young lumber industry was booming, with more than four hundred million board feet of lumber being rafted downriver to sawmills each year. There was pressure from the South as well, as boats bound upriver found their progress stymied by the swift shallow water at Keokuk and Rock Island. For some of the larger boats, Keokuk was the head of navigation because they feared the two rapids above.

At the end of the Civil War, Keokuk, at the foot of the Des Moines Rapids, was a busy place with over a thousand steamboat landings at its levees every year. More than three hundred steamboats were engaged in commerce on the Upper Mississippi, yet they still had to transfer cargo to lighters or to the new railroad that flanked the rapids between Keokuk and Montrose.

Anxious to heal the wounds of the Civil War and aware of the need to improve the Mississippi for trade between North and South, Congress authorized a four-foot channel north of St. Louis. As part

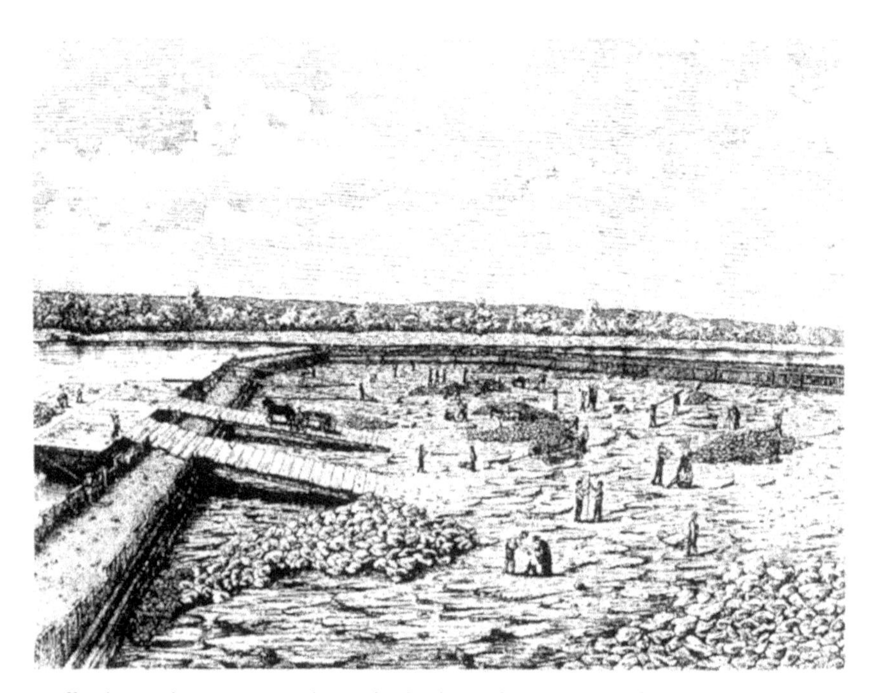

Coffer dam and excavation at the Rock Island Rapids. Once the coffer dam had been built and the contained water pumped out, obstructive rocks could be blasted with black powder charges. Earlier excavators shattered rocks by heating them with bonfires and dousing them with water (sketch by draftsman Henry Bosse, U.S. Army Corps of Engineers, Rock Island District).

of the Rivers and Harbors Act, money was provided for improvement of the Des Moines Rapids and the Rock Island Rapids. Engineers concluded that if it wasn't practical to create a channel through the Des Moines Rapids, the solution was to build a canal around them.

About a thousand men were employed at the peak of construction of the Des Moines Rapids Canal that opened to traffic in 1877. The lateral canal ran along the Iowa shore and was 7.6 miles long, three hundred feet wide, and five feet deep. Its three locks provided a total lift of 18.75 feet. The locks were constructed of limestone quarried from the adjacent bluff (Tweet 1980).

Work on the Rock Island Rapids began in the fall of 1867. The Corps' contractors used several methods to excavate the rapids. In shallow water, they constructed cofferdams around the areas to be excavated, dewatered the areas, and blasted the exposed rock. In deeper areas they used chisel boats, dredges, even submarine blasting. By 1886, the work was virtually complete with a channel with a minimum width of two hundred feet and a minimum depth of four feet having been cut. The fast currents in the twisting channel still made the rapids dangerous, and "rapids pilots" were employed to guide craft through the rapids. A significant improvement to the rapids was the removal of the awkwardly placed Rock Island railroad bridge and its replacement in 1872 with a new, safer bridge that could accommodate both trains and wagons (Tweet 1980). Nevertheless, the Rock Island Rapids remained a major obstacle until the Moline Lock, completed in 1907, enabled boats to bypass the worst of them.

Rivers Carry the Frontier

Despite hazards and obstacles, steamboats played a vital role in the settlement of the Upper Mississippi Valley. The majority of European immigrants who came to the Upper Mississippi Valley between 1830 and 1870 used the steamboat for all or part of their journey.

Steamboats carried the American frontier both west and north. In 1820 the area around St. Louis was well populated, and by 1830 the frontier line had reached Keokuk. By 1840 the frontier included the lead-zinc region around Dubuque and Galena, and by 1850 it had reached the northern boundary of Iowa. During the 1850s, steamboats shoved the frontier a short distance beyond St. Paul.

By 1847 St. Louis had become the base of navigation for the Upper Mississippi River and its tributaries, and also the head of navigation for the larger steamboats from the Ohio and Lower Mississippi Rivers. Its population had grown to about fifty thousand, and it was the world's greatest steamboat port next to New Orleans (Hartsough 1934).

In 1856 the total steam tonnage operating on the Mississippi and its forty-four navigable tributaries was said to exceed the total steam tonnage of Great Britain. In 1862 1,015 steamboats landed at St. Paul, which had become the great entry point for the undeveloped country to the Northwest (Russel 1928).

Steamboats not only plied the main stem of the Mississippi, they brought settlers and supplies to some remote reaches of the big river's tributaries.

For almost a century the Missouri River was the most important means of entry into the great expanse of land between the Mississippi and the Pacific Ocean. Men had followed it searching for a water route to the Pacific Ocean and the riches of the Orient (spices, silk, gemstones, and tea), mythological kingdoms, beaver pelts, gold, homesteads, adventure, and glory. Most Americans today believe that the West was opened solely by wagon trains and railroads, but prairie schooners carried little cargo and the "iron horse" did not reach the Rockies until the late 1880s (Time-Life Books 1975).

For boatmen, the Missouri was one of the greatest challenges of all. Its tortuous bends and shoaled, muddy waters challenged canoeists, keelboaters, and eventually steamboatmen. Even so, the Missouri provided a water highway deep into the heart of the West. On July 17, 1859, the *Chippewa* made it to within fifteen miles of Fort Benton (Montana), 3,560 miles from, and 2,565 feet above, the ocean—the whole distance made by steam on a river unimproved by artificial works (Petersen 1968)

Elsewhere in the upper Midwest, steamboats operated regularly on streams now relegated to canoes and fishing boats. In Iowa, steamers traveled the Des Moines River all the way to the trading post located at the site of Iowa's capital city, Des Moines. They traveled, by schedule, on streams like the Iowa, Cedar, and Maquoketa. In Wisconsin, they ran 150 miles up the Rock River and up the St. Croix River to Taylor's Falls.

In the early days, fair-sized steamboats ran up the Chippewa River as far as Eau Claire, Wisconsin (Fugina 1945). An excursion boat, the first steamer in twenty-six years, made it upstream seventeen miles to Durand in 1912. The trip was arranged by businessmen to demonstrate the navigability of the Chippewa, but it was the last steamboat trip ever made on the river (Russel 1928).

The steamboat *Enterprise* was built on the Wisconsin River at Wausau, in north-central Wisconsin, and ran about 225 miles down to the Mississippi, negotiating a significant waterfall at Stevens Point, many rapids, and innumerable shoals (Russel 1928).

In southern Minnesota, in 1858, the *Little Frank* ran up the Root River for fifty miles to Rushford, carrying lumber upstream and grain and flour downstream.

To the north, the *Governor Ramsey,* built and operated by my great-grandfather John Rollins, made her trial run above St. Anthony Falls in May 1850 and found it feasible to run up another seventy miles to St. Cloud and Sauk Rapids. Rollins's second boat, the *North Star,* ran from St. Anthony to Sauk Rapids until 1857. Later, diminutive steamers ran north as far as Bemidji, on the Mississippi Headwaters, about a hundred miles from the Canadian border (Merrick 1987).

Cargoes

The first steamboats on the river carried trade goods, supplies for the military and Indians, then settlers, then soldiers who crushed Indian resistance, and finally they carried the vanquished Indians to distant reservations.

After 1823, the primary stimulus for expanded steamboat traffic on the Upper Mississippi River came from the lead district centered around the ports of Galena and Dubuque. Between 1823 and 1848, 200 of 365 boats that sailed above the Des Moines Rapids were primarily in the lead trade, handling shipments from Galena (Tweet 1983). As early as 1843, steamboat arrivals from St. Louis alone totaled 243, with 50 keelboats in tow (Hartsough 1934). During the 1850s, Galena was the western terminus of the Illinois Central Railroad, and was the most important and populous point above St. Louis.

But the lead boom would be relatively short. By 1840, lead was being hauled overland to Milwaukee and then via the Great Lakes to eastern markets. As traffic increased on the Great Lakes and eastern canals, the overland route grew in importance, and by about 1857 the lead trade on the Mississippi had virtually disappeared (Hartsough 1934).

By the 1850s, downstream traffic was chiefly agricultural produce—wheat, potatoes, barley, oats, flour, and cranberries. In later years, cargoes included Minnesota ice destined for the Deep South, vitreous white sand (St. Peter Sandstone) from St. Paul for glass making, and even circus menageries.

A typical upstream boatload might include four hundred tons of groceries, dry goods, and farm machinery, 465 cabin passengers, and ninety deck passengers (Hartsough 1934; Merrick 1987). Upstream boats also carried herds of cattle, horses, mules, sheep, and pigs as well as flocks of turkeys, ducks, geese, and chickens. They carried sawmills, gristmills, and flourmills, bricks, church organs, threshers, reapers, fire engines, railroad locomotives, railroad rails, liquor, and kegs of nails. In 1857 coal was shipped from Pittsburgh down the Ohio and up the Mississippi to St. Paul, a distance of about two thousand miles.

In the early days, steamboats frequently towed keelboats full of additional cargo, but the keelboats were replaced by wooden barges beginning in the 1850s. By the end of the Civil War, flat-bottomed, shallow-draft barges had emerged in large numbers. They offered increased cargo capacity, ease of loading, and flexibility like railroad cars. Grain didn't have to be sacked as it did when carried on boats, and like rail cars, barges could be picked up or dropped off during a trip up or down the river (Tweet 1983).

All barge cargo was carried on deck. Goods that could be damaged by moisture were stored in a shed atop the barge. Flour and other perishables were packed in barrels that could be rolled on and off. Later, flour was packed in bags that could be carried.

By the 1870s the grain trade had grown to immense proportions, with steamboats towing barges of grain to Upper Mississippi railroad elevators as well as to St. Louis and New Orleans, but it was the upstream swarming of immigrants that made the grain trade possible, staving off for almost twenty years the decline and ultimate extinction of steamboating on the Upper Mississippi River.

Passengers

The *Virginia*'s historic voyage in 1823 marked the beginning of a new era for the Mississippi. Until this time, the fastest means of travel in the river corridor was a man on horseback. The steamboat made it possible for people to travel to the frontier without bearing the hardships that early explorers had experienced.

The "Fashionable Tour," endorsed by George Catlin in 1835, brought about important changes in steamboat accommodations and introduced tourism as a factor in the river economy. Realizing that tourists would patronize those boats with the best facilities, captains created luxurious accommodations—staterooms with a washstand and beds with spring mattresses. Few steamboats dared compete without a band or orchestra. Excursionists soon learned which craft served the best meals.

During the 1850s, passenger traffic became the most lucrative branch of the steamboat business as settlers began to pour in. Just as lead dominated river commerce before 1850, the value of passenger traffic exceeded that of all other cargoes between 1850 and 1870.

The creation of the Territory of Minnesota in 1849 brought about the "Golden Age of Steamboating" on the Upper Mississippi. The number of immigrants boarding at St. Louis and traveling upriver to St. Paul dwarfed the 1849 gold rush to California and Oregon (Tweet 1983).

A deluge of immigrants from states farther east and from foreign countries poured into the Upper Mississippi Valley during the mid-1850s and 1860s. Each train brought hundreds of immigrants to the banks of the Mississippi where they boarded steamboats for the trip upriver to the "promised land."

They came singly, as families, and sometimes in groups of a hundred or more to found settlements in the new country. The nascent settlements were often named for the cities in the "old country" from which the settlers came—New Albion, New Trier, New Prague, New London, New Lisbon. Some of the immigrants could afford the steamboats' cabins, but most lived on deck at half the cabin rate, supplying their own beds and food and assisting the crew in wooding up (Hartsough 1934). A boat two hundred feet long and thirty feet wide could accommodate two hundred cabin passengers comfortably, one hundred second-class passengers on deck, and a crew of as many as eighty (Merrick 1987).

Deck passage on a steamboat was often horrific, with crowds of immigrants sharing the deck with livestock. Washrooms were filthy, and there were polluted water, mud, and offensive odors everywhere. Tobacco chewing was common, and the decks were fouled with spit. Due to the crowded, unsanitary conditions and ignorance of how communicable diseases were spread, cholera took a heavy toll of lives. Passengers who died were usually hastily buried on shore but sometimes just thrown overboard.

At the end of the 1850s, most steamboat passengers reached the

The Grand Excursion of 1854

The Grand Excursion of 1854 was by far the most brilliant event of its kind that the West had ever witnessed. President Millard Fillmore, one of the honored excursionists, declared that it was one for which "history has no parallel, and such as no prince could possibly undertake."

The first railroad to unite the Atlantic Ocean with the Mississippi River had reached Rock Island on February 22, 1854. To celebrate the event, the railroad contractors invited leading citizens of the country to participate in a joint railroad and steamboat excursion to the Falls of St. Anthony. The response was so enthusiastic that the number of chartered steamboats was increased from one to five. The preparations were so lavish that an eastern newspaper declared the affair "could not be rivaled by the mightiest among the potentates of Europe." Almost every metropolitan newspaper sent a writer to accompany the excursion (Petersen 1968).

Shortly after 8 A.M. on the fifth of June, two gaily decorated trains of nine coaches each left Chicago filled with excursionists. They were met at each stop by bands, fireworks, and parades and reached Rock Island at four o'clock that afternoon, filling the five chartered steamboats and two additional craft. A shortage of berths caused about one-third of the guests to renounce the steamboat trip and return to Chicago, but at least twelve hundred remained on the boats and were served a "sumptuous feast" equal to any afforded by the country's best hotels.

After speeches at Rock Island and Davenport, the passengers were entertained by a brilliant display of fireworks from Fort Armstrong. Decorated with prairie flowers and evergreens, the steamers left Davenport at 10 P.M. with bells and whistles sounding. With music on deck, they sailed by moonlight up the Mississippi, arriving at the entrance to the Fever River at dawn. The boats cruised up the Fever to Galena for field trips to the lead mines and a picnic dinner in the woods, complete with wines from France and Ohio. Leaving Galena, passengers were free to go ashore whenever the boats made landings at towns and scattered settlements and whenever the boats stopped to "wood up." When underway, the boats were often lashed together, enabling folks to socialize, enjoy the music, promenade, and dance from boat to boat (Petersen 1968).

Only a little more than six years old, St. Paul boasted six thousand inhabitants and made a fine appearance from the decks of the arriving steamers. Shortly after arrival, excursionists were transported overland

Upper River by rail at Rock Island, Dubuque, Galena, or Prairie du Chien. Prairie du Chien was the terminus of the railroad nearest St. Paul and the Upper River. It had the promise of being a big city, the outlet and entry point for the trade of a great territory. A dozen steamers might be seen loading merchandise from the railroad or transferring wheat and produce to railroad cars destined for Milwaukee or Chicago (Merrick 1987).

By 1860, the 2,028,948 inhabitants of Missouri, Iowa, and Minnesota exceeded by almost four hundred thousand the combined populations of Arkansas, Texas, New Mexico, California, Oregon, Washington, Nevada, Utah, and Colorado; by 1870 a fairly dense population extended for many miles inland along both banks of the Mississippi. The 1870s saw a precipitate decline in passenger traffic as north–south railroads afforded more dependable year-round transportation (Petersen 1968).

---

to the Falls of St. Anthony in whatever carts, wagons, and buggies were available. After more field trips, the fleet cast off from St. Paul for the return trip.

Meals on board rivaled those of modern cruise ships. Meats and vegetables were prepared in one kitchen, while pastry and deserts were readied in another. When needed, trout, river fish, wild game, lambs, pigs, chickens, turkeys, ducks, eggs, and vegetables were bought along the way. Oysters and lobsters were served daily, rushed by rail from two thousand miles away and packed in ice. Cows on the lower deck provided fresh milk. From the bakery came fresh breads, cakes, fancy desserts, and decorative candied sugars. Excursionists were invited to inspect the pastry kitchen at any time, but they were never allowed in the meat and vegetable kitchen, where animals were dispatched, plucked, scalded, butchered, and cooked. Scenes there could spoil appetites. As on all steamboats, all wastes went into the river (Petersen 1968).

"Fashionable Tours" to the Falls of St. Anthony formed a unique phase of steamboating that did much to advertise the Upper Mississippi. Influential eastern newspapers encouraged tourists, various organizations, foreign travelers, and honeymooners to take the tours. In the following years, hundreds of excursions were made to this garden spot of the West. The cooperative rail–steamboat ventures were wonderful but short lived. The honeymoon didn't last long, and a bitter divorce ensued.

Railroads Bridge and Flank the River

The railroads were undoubtedly the principal factor in the decline of Mississippi River transportation. At first they were welcomed as feeders and supplementers of water routes. They stimulated river traffic in the 1850s and 1860s with a mutual exchange of freight and passengers, but the cooperation lasted only until the railroads bridged the river.

The Chicago and Rock Island Railroad reached the Mississippi at Rock Island in February 1854 and bridged the river over to Davenport, Iowa, in 1856, crossing the narrowest part of the treacherous Rock Island Rapids where the current was fastest.

The Rock Island bridge was followed by fourteen more on the Upper River by 1887. Not only did the railroads extend westward but soon paralleled the river, capturing north–south trade as well as east–west trade. In 1871 the Lake Superior and Mississippi Railroad connected the Twin Cities with Duluth, diverting additional traffic from the upper valley to the Great Lakes route (Hartsough 1934).

The Mississippi River was a major railroad obstacle, but it was met with imaginative engineering feats. In 1862, beginning on the west side of the Mississippi, tracks were laid westward for the Winona and St. Peter Railroad Company, and a wood-fired locomotive dubbed "Old Tiger" was shipped to Winona by its now-faltering competitor, the steamboat. During the early years, before bridges, railroad cars were laboriously floated across the Mississippi by ferries. Traffic was speeded when the river froze each winter, allowing temporary rail beds to be laid across the ice. Temporary winter bridges spanned thin spots, were high enough to drive a team and wagon underneath, and were removed in the spring.

Permanent railroad bridges spanned the Mississippi at Hastings, Minnesota, in 1869; Winona in 1871; and at La Crosse in 1878. Westward progress had been slowed by the Civil War, but by 1873 the Winona and St. Peter Railroad had breached the western border of Minnesota and plunged into the Dakota Territory. By 1873 Minnesota had 1,906 miles of completed railroads, of which 70 percent had been built in four years. By 1900 Winona residents had daily rail service to Chicago, Kansas City, Green Bay, Minneapolis, and St. Paul.

Direct west to east railroad routes had great advantages over the Mississippi River route. The railroads offered quicker, more direct transportation, usually without transshipment. They were also more reliable, not being tied up by ice, low water, or floods. Their rates were higher at first but were partially compensated by lower insurance rates (Hartsough 1934).

Rail routes were always shorter than water routes between the same two points. St. Louis and St. Paul were 683 miles apart by water, but only 585 miles by rail (Tweet 1983). From Galena it was a few miles farther to New Orleans by river than it was by rail to New York. Continuing on to Liverpool, England, the New York route was over fourteen hundred miles shorter than the New Orleans route. Also, New Orleans had never produced the manufactured goods that the West needed, making it difficult to secure return loads. All of this was exacerbated by the economic somnolence of the South in the immediate post–Civil War period (Hartsough 1934).

Railroad competition with the steamboats was by no means passive. As soon as railroads could exert their independence they became aggressive and ruthless competitors, cutting their rates at all competitive points, often carrying goods at a loss during the navigation season and recouping their losses by raising rates in the winter. The railroads could have cooperated with water routes by making through rates such as the rail lines commonly made with one another, but they refused to do so. They even charged more for short hauls than long hauls, further decreasing the likelihood of cooperation (Hartsough 1934).

Rail lines squelched river competition even more with the purchase of waterfronts, terminal facilities, and even steamboat and barge-line companies for the purpose of preventing use and development of water routes. By 1910 commercial waterfronts not publicly owned were largely controlled by railroad interests (Hartsough 1934).

Declining river traffic was also the fault of the steamboat business itself. While the railroads were built and managed by some of

The St. Louis waterfront was a busy place in 1896 when the packet steamboat *Dubuque (B)* loaded coal below the famous Eades Bridge from which this photo was taken. River level was fairly low, as shown by the amount of floodable, sloping levee (complete with railroad tracks) available for cargo handling and sight seeing. Today, much of the area shown in the right side of the photo has been developed as the Jefferson National Expansion Memorial, which includes the landmark Gateway Arch (courtesy of Winona County Historical Society).

America's shrewdest and most aggressive entrepreneurs, the steamboat business lacked expert organization and management. Steamboat men tended to be a fiercely independent breed, running their own boats, establishing their own schedules, setting their own rates, and competing with each other.

Facilities for handling freight and passengers along the waterfront also failed to meet the needs and rising expectations of the public. When the Mississippi was busiest, a typical waterfront at a river town was a dirt bank sometimes paved with rock. Most river ports lacked adequate docks, loading and unloading equipment, and warehouses. Freight was often piled in a heap and covered with a tarp to protect it from rain, mud, and sun. Warehouses were located near the river where they were subject to flooding.

Disunited and disorganized, the steamboat interests could not compete with a natural monopoly like the railroads (Hartsough 1934).

The End of an Age

The decade of the 1850s was marked by depression and the beginning of a trend that would relegate steamboats to a minor role in the economic life of the Mississippi Basin. The decline was caused by many factors, but mainly they were the loss of traffic to railroads, changing direction of traffic from west to east instead of north to south, the growing severity of competition among steamboatmen, steamboat disasters, low water years, and the deteriorating conditions of the river itself (Hunter 1969).

For St. Louis the Golden Age of Steamboating ended abruptly in 1860 with the beginning of the Civil War that closed southern ports. By 1860, nine Upper Mississippi ports had been united with the Atlantic seaboard by rail—Cairo, St. Louis, Rock Island, East Burlington, Dunleith (East Dubuque), Fulton (opposite Clinton), Quincy, Prairie du Chien, and La Crosse. In 1867 St. Paul was linked by rail with La Crosse. At the same time only three railroads had reached the Mississippi south of Quincy—at St. Louis, Cairo, and Memphis.

After the war, the St. Louis to New Orleans trade never fully recovered. Recovery was also hindered by close identification with the South. St. Louis had a number of railroad connections and was the chief market and wholesale center for the Upper River. River trade was still the chief basis of prosperity for St. Louis, but the "writing was on the wall" because the Great Lakes and rail lines to the East were depriving it of trade. St. Louis had bitterly fought the building of the railroad bridge between Rock Island and Davenport, and now Chicago threatened to depose St. Louis and Cincinnati for commercial supremacy (Hartsough 1934; Merrick 1987).

Commercial interests had always perceived St. Louis as southern and Chicago as northern. Shippers got onto the habit of moving grain by rail to Chicago during the war and retained the habit after the war (Tweet 1983).

At St. Louis the river had lost to the railroad. St. Louis became a rail center in the 1870s, but Chicago became the great link in the first transcontinental railroad, completed in May 1869 at Utah's Promontory Summit. Chicago went on to become the rail center of the Midwest if not the nation.

13

"Improving" the River, 1878–1930

> The military engineers of the Commission have taken upon their shoulders the job of making the Mississippi over again, a job transcended in size only by the original job of creating it.
>
> Mark Twain, *Life on the Mississippi,* on management of the Mississippi

The Four-and-One-Half-Foot Channel Project

The railway boom that followed the completion of the first transcontinental railway in 1869 threatened to strangle river commerce. Not only did railroads carry freight quickly and efficiently inland from river ports to markets in the east and west, they also ran when the river was frozen or impassable due to extreme low flow.

While the country enjoyed the benefits of a more efficient transportation system, the fear arose that, unhindered by competition, the railroads could impose a transportation monopoly on the country. In the Congress, representatives concerned with the growing power of the railroads came to look at a revitalized river transportation system as a competitor to the railroads that would act as a force to hold down freight rates. However, the burgeoning rail lines and shrinking number of boats on the river made it all too clear that action would have to come quickly if a rail monopoly was to be offset.

Acting on this impulse, on June 18, 1878, Congress authorized the

An October 22, 1922, view downriver from Schamaun's Bluff (RM 738.5), about ten miles upstream from Winona, Minnesota. The wing dams on the left (Wisconsin) shore were built in 1893, and new wing dams are being built along the Minnesota shore by the construction fleet in the distance. Note the sand accumulation between the older wing dams. In subsequent years they were buried by dredged sand as part of the 9-Foot Channel Project. The area is now forested (U.S. Army Corps of Engineers, St. Paul District).

creation and maintenance of a navigational channel four-and-one-half-feet deep on the Upper Mississippi River between St. Paul, Minnesota, and the mouth of the Ohio River, to be implemented by the U.S. Army Corps of Engineers. Funding for the project was provided in the initial bill and renewed annually, but the magnitude of the project was such that it was not substantially completed until 1907.

The technologies used by the Corps were not new. The plan called for using the power of the river itself to create and maintain a channel of the specified depth. To do this, the force of the river current would be directed to the designated main channel through the construction of wing dams (called wing dikes in southern reaches), closing dams, shoreline protection, and dredging where necessary.

The purpose of wing dams and closing dams is to constrict the area through which the river flows during low flow, thus forcing it to

Wing dams were made by sinking successive willow mattresses with "one-man rocks" quarried from nearby bluffs. While rock was relatively expensive, hand labor was cheap, allowing crews to balance the barge load by neatly piling the rocks. This also made inventory and volume estimation easier. Because the wooden barge carried its cargo atop its deck, it had a high center of gravity and had to be carefully unloaded to maintain its balance (U.S. Army Corps of Engineers 1891 photo by Henry P. Bosse, courtesy of Winona County Historical Society).

scour itself deeper. By using wing dams to direct the river's flow down a single narrow channel, and closing dams to cut off the flow in alternate channels, a swift current was created that prevented deposition of sediments in the main channel. Instead, most sand accumulated on the downstream side of the wing dams where it filled in the area from the channel border to the shore.

Wing dams had been used for a thousand years in China and on the Rhine and other European rivers. American engineers used them successfully to keep traffic moving on the Ohio and Illinois Rivers, and loggers on the Chippewa River had used brush wing dams in the 1850s to maintain an open channel. Henry Shreve had used brush dams to prevent shoaling on the Lower Mississippi River beyond New Orleans.

Most wing dams and closing dams were constructed not only in the summer but also in winter when loads of brush and rock were hauled onto the ice with horse- or mule-drawn sleighs. After layers of brush and rock had been laid on the ice, workers used special saws to cut through the ice around the dam, letting it fall into place. Additional layers were added until the dam reached the required height (courtesy of the Winona County Historical Society).

Brush dams are not effective on rocky, clear-water rivers. For the dams to be effective, the river must transport enough sediment to fill the interstices of the brush mats, making the dam a solid structure, entombing the wood, and preserving it. The Mississippi and Missouri Rivers were ideal candidates.

Wing dams were constructed of readily available materials—willow saplings cut from the river bottoms and limestone or dolomite quarried from adjacent bluffs. The willows, tied in bundles twenty feet long and twelve inches in diameter, and rocks were barged to the dam site. There the bundles, called *fascines,* were loaded aboard a building barge and tied together to make mattresses. They were skidded into the river and held in place by ropes until sufficient rock could be loaded on them to sink them, layer after layer, into the depths. The dams were built in from five to forty feet of water and were constructed so that they projected as much as six feet above the

This rare 1927 aerial photo shows how wing dams and closing dams were positioned just downstream from Fountain City, Wisconsin (RM 728.5 to RM 730.5). The river is flanked on the Wisconsin (lower) shore by the Burlington Northern Railroad. River flow is from right to left. Dendritic white lines are caused by crazing of the aged photographic emulsion. Note that the bluff-top Wisconsin farm fields are not contoured or stripped and that sheet and gully erosion are obvious. Bluff faces are severely grazed. Most of the floodplain in the photo was subsequently inundated in 1935 by the Nine-Foot Channel Project to form Pool 5A (U.S. Army Corps of Engineers, St. Paul District).

surface. Quarried rock was brought down to the river by horse cart originally, but some quarrymen developed tram systems that worked much more efficiently.

Since rock had to be piled, picked up, and dropped onto the mattresses by hand, the boulders were hewn to be "one-man rocks." In an era before workmen's compensation or health insurance, no one calculated the cost in smashed fingers and toes, muscle strains, inguinal hernias, and assorted back problems.

The usual practice was to build wing dams in a series with the shorter ones on the upriver end. The spaces between most wing dams rapidly filled with sand as high as the dams. Willows soon sprouted on these sandbars, creating new islands in a few years (Fugina 1945). There was, however, a downside to this effect—they also collected sewage and garbage in an era when the river was treated like a sewer.

The most difficult rock and brush structures to build were the closing dams to cut off backwater channels and keep water flowing in the primary channel. As the dam was built, the velocity of the current through the shrinking opening steadily increased. This high-velocity water flowing through a partially completed closing dam could sweep away the works that had been completed.

Riprap and wing dams kept the river from carving a new channel through the floodplain, the normal riverine process of meandering. The structures "hardened" the river or "fixed" it in position.

Farther south where river bluffs no longer flank the river and rock is less accessible, wing dams were mainly brush, with just enough rock to sink them. To address this problem, hurdles (piling dikes) were first employed in the St. Louis area in the 1880s in conjunction with stone dikes (wing dams).

To construct a hurdle, a row of piles five feet apart were tightly interwoven with brush. The hurdle worked by letting the river itself provide the materials necessary for channel constriction. These permeable structures slowed the water flowing through them, causing much of the tremendous silt load contributed by the Missouri River to settle out, narrowing the channel, filling sloughs, and forming new river banks. Solid dikes were virtually abandoned in the St. Louis area in favor of the cheaper, more efficient hurdles (Dobney 1978).

Riverbanks tend to erode and cave away, especially on the outsides of bends. To prevent this, willow mats and rocks were used. The river bank was divided into three zones: the first was beneath the water surface extending from the lowest point of erosion up to the historic low water mark; the second from the low water mark up to a level where willows would grow; and the third from the latter point to the top of the bank. The lower zone was protected with a brush mattress weighted with rocks. Because rotting was prevented by constant inundation, the mats lasted indefinitely. In the second zone where rotting would occur, rock riprap was used. Willows were planted in the upper zone. This system of bank protection was effectively used from St. Louis to Cairo and was used in modified form throughout the Upper Mississippi (Dobney 1978).

Water rushing around the ends of the wing dams intensified erosion on the opposite shore. Consequently, the shore was usually fortified with mattresses and rock to keep it from being washed away.

Weighted mats were also placed on the shore from which a wing dam extended. In the spaces between the dams, timber, leaning trees, and stumps were removed from the shoreline, which was then graded to a three-to-one slope, faced with layers of brush, and covered with stone (Hill 1961). Without protecting this shore, the river might cut behind the dam and negate its function.

Up and down the channel, a dipper-type dredge removed troublesome sandbars.

The Six-Foot Channel Project

While construction proceeded on the Four-and-One-Half-Foot Channel, riverboat design evolved—and the larger, more powerful boats needed a deeper channel to carry greater payloads. To address this need, additional funds were appropriated by Congress in 1907 to deepen the navigable channel to six feet by constructing two thousand additional wing dams, additional shore protection, additional dredging, and the construction of two new locks.

The first of the two locks was completed in 1913 as part of the Keokuk and Hamilton Water Power Company's waterpower dam and lock at Keokuk, Iowa. The dam inundated the entire Des Moines

Rapids and its canal under a fifty-mile-long pool of water as deep as forty feet. The second project was the LeClaire Canal and Lock completed in 1922. Its six feet of water permitted boats to bypass the upper 3.6 miles of the Rock Island Rapids (Tweet 1980).

Powerplant and Lock and Dam 19 at Keokuk

The Des Moines Rapids Canal that had opened in 1877 performed well for many years, but it had limitations. The massive log rafts that floated downstream wouldn't fit in the canal and had to be broken up and reassembled, a procedure that could take forty to fifty hours. Almost all boats going upriver used the canal rather than fight the current, but during high water, about 15 percent of downbound steamers chose to run the rapids and bypass the canal, saving themselves over an hour of travel time.

The beginning of the end for the Des Moines Rapids Canal came when the Rivers and Harbors Act of 1902 authorized a survey at Keokuk to determine if building a dam at the foot of the rapids would benefit navigation. The report favored a dam that would flood the entire rapids. Negotiating the single lock associated with the dam would cut travel time and operating expense, compared to negotiating locks at both ends of the canal. Raft traffic would suffer, but it was already nearly dead, with only one sawmill remaining on the Mississippi south of Keokuk.

The plan was not without its critics. Some believed that the river had outlived its usefulness. Waterways were decried as commercially obsolete. Public money spent on their maintenance and improvement was spoken of as no more than "pork barrel" contributions to the demands of local politics.

Yet, in 1905, the Keokuk and Hamilton Water Power Company (now the Union Electric Power Company) was authorized to construct a dam with a hydroelectric plant, a lock, and a drydock at Keokuk. These structures, with the exception of the dam and powerhouse, were turned over to the United States upon their completion in 1913. The Keokuk dam was the only dam on the Mississippi below St. Anthony Falls, and its hydroelectric plant was one of the world's largest. Like all other projects of the time, it was built with virtually no public input or concern for the environment.

The dam severely impacted the ecology of the Mississippi River. Water that did not go through the power plant's massive turbines had to go over the top of the forty-foot-high dam. Consequently, the structures became a nearly impenetrable barrier to migrating fish. Ebony shell mussels (clams) disappeared above Lock and Dam 19 because the dam blocked migrations of skipjack herring, which were the prime hosts to the larvae (glochidia) of the ebony shell clams. Without the skipjacks to distribute the clam larvae to good habitat, the ebony shells were extirpated above the dam. They had furnished valuable raw materials for the pearl button industry.

The huge hydroelectric plant generated electricity far in excess of demand. To market this surplus power, drainage districts were formed that used electric pumps to dewater floodplain land for agriculture. This scheme resulted in the destruction of large expanses of wetlands and floodplain forest.

Limitations of the Six-Foot Channel

The magnitude of the early channelization projects is mind-boggling. In constructing the Four-and-One-Half-Foot and Six-Foot Channels, the Corps of Engineers built thousands of wing dams and closing dams between St. Louis and Minneapolis. In addition, hundreds of miles of shoreline were protected with riprap. Most of the limestone and dolomite bluffs that abut the river have quarries in them where rock was excavated for constructing the channelization structures.

While the construction of wing dams and other rock channel-training structures allowed a more continuous flow of river commerce by keeping the river open during a longer part of the season and during more times of low water, their capabilities were limited. Maintaining a navigable channel in the free-flowing river depended on a minimal volume of water; and when that water was not forthcoming, river traffic stopped.

This limitation became starkly apparent when the transport of freight on the Upper Mississippi declined precipitously after the lumber industry peaked out in 1892. The Four-and-One-Half-Foot and Six-Foot Channels limited movement of other types of freight that required larger, deeper draft barges to be transported economically.

The final blow came in 1925, when, with the Six-Foot Channel only half completed, the Corps determined that the six-foot depth would not be possible for the entire length of the river with the techniques then being applied. In response, the River and Harbors Act of 1927 abandoned the six-foot proposal and authorized an eventual Nine-Foot Channel.

Requiem for Steamboats

By the early 1900s, the river that had bustled with traffic lay virtually deserted. The government fruitlessly continued to spend money on river improvement—wing dams, closing dams, and snag removal. Hundreds of millions of dollars were spent, chiefly by the federal government, to revitalize commercial river traffic. Ironically, most of the money was spent after the river had ceased to be an important transportation route. The channel was so carefully marked by government buoys and lighted markers that even inexperienced navigators ran little risk. But except for an occasional construction barge or recreation boat, there was no one to use the channel markers in the 1920s. The 359 crossing lights between St. Paul and Cairo burned religiously every night, but except for an occasional excursion boat there were no steamers to see them (Russel 1928).

World-class Smallmouth Bass Fishing

A modern fisheries manager with an unlimited budget would have been hard pressed to design and implement habitat as good as that created by the early channelization projects of the 1878 to 1925 period, which produced incredible fishing for smallmouth bass and walleyes in areas not grossly polluted.

The rock rubble and brush of wing dams, closing dams, shoreline protection, and the rock foundations of channel markers provided excellent habitat for the growth of invertebrate animals, prime food of fish and other wildlife. Their location in the current—and because the water that flowed over them was nutrient-rich and well-oxygenated—meant the wing dams were rich in species and total numbers of invertebrates. Crayfish, a favorite food of smallmouth bass, walleyes, and turtles, proliferated in the nooks and crannies of the wing dams, grazing on abundant algae, other invertebrates, and decaying plants and animals.

The first two decades of the twentieth century saw the Upper Mississippi almost cease to be an avenue for long-haul traffic. By 1924 only five freight boats were still making runs in the St. Louis to St. Paul trade. Raw statistics accurately depict the decline in long-haul traffic, mainly the rapid loss of rafts and raft boats, but they do not show the shift to less glamorous short-haul work boats hauling sand, gravel, rock, and other low class bulk commodities. After 1914 until 1940, the largest user of the river was the Corps of Engineers engaged in river improvement projects. In 1921, 94 percent of the tonnage carried on the river was sand, gravel, and rock. After 1929, operations were an almost continuous struggle against low water

---

At the same time that wing dams set out an unmatched underwater smorgasbord, they furnished underwater "structure" so critical to some species of fish, especially the popular smallmouth bass, among the scrappiest of freshwater fish, known for its spectacular jumps.

The smallmouth bass fishing was world-class, and life was indeed simpler in the era preceding the building of the nine-foot channel dams of the 1930s. Because outboard motors were scarce, most boat fishermen employed the lost art of rowing, often for several miles to get to their "secret spots." Other anglers walked the railroad tracks that flanked the river and then walked out on wing dams or along the shoreline riprap.

When feeding heavily in the summer, smallmouth bass and walleyes preferred the upstream edge of the dam or the outstream end where current was fastest. Between the dams there were areas of lessened current where fish could laze when they weren't in the feeding mode.

Equipment was simple, usually a cane pole, often outfitted with piano wire to withstand the inevitable snags. Frogs, now scarce, were abundant then and were a popular bass and walleye bait. Willowcats (diminutive, venomous members of the catfish family) were especially deadly for walleyes (and they still are).

The "compleat angler" of the day used fancier equipment—steel casting rod, level-wind reel, split bamboo fly rod, and artificial lures. Kids, strapped for cash, used a willow pole, bobber, and worms to catch bass, sunfish, crappies, channel catfish, flathead catfish, drum, suckers, and an occasional eel. And there wasn't a water-skier, personal watercraft, yacht, or bass boat in sight!

and poor channel conditions (Tweet 1983). The prolonged drought and economic depression of the 1930s were double disasters. The river was so low in the summer of 1932 that people at Winona could wade, chest deep, across the river.

In retrospect, it should have been obvious that steamboat trade on the Upper River was doomed. If the railroads hadn't killed the steamboat freight business, trucks and airplanes would have. Cars, buses, and airplanes would have killed the steamboat passenger service just as they have killed it on most rail lines. The railroads have survived to this day as the nation's prime freight haulers, cooperating with trucks and ocean-going ships by hauling containerized cargoes.

Concurrent with the twilight of commercial traffic on the Upper River, a new type of towboat evolved that would have momentous impact. Two new diesel-powered boats went into service on the Illinois River in midsummer 1933. They each developed one thousand horsepower, yet the tunnel-type, twin-screw boats only drew four and one-half feet of water—no more than paddleboats. They were more powerful and maneuverable than stern-wheel towboats of the same size, yet operated with a smaller crew (Hartsough 1934). In combination with the Nine-Foot Channel Project, also begun in 1933, the diesel towboat would bring about a renaissance in river commerce.

Floodplain "Reclamation" Schemes

Early channelization projects had manipulated the river to improve it for navigation, but subsequent schemes were aimed at improving the river's floodplain for agriculture. According to the prevailing attitude of the time, wetlands were wastelands.

During the early 1920s, owners of floodplain land between La Crosse and Prairie du Chien proposed draining their land to make it suitable for agriculture. The proposed reclamation project included clearing timber, constructing dikes to protect the land from high water, and digging internal ditches to carry water toward pumping stations where it would be pumped over the dikes. The landowners urged that the drainage districts be created under state law, publicly financed with the drainage costs charged back to the landowner who benefited.

Opponents of the scheme insisted that floodplain areas should be preserved for recreation and for the conservation of plant and animal life. The Izaak Walton League of America, which strongly supported the parties opposing drainage, requested the U.S. Department of Agriculture to investigate the practicability of reclaiming floodplain land between St. Paul and Rock Island (Marsden and Shafer 1924).

The resulting survey revealed that there were about 343,000 acres of floodplain land between St. Paul and Rock Island and that the principal agricultural use of the land was for pasturage for cattle in dry seasons. Less than a fourth of the land was mowed for hay and only a very small part was cultivated.

About ten thousand acres of the land had already been drained by 1924. Most of this early land reclamation was done in Wisconsin where 6,600 acres of bottomland in Buffalo and Trempealeau Counties had been drained by 1912. A break in a dike in 1913 left most of the area flooded, and no pumping had been done between 1913 and 1924. Ultimately that land reclamation program was abandoned and most of the land became the Delta Fish and Fur Farm, later to be incorporated into the Trempealeau National Wildlife Refuge.

A second drainage district of 3,600 acres was completed just below Savanna, Illinois, in 1925. The survey reported that another 86,000 acres could be reclaimed at an average ditching and diking cost of forty-five to seventy-five dollars per acre. Operation and maintenance of the drainage pumping plants were to be provided by an additional annual assessment. Farm land thus created was to be utilized for growing corn, the report continued, because the dairy farmers on the hills bordering the Mississippi were reported to have insufficient land suited to the growing of corn and were forced, therefore, to import cattle feed from other states. This rationale seems incredible today when most commercial river traffic is geared to the export of surplus grain, mainly corn.

Agricultural development of the Upper Mississippi River floodplain is most prevalent below Rock Island where it has an average width of four to six miles but may exceed ten miles in some areas.

Floodplain agriculture depends upon levees. From Minneapolis to Rock Island the floodplain is narrow, and only about 3 percent has been leveed (about 15,000 acres). Between Rock Island and St. Louis, where the floodplain is wider, about 53 percent (about 530,000 acres) has been leveed. From St. Louis to Cairo, about 82 percent (about 543,000 acres) has been leveed (Fremling et al. 1973).

The Canal Craze

The Portage Canal

When nature didn't provide a water link between two points, nineteenth-century Americans often attempted to rectify the oversight with pick, shovel, and determination.

Sparked by the successful 1825 completion of the 363-mile-long Erie Barge Canal that connected the Hudson River with Lake Erie, Wisconsin business leaders and politicians promoted a canal that would ultimately connect New Orleans with the Atlantic seaboard. They proposed digging a canal between the Fox and Wisconsin Rivers at Portage, Wisconsin, where the rivers flow within a mile and a half of each other. When complete with locks and dams on the two rivers, the canal would enable steamboats to navigate from Buffalo, New York, through the Great Lakes, up the Fox, and down the Wisconsin and Mississippi to New Orleans.

Digging on the Portage Canal began in 1835 and the first steamboat traveled the Fox–Wisconsin waterway in 1858. By the 1920s as many as 250 steamboats had used the system, but it was doomed—primarily by shifting sandbars and fluctuating water levels on the Wisconsin River and competition from railroads offering cheaper, faster, year-round transportation. For the last half-century of operation, the main users of the canal were recreational boaters. The federal government officially closed the Portage Canal in 1951.

Canals Connecting Lake Michigan with the Illinois River

As early as 1836, ambitious Chicago businessmen developed a plan to connect Lake Michigan with the Illinois River and the Mississippi, a project first suggested by early explorers and traders in the region. The Illinois and Michigan (I&M) Canal was completed in 1848, helping establish Chicago as a national transportation hub. So successful was the canal that soon after its completion, the number of steamboat arrivals at St. Louis from the Illinois River was almost equal to those arriving from the Upper Mississippi (Petersen 1968; Hill 2000).

A more ambitious project, the Chicago Sanitary and Ship Canal, also connected Lake Michigan with the Illinois River. The project, begun in 1892 and completed in 1900, was intended to bolster both the economic health of the city and the physical health of its people. As Chicago's population grew and industry flourished, Lake Michigan, the city's source of drinking water, became increasingly contaminated by domestic and industrial sewage. The Chicago Sanitary and Ship Canal project reversed the flow of the Chicago River, allowing the city to use Lake Michigan water to flush its waste down the canal, into the Illinois River, and into the Mississippi. The canal was successful in reducing the threat of cholera and other waterborne diseases in Chicago, but at a tremendous cost to the health of the Illinois River environment.

A third shipping canal, the Cal Sag Channel, completed in 1922, further connected Lake Michigan with the Illinois River. The old I&M Canal was closed in 1933 (Capano 2003).

The impact of these canal projects has been profound and continuing. In addition to severely degrading the Illinois River, the canals also provided avenues for exotic biota, such as zebra mussels, brought into the Great Lakes via ocean-going ships, to enter the Mississippi and its tributaries. Likewise, undesirable exotic fishes from the Mississippi threaten to enter Lake Michigan via the Illinois River and connecting canals. In 1955 the American Society of Civil Engineers declared the Chicago Sewage Disposal System one of the seven wonders of the modern engineering world.

The Proposed Lake Superior–Mississippi River Canal

During the late 1800s and early 1900s, navigation interests agitated for a canal from Lake Superior to the Mississippi River. In 1912 Congress directed the Corps of Engineers to determine the feasibility and commercial potential of the idea. The engineers reported in 1913 that a canal up the St. Croix River, over the Continental Divide, and down the Brule River to Lake Superior was practical but not feasible commercially. The canalized section between Taylor's Falls on the

St. Croix and Allouez Bay on Lake Superior would have been 160 miles long. The canal commissions of Minnesota and Wisconsin appealed the decision, but the canal was never built and the plan was never revived (Fugina 1945). Such a canal would have been an ecological disaster, destroying two of North America's most pristine rivers and enabling movement of fish and other organisms between two continental drainage systems that had been isolated from each other for ten thousand years.

Headwaters Reservoirs

In its Headwaters, the Mississippi flows through nine glacial lakes, including Lake Winnibigoshish and Lake Pokegama, both of which were dammed in the late 1800s as part of a U.S. Army Corps of Engineers navigation and flood-control system that includes four other lakes in the Headwaters watershed.

The project's original purpose was to store spring runoff in order to augment low summer flows for commercial navigation between St. Paul and Prairie du Chien, but the Nine-Foot Channel Project of the 1930s made that function unnecessary. Because the Headwaters dams raised water levels and stabilized them, lakeshore on the beautiful, enlarged reservoir lakes became precious. Property owners built lake homes in accordance with the artificially maintained lake levels, and the owners became so numerous and politically influential that tinkering with the lake levels for any purpose was out of the question. The reservoir dams are now mainly used for flood control, recreation, residential amenities, and conservation.

14

Mission Impossible

The Nine-Foot Channel Project

Who Needed a Nine-Foot Channel?

Like the construction of the Erie Canal early in the nineteenth century, construction of Chicago's Sanitary and Ship Canal trained a generation of engineers. Many veterans of the "Chicago School of Excavation" went to work on the Panama Canal, made possible by techniques developed in Chicago (Capano 2003). The opening of the Panama Canal in 1914 awakened the Midwest to the potential of a restored transportation system on the Mississippi.

With the opening of the Panama Canal, freight could suddenly be shipped coast to coast by water at dramatically lower cost than by rail. It became cheaper, for example, to send farm machinery manufactured in New York to San Francisco by water than it was to ship the same equipment directly from the John Deere factory at Moline, Illinois, to San Francisco by rail (Tweet 1983). If a viable river navigation system were in place, Deere could ship its machinery to San Francisco by water via New Orleans and the Panama Canal at less expense than from New York.

The economic impact on river cities of the shift from shipping by river to shipping by rail could be seen in the Minneapolis flour industry. In the early 1920s, Minneapolis was still the nation's flour milling center, producing nearly twenty million barrels a year. But by the mid-1920s it was less expensive to ship unprocessed wheat to Duluth by rail for Great Lakes shipment to eastern markets than it was to mill the grain in Minneapolis and ship the flour east by rail. By 1935 flour production in Minneapolis had dropped to less than eight million barrels a year (Tweet 1983). With a nine-foot channel, flour could be barged less expensively to the East via the Mississippi and Ohio Rivers.

As the population and industry of the Upper Midwest region grew, there was an increasing demand for cheap coal for power generation. The need for coal in the Upper Midwest was complemented by the need to ship the Midwest's surplus grain south to other centers of population. Thus, even greater economies could be realized by having cargoes going both directions on the upper reaches of the river.

The steel barges and diesel towboats, which gradually replaced the steamboats on the river below St. Louis, were capable of carrying the freight upriver—if only there were a nine-foot channel above St. Louis.

In the Rivers and Harbors Act of 1927, Congress authorized surveys of the Upper Mississippi River looking toward the creation of a nine-foot channel. The preliminary report by the U.S. Army Corps of Engineers was adverse, but Minnesota's Senator Shipstead succeeded in having the proviso for a nine-foot channel written into the bill that, when signed by President Hoover, became law in July 1930.

The bill authorized a nine-foot navigation channel with a minimum width of four hundred feet to accommodate long-haul, multiple-barge tows. This was to be achieved by the construction of a system of locks and dams, supplemented by dredging. But the project was only authorized, not funded. Senator Shipstead advocated a bond issue of a half billion dollars, based on the plan followed in building the Panama Canal, but the bill embodying this plan failed in the Hoover administration (Hartsough 1934).

Although authorization for the project had come in 1930, it

received minimal funding during the early years of the Great Depression and the last years of the Hoover administration. With the advent of the Franklin D. Roosevelt administration in 1933 and its New Deal, the Nine-Foot Channel Project was resurrected as a way to put people back to work. Within a few months, contracts had been let for several dams. Incredibly, work was underway before winter set in.

From today's perspective, those were unusual times. Cumbersome, expensive, time-consuming Environmental Impact Surveys were not required; the nation was in severe crisis, the worst economic depression the nation had known. Equally important, there were thousands of skilled construction workers desperate for employment. The project moved ahead at a breathtaking pace. Four dams were completed in 1934 and sixteen more were under construction. It was estimated that the Nine-Foot Channel would be accomplished by 1936 (Hartsough 1934).

The scope of the project itself was enough to take the breath away. The Corps had been authorized to build and operate one of the largest public works projects in the history of the United States. The effort ultimately resulted in the construction of twenty-nine locks and dams on the Upper Mississippi River. The system created a navigation channel that allows modern towboats to traverse the 400-foot elevation gradient and 670 miles of river between St. Louis and Minneapolis.

Environmental Concerns about the Proposed Nine-Foot Channel Project

Channelization projects prior to 1930 had employed wing dams, closing dams, shore protection, and auxiliary dredging instead of locks and dams for maintaining the navigation channel. These methods were not only less costly and did not completely alter the river environment, but they also permitted open-channel navigation, preferred by those who ran log rafts and packet boats. However, the short-lived logging boom that began in 1878 peaked out in 1892 and expired in 1915. Traffic by six-foot-draft steamboats had dwindled because the obsolete craft could not compete with rapidly expanding railroads.

The lock and dam system required by the Nine-Foot Channel Project would be the end of the free-flowing river, irrevocably transforming it.

Not everyone was thrilled about the proposed project. There had been early concerns about possible adverse biological impacts. Writers of outdoor columns in newspapers had been vocal in condemning the project. For example, the Voice of the Outdoors stated, "We are still against the alleged nine-foot channel under the dam form of construction. We are now more convinced than ever that it will be a gigantic commercial failure and will be impossible to maintain without spending millions of dollars each year in dredging operations. It will completely destroy bass fishing on the river and will look like a lot of link sausages on a map and smell worse than said sausage if they were left exposed to the present heat for a week. The scenic attraction of the river will be completely wiped out" (*Winona Republican Herald,* July 26, 1930).

In numerous pronouncements, the Izaak Walton League condemned the Nine-Foot Channel plan as detrimental to the environment. The league was especially concerned that soil erosion and pollution had to be controlled before the project began.

Many observers had expressed concern that soil erosion would constitute a severe problem in the proposed navigation pools. C. G. Bates, a forestry engineer, was quoted by the Voice of the Outdoors as predicting that the proposed pools would be completely filled with sand in a period of twenty years (*Winona Republican Herald,* July 23, 1930).

The U.S. Bureau of Fisheries had viewed the Nine-Foot Channel Project with serious misgivings. The Bureau's views were expressed in written testimony by Captain Culler, presented at a hearing in Wabasha, Minnesota:

> The Bureau of Fisheries views with much concern the establishment of a series of slack-water pools along the Upper Mississippi River until the problem of pollution and erosion as they affect this upper section of the Mississippi River are solved. If the lake formed by the Keokuk Dam may be taken as a criterion, the creation of similar pools may mean the eventual elimination of all fish life

> inasmuch as the production of fish in Lake Cooper, which is formed by the Keokuk Dam, has declined according to the official statistics of the Bureau of Fisheries from 701,181 pounds in 1922 to 350,750 pounds in 1929.
>
> The construction of slack-water pools such as the one that is contemplated at this time and in this particular section north of Winona, will mean the eventual elimination of the smallmouth black bass for which this section is so widely known. (U.S. Bureau of Fisheries 1931)

But biologists were not unanimous in their condemnation of the project. The U.S. Bureau of Biological Survey maintained that the

---

The Izaak Walton League Leads the Fight for an Upper Mississippi River Wildlife and Fish Refuge

During the time that the 1920s "reclamation feasibility" study for the Upper Mississippi floodplain was being done, the Izaak Walton League of America intensified its efforts to reserve vast areas of the floodplain for a wildlife and fish refuge.

Primarily because of the enthusiastic sponsorship by the "Ikes," the U.S. Congress on June 7, 1924, authorized appropriations aggregating $1.5 million for purchase of Mississippi bottomlands on a "willing seller" basis to be administered as the Upper Mississippi River Wildlife and Fish Refuge. The refuge, which was originally intended for protection of smallmouth bass, extended from the foot of Lake Pepin to Rock Island.

By 1930 the Upper Mississippi River Wildlife and Fish Refuge encompassed about 87,000 acres of floodplain land. Similarly, the Nine-Foot Channel Project authorized the Corps of Engineers to condemn land to obtain flowage rights. It soon became obvious that it was redundant for federal wildlife interests and federal navigation interests to compete for land. In a remarkable instance of interagency cooperation, the U.S. Bureau of Sport Fisheries and Wildlife gave the Corps of Engineers flowage rights on refuge land in return for wildlife management rights on land flooded by the Corps. Fortuitously, the size of the Upper Mississippi River Wildlife and Fish Refuge was increased to about 195,093 acres, creating a vast public recreation area. The word "National" was inserted into the refuge name as a result of the National Wildlife Refuge System Administration Act of 1966.

Nine-Foot Channel Project could be beneficial to waterfowl and muskrats if water levels were stabilized. The Bureau's conclusions were based on a comprehensive study of the biological effects of Lock and Dam 19 on the Mississippi River. According to those findings:

> It is very probable that considerable portions of the Upper Mississippi River Wildlife and Fish Refuge would be benefited by the construction level above a maximum of five feet in depth over the newly flooded bottomlands, provided that stable water levels are maintained throughout the year. The construction of these dams will undoubtedly make an entirely different type of Refuge, for most of the bottomland timber will be destroyed and the percentage of land unaffected by the flooding will be relatively small. Immediately following the construction of any system of dams flooding the lowlands, an adverse period must be anticipated, but following the re-adjustment and re-establishment of the aquatic and marsh vegetation, the Refuge should be an improved place for waterfowl and probably also for muskrats. (Henderson 1931)

Construction Begins

Early concepts of a nine-foot channel were grandiose. Some thought that the dams could generate electricity and control flooding on the Upper River. The eventual reality was not so ambitious.

The dams were designed for navigation purposes only, except for some power generation at Upper St. Anthony Falls and Dam No. 1. Lock and Dam 19 at Keokuk, of course, was an integral part of a major hydroelectric facility completed in 1914. The low dam elevations and small pool capacities relative to flood volume precluded operation of the dams for flood control. All the gates in each dam are removed from the water long before flood stage is reached so that natural open river conditions exist during the flood period.

The locks designed for the nine-foot channel dams were fairly conventional, but the design of the dams had to take into account special conditions on the Upper Mississippi River. Floods demanded moveable gates that could be raised completely out of the water during flood stage. The gates also had to be able to withstand ice jams and allow ice floes to pass through during the spring breakup.

To minimize environmental problems that the dams might cause, roller gates were selected because they would permit fish

Lock and Dam 4 at Alma, Wisconsin. The six-hundred-foot lock chamber is at the top of the photo with an uncompleted auxiliary lock adjacent to it. River flow is from left to right. Five roller gates are closest to the auxiliary lock, and a long series of tainter gates runs to the earthen dike that extends for a mile across the river bottoms to the Minnesota shore. The T-shaped structure is a popular commercial fishing float where fishermen catch a wide variety of fish species. A launch transports fishermen to and from the float.

migration, aerate the water passing under them, and allow free passage of silt, sewage, and debris. This was a new idea imported from Germany where roller gates were common, but in 1930 there were only nine other roller dams in the United States (Tweet 1980).

Because Dam 15 was located in an extremely narrow part of the river and subject to frequent ice jams, it was designed with nothing but roller gates. Most of the other dams have combinations of roller and conventional tainter gates (which also permit water to flow beneath them), earthen dikes, and concrete spillways (Tweet 1980). The Keokuk Dam constructed in 1913, a year before the powerhouse, is a notable exception. It has conventional gates that allow excess water to overflow them.

At each location for a lock and dam, the lock was built first so that navigation would be impeded as little as possible. Originally, each main lock was to have a smaller auxiliary lock built along side to accommodate future projected traffic increases. The only one actually completed was at Lock and Dam 15. The navigation dams transformed the free-flowing river into a series of "lakes" or slack-water navigation pools from St. Louis to Minneapolis.

The pool areas of the proposed dams contained expanses of floodplain forests. In some projects, the forests were simply inundated, killing the trees, which eventually rotted off at the water line. In other projects, the trees were felled, logs hauled away, and the brush burned. In either case, the stumps are still there, providing fish habitat, but creating a hazard for the propellers and lower units of motorboats.

Construction on the Nine-Foot Channel Project began in 1931, at the foot of the troublesome Rock Island Rapids, with the construction of a dam with two locks. Completed in 1934, it submerged the Rock Island Rapids under nine feet of water all the way to the LeClaire Lock. Because the LeClaire Canal and Lock already provided a fairly good channel over the upper section of the Rock Island Rapids, Lock and Dam 14 adjacent to the LeClaire Lock was the last of the locks and dams to be built in the Nine-Foot Channel Project. Completed in 1939, it was the death of the troublesome Rock Island Rapids.

The locks and dams are numbered 1 to 27. There is no Lock and Dam 23, but there is an "extra one" at Winona (Lock and Dam 5A). With the Upper and Lower St. Anthony Locks and Dams they total twenty-nine. The northernmost lock is Upper St. Anthony, completed in 1963. The southernmost lock is number 27 at St. Louis. It lies one mile up a ten-mile long canal that bypasses the Chain of Rocks Rapids. A vessel traveling from a point below St. Louis to north Minneapolis would have to "lock through" twenty-nine times.

Because twenty-six of the locks are too small to accommodate towboats pushing more than nine barges, larger tows must be "broken" and locked through in two segments. Tows with inexperienced crews may take as long as two hours to make a double lockage. If two tows arrive at a lock at the same time, they may tie up the lock even longer—to the extreme displeasure of pleasure boaters. Presently, only Locks 19, 26, and 27 are long enough to handle a fifteen-barge tow in a single lockage.

This 1980 photo shows Lock and Dam 19, located at Keokuk, Iowa (RM 364). From the hydroelectric plant, completed in 1913, the dam, with its 119 gates, extends across the river to Illinois. Old Lock 19 lies below the powerhouse. New Lock 19, completed in 1957, lies along the Iowa shore. Note that it has a vertical lift gate at the upstream end and miter gates at the downstream end. Slipping a fifteen-barge tow into Lock 19 is an extreme test of skill. Downbound, the tow must pass between the Iowa shore on the left and the tip of the long ice fender jutting out from the power plant. Flow of water through the turbines of the hydroelectric plant, situated at right angles to the lock, creates a strong undertow. The tow, which is 105 feet wide, must be aligned exactly to slip into the lock which is only 110 feet wide, leaving only 2.5 feet of clearance between the sides of the tow and the lock walls. If the pilot is timid, the power plant can suck the stern of the tow sideways, sometimes right into the powerhouse. Yet the tow cannot proceed too fast or it could crash into the miter gates at the lower end of the lock. The swing span of the railroad/highway bridge opens to let tows through. A new highway bridge was constructed in the late 1980s just downstream from the lock, and the old bridge is now used only by the railroad. The hazardous ice fender was shortened to make tow traffic easier and safer (U.S. Army Corps of Engineers, Rock Island District).

Locks and Dam 26 (Melvin Price Locks and Dam) is located at Alton, Illinois (RM 201), where it must handle the combined traffic of the Upper Mississippi and Illinois Rivers. Completed in 1990, it replaced old Lock and Dam 26, which acted as a bottleneck, sometimes delaying tows for days. The new facility, shown here looking upstream during construction, was born amid a storm of controversy, pitting environmentalists and railroads against the Corps and the towing industry. There are two locks, a 600-foot one on the extreme right and a 1,200-foot one in the left center of the photo. The 1994 Clark Highway Bridge is visible upstream.

Most of the twenty-nine locks and dams were constructed during the 1930s. An historic exception is Lock and Dam 19 at Keokuk, constructed as part of the hydroelectric facility completed in 1914. A 1,200-foot lock was added at Keokuk in 1958. The newest facility is Lock and Dam 26 at Alton, Illinois, completed in 1990 just below the mouth of the busy Illinois River.

Early Environmental Impacts of the Nine-Foot Channel

Submerged Channel Training Structures

By the end of the 1930s, the Nine-Foot Channel and the lock and dam system had inundated the rock wing dams and closing dams of earlier channelization projects, drastically altering their long-term impact on the river environment. If not for the Nine-Foot Channel, sedimentation behind the emergent wing dams would probably have created a river similar to the lower unimpounded reaches of the Missouri River, with a narrower, faster channel. Instead, once submerged, the ability of the old entrainment structures to direct flow to a narrow channel and to trap sediments was greatly reduced. In fact, rather than holding large volumes of sediment, some wing dams developed large scour areas.

Despite their reduced effectiveness, most wing dams are still partially functional, and they provide rocky corrugations of the river bottom, in effect increasing the total surface area of the river bottom. This, in turn, increases the carrying capacity of the river for fish-food invertebrate animals and algae. Unfortunately, the effectiveness of most wing dams as fish habitats and producers of fish food has decreased over the years. Most dams have slumped to some degree, and sedimentation has taken its toll, filling the spaces between dams and even overtopping them. Many dams were buried by

"Hello, Hawkeye Hotel"

The original 358-foot Keokuk Lock (later named Lock 19) became an impediment to river traffic when the Nine-Foot Channel Project was completed in the 1930s. The lock couldn't accommodate the newer, bigger tows operating on the river, forcing them to wend their barges through multiple lockages. Located on the Iowa side of the river adjacent to downtown Keokuk, the delay at the old lock was not as unpopular with boat crews as might be thought. The lockage time was sufficiently long that idle crewmembers could make a quick visit to one of the brothels conveniently operating just up the street, the most notorious of which was the Hawkeye Hotel.

The obsolete lock was replaced by a lock 1,200 feet long and 110 feet wide, completed in 1957. Filling time for the new lock is only about ten minutes and emptying time is about nine minutes. A fifteen-barge tow can pass through the lock in a half hour if the lockage goes smoothly, hardly leaving time for a trip downtown. Nostalgically, as late as 1959, when the author was stationed there, veteran towboat captains making radio contact with lockmasters of the new lock often began by saying, "Hello, Hawkeye Hotel, this is the (name of towboat)."

Corps' dredge spoil prior to 1973, when the practice was made unlawful (Fremling and Claflin 1984).

For the recreational boater and fisherman, the submerged rock structures are very important. Above St. Louis, they usually lurk, unmarked, about propeller deep. Most serious boaters have accidentally hit them—usually with dire consequences such as a mangled propeller or a damaged lower propulsion unit. Below St. Louis, where the river is not impounded, wing dams still rise above the water during normal flow.

The navigation dams transformed the Upper Mississippi River, which was formerly a braided stream, into a series of large, well-fertilized, silted impoundments through which an appreciable current still flows. Because navigation pools occupy most of the floodplain, the river is much wider (but much shallower) above most dams than it is at New Orleans where the river is undammed, narrow, swift, and very deep. The main stream of the river is punctuated by navigation markers and is flanked in many areas by extensive deposits of dredge spoil excavated from the main channel. Railroad beds, highways, landfills, and municipal flood dikes have constricted the flood plain in many areas and intercepted historic channels.

Drawdowns

Whenever flooding threatens in the Mississippi River Valley because of high water content of the winter's accumulation of snow, some people believe that the navigation pools should be drawn down to provide storage capacity for the coming floodwaters. This was done in earlier years, but the result was devastating losses to fish and wildlife populations (Greenbank 1946), resulting in the 1934 "Anti Drawdown Law." It directs the Corps of Engineers to operate and maintain pool levels as though navigation were carried on throughout the year in recognition of the needs of fish and other wildlife resources and their habitats. Also, the storage capacity of the navigation pools is so small in comparison with the magnitude of spring flood flows that a drawdown would be refilled in a matter of hours and would not appreciably lower the stages reached by the flood. Recently, summer drawdowns have been employed experimentally to revitalize senescent wetlands.

Immediate effects of impoundment included an increase in the water surface area as marshes and floodplains were flooded, a general stabilization of water elevations within the pools, and an increase in production of aquatic plants and animals. In lower pool areas, backwaters that had been intermittently flooded were transformed into permanent lakes.

The inundation of the floodplain created an immense, highly productive system with a great diversity of habitats. Some species, such as the paddlefish that relied upon open river habitat, were adversely affected by the creation of the lakelike environments. These adverse effects, however, were partially offset by the creation of new, and enlarged, habitats for such species as largemouth bass, crappies, and sunfishes.

Each impoundment consists of three distinct ecological areas. The tailwater areas just downstream from the dams look about the same as they did before impoundment—typified by deep sloughs and wooded islands. The middle and downstream reaches of most pools contain large open areas with few large trees, because stands of timber were usually cut prior to impoundment, or if left uncut, rotted off at the water line. The inundated floodplain prairies and hay

Old Faithful: The Hydraulic Dredge *Thompson*

The Nine-Foot Channel Project raised water levels in most reaches of the Upper Mississippi, but in areas where there is less depth than programmed for, it is necessary to dredge. The Nine-Foot Channel is actually kept at least ten-and-one-half-feet deep to reduce water resistance between the river bottom and the bottom of barges that have nine-foot draft. Problem areas are dredged to eleven or eleven and one-half feet. Most channel deepening is accomplished by using a hydraulic suction dredge and discharging to channel-side higher ground through pipes floated on pontoons. The Corps of Engineers' dredge *William A. Thompson*, which performs most of this function, has been in operation since the 1930s. The *Thompson* isn't as powerful as it looks. Much of its bulk is composed of living accommodations no longer needed by today's commuter crews. Newer dredges are "leaner and meaner," able to be disassembled for transport and reassembled on site.

Dredging is necessary perennially to maintain a nine-foot navigation channel. In this 1973 photo, the hydraulic dredge *Thompson* (right foreground) is pumping sand from the river bottom via a floating pipeline to a booster dredge that relays it to shore. Using this technique, sand can be pumped for several miles. Sand dredged in previous years is encroaching on a valuable wetland, destroying it. Such indiscriminant placement of spoil has been unlawful since the mid-1970s.

Chain of Rocks Canal, Locks and Dam 27

Between St. Louis and the mouth of the Missouri River, a seven-mile stretch of rock ledges, known as Chain of Rocks, had been a low-water hazard for over one hundred years. After the Nine-Foot Channel was completed in 1940, the Chain of Rocks became an increasing annoyance because deep draft tows often had to break up to get through. In the late 1950s, a ten-mile-long canal and two locks were built to bypass the Chain of Rocks. The main lock is twelve hundred feet long, and the auxiliary lock is six hundred feet long, both located just over a mile from the lower canal entrance. Then, because of difficulty maintaining a nine-foot depth below Locks and Dam 26 at Alton, the Corps constructed a low water, fixed-crest dam, from shore to shore, across the river at the canal site. Dam 27 was the first complete barrier across the Mississippi below Minneapolis. The Chain of Rocks Locks and Canal were opened to traffic in 1963 (Tweet 1983).

meadows of the midpool areas now provide the best marsh habitat and are among Earth's most productive ecosystems. The downstream reaches of pools are deeper, mainly open water, with heavily silted bottoms. Generally, marsh vegetation creeps downstream as pools fill with sediment. Marsh vegetation of the middle pool areas is replaced, in turn, by trees and other terrestrial vegetation.

New Locks and Dam 26 at Alton

Located just below the mouth of the Illinois River, the old Locks and Dam 26 at Alton had to handle the combined long-haul traffic of the Upper Mississippi and Illinois Rivers. They were unquestionably the busiest locks on the Mississippi, a real bottleneck. When not broken down, the two locks (six hundred feet and twelve hundred feet) could handle about thirty tows a day, but breakdowns were frequent, causing backups of traffic for miles in both directions. Delays of three days were not uncommon, and for the towing industry, time is big money.

Operating under the authority of the 1909 Rivers and Harbors Act, which allowed replacement of deteriorated facilities without specific Congressional approval, the Corps proposed a new dam with two 1,200-foot locks, capable of passing boats with eleven-and-one-half-foot draft (Tweet 1983).

Congress approved $350,000 for planning and the design work was about 90 percent complete in 1974 when the Izaak Walton League, Sierra Club, and twenty-one railroads filed suit in Federal Court to halt the project. The court ruled, late in 1974, that the project would require congressional approval. Four years of intense litigation ensued, with the railroads worrying about competition, and environmentalists worrying about the project's impacts on the health of the river (Tweet 1983). The new facility, named the Melvin Price Locks and Dam, became operational in 1990.

15

The Glory Years

Prior to their inundation in the 1934 to 1940 period, the river bottoms were primarily wooded islands separated by deep, running sloughs. Hundreds of small lakes and ponds were scattered through the wooded bottoms. Bay meadows and small farming operations, mainly haying and grazing, occupied some areas on larger islands. Marshes were limited to the shores of lakes and guts leading off the sloughs. Marsh vegetation was also limited, dominated by river bulrush. The bottomlands were subject to wide fluctuations of water levels, ranging from flooding in the spring to drying out in late summer when most wetlands usually dried up. During dry years, the entire 284-mile length of the Upper Mississippi River Fish and Wildlife Refuge became a virtual tinderbox. Wildfire was a constant threat.

In summer, hay was cut and stacked in the bottoms and after freeze-up the farmers would go out with large sleds, wagons, and horses to bring the hay home. In the early 1900s, up to 3,300 head of cattle grazed the expansive Lost Island area of what is now Pool 5. Each owner put his own bells on a few of the dairy cattle in his herd. The bells, brought by immigrants from Germany and Switzerland, had distinctive sounds, and their bucolic tintinnabulation was musical as the cows waded belly-deep across Belvedere Slough in the evening, coming home to be milked (interview with Lloyd Fetting, Cochrane, Wisconsin, 1976).

The creation of the Nine-Foot Channel changed the human and natural environment of the Upper River dramatically and irrevocably. Along with the stabilization of water levels and flooding of great expanses of floodplain, ownership and control of virtually all the river landscape passed from private to public hands,

By dedicating almost 100 percent of the lands in the river bottoms to public ownership and control, the Nine-Foot Channel Project brought to fruition a long-sought dream of conservationists from all walks of life. In an era when "no trespassing," signs were becoming increasingly prevalent, it made the lands available, in perpetuity, for public use. Complete federal ownership of bottomlands also permitted efficient designation of sanctuaries and open hunting areas to the welfare of migratory waterfowl populations during the hunting season.

Together, the Upper Mississippi River Wildlife and Fish Refuge Act of June 1924 and the Nine-Foot Channel Project authorized the purchase of bottomlands from St. Paul to St. Louis, thus removing marginal farming operations from high-risk areas. Crop production, haying, and grazing had always been subject to flooding, and access was often difficult or impossible in high water.

By 1958 the Upper Mississippi River contained three National Wildlife Refuges: Upper Mississippi River Wildlife and Fish Refuge—305 square miles (1924); Trempealeau National Wildlife Refuge—17 square miles (1943); and Mark Twain National Wildlife Refuge—51 square miles (1958). Their major emphasis became migratory waterfowl management rather than fish management as envisioned by the Izaak Walton League, which had fought for their establishment. The U.S. Department of the Interior, in cooperation with adjacent state governments, is responsible for their management.

The Nine-Foot Channel Project enhanced the opportunities for recreational boating on the river, making it possible for a variety of craft to use it, from kayaks and canoes to houseboats and seagoing cabin cruisers. The flooded bottomlands offered a labyrinth of channels and backwater lakes, available to pleasure boaters, anglers, trappers, and hunters.

Dredging, required in approximately 20 percent of the navigable channel, entailed the disposal of several million cubic yards of sand and silt each year. Prior to 1973, dredge disposal sites were at the Corps' discretion—usually at the handiest spot. Varying in extent from less than one acre to more than one hundred acres, these deposits ranged on either side of the navigable channel, creating beautiful sand beaches that could be used, free of charge, by the public for camping, picnicking, swimming, and other water sports. Unfortunately, dredged sediments were often pumped directly into wetlands, destroying them.

The huge expanses of water in the impoundments improved the scenery of the area by opening new vistas to river residents and motorists. To most of them, the sight of a modern towboat with a full complement of barges added beauty, interest, and contrast to the naturalness of the river setting.

The pools were new and dynamic, providing unlimited habitat and little competition for hundreds of plant and animal species. Prior to the project, large-scale fish rescue programs were carried out each year necessitated by fluctuating water levels that left fish stranded in floodplain pools. Stabilization of water levels made this work unnecessary.

It is often said that nature abhors a vacuum, and that if a new environment is created nature will fill it. This was certainly true of the new navigation pools where environmental resistance was low.

The Nine-Foot Channel Project magnified the river's surface, increasing its exposure to the sun's energy. Because of increased photosynthetic activity, the river produced many times more pounds of fish-food organisms and fish per linear mile than it did before impoundment. Fish grew rapidly, exhibiting small heads and plump, robust bodies.

Moreover, the tailwaters of the dams became virtual feedlots for fish that congregated there, receiving food produced in the huge expanse of the impoundment above. Because fish congregated below dams, anglers could harvest them more efficiently. The river became so productive that anglers in most reaches were able to fish year-round, with two or more lines.

The navigation dams increased habitat for waterfowl, muskrats, beaver, mink, otter, and other marsh animals. In addition to being valuable monetarily, the animals provided a distinct recreational resource for trappers and wildlife observers. Significant portions of the world populations of canvasback ducks and tundra swans utilized the river for resting and feeding during fall migrations.

The formation of the navigation pools led to greater cooperation between the natural resource departments of states bordering the river, enabling them to manage fish and wildlife resources more efficiently. The impoundments usually extended to the railroad tracks that flank the river on either side. The tracks served as easily recognized boundaries to the area of fishing reciprocity that lay between states on opposite sides of the river. For example, an angler with a Wisconsin fishing license could fish to the Iowa railroad tracks and an Iowa angler could fish all the way to the Wisconsin tracks. Reciprocity was not extended to hunting and trapping.

The closure of the Nine-Foot Channel dams and the creation of additional wetlands was viewed by most hunters, fishermen, and other outdoor enthusiasts as a boon to the Mississippi River system.

"River rats" had to develop a new vocabulary and *modus operandi* for their "new river." The river reach between two dams is called a pool. Pool 9, for example, included the extensive area from Lock and Dam 9 upstream to Lock and Dam 8. When an experienced river fisherman told a neophyte that he had caught his limit of walleyes by the big stump in Pool 9, he wasn't revealing any secrets.

Lucky was the kid who could prowl the river during the war years of the 1940s when most of the "old guys" were in the military, leaving behind a wonderland of biological diversity to be enjoyed by all. Duck hunting was phenomenal. Shotgun shells and gasoline were rationed, but that lessened hunter competition. A youngster was in business if he had a single-barreled shotgun and a few handmade wooden decoys. Luxuries included hip boots, a dog, a wooden boat, and a little outboard motor, usually called a "kicker."

Ducks were proudly brought home, plucked, singed or waxed to remove the last traces of down, gutted, usually stuffed, and roasted.

Today's wantonly wasteful practice of "breasting out" waterfowl was rare. Old-timers knew that the flavor of a duck was mainly in the fat layer under the skin. Grandmother preached, "Waste not, want not," and mother was adamant, "If you're not going to eat it, don't shoot it!" Many of today's affluent "sport" hunters have a hard time with those philosophies.

Unfortunately, the early, dynamic, postimpoundment environmental changes weren't well documented. Hitler's Panzer divisions were roaring across Europe, and young men were being mobilized, even before America's entry into World War II in 1941. Most able-bodied young men were in the armed forces until Japan's surrender in 1945. State and federal natural resource agencies were understaffed, yet seldom employed women as biologists. State departments of natural resources, usually called conservation departments, paid little attention to rivers, devoting funds to fish culture, rough fish removal, and law enforcement. The Corps of Engineers had no interest or expertise in river ecology. Research funds were virtually nonexistent. Even after the war, when American colleges and universities were replete with veterans studying under the GI Bill, the Mississippi River went unstudied.

Dr. William E. Green of the U.S. Fish and Wildlife Service was a notable exception. Stationed at Winona, headquarters for the Upper Mississippi River National Wildlife and Fish Refuge, Bill was Refuge Biologist from 1940 until 1945 when he entered the military. "Doc Green" documented marsh conditions during the "glory years," plotting major beds of aquatic plants throughout Pools 4, 5, 5A, 7, and 8 that had been created in the 1935 to 1937 period. He resumed his studies after discharge from the navy in 1947.

Bill reported that in the first few years after impoundment the river bottoms turned red in late June and early July due to the explosion of aquatic smartweed, a pioneer marsh plant that produced immense quantities of seeds relished by waterfowl. But smartweed, like lots of other aquatic plants, needed to have its "feet" dry out occasionally. After less than a decade of constant inundation the smartweed quit producing seed and then disappeared. Yet, the bottoms were so rich in aquatic plants that an armful could include ten or fifteen other species.

Similarly, *Phragmites communis,* a tall, slender, reedlike grass, became so abundant in the early years that duck hunters didn't have to build blinds. A hunter could get lost on a foggy morning in the famous Weaver Bottoms at Minneiska, Minnesota, in a virtual, impenetrable sea of *Phragmites* over seven feet tall. But by 1960 the mammoth, unbroken stands of *Phragmites* were history. Their demise was due to the interplay of several factors.

The outboard motors of postwar duck hunters had torn a maze of trails through the shallow root masses of dense *Phragmites* stands, dissecting them, creating smaller and smaller patches. Ice, during the spring breakup, lifted root masses and carried them downstream or piled them up on shore. Like aquatic smartweed, *Phragmites* needed dry feet once in a while, and constant submersion stressed the plants. As they weakened and died, areas of open water became larger, increasing wind fetch. Final obliteration was caused by wave action that riled the water and dislodged soil from around the root masses. By the mid-1960s, the lush Weaver Bottoms had become a broad expanse of open water, virtually devoid of ducks and duck hunters. But *Phragmites* is a tough plant, still existing today in the once-glorious Weaver Bottoms as tiny, isolated islands in a sea of wave-churned, muddy water.

Muskrats did exceedingly well in the early years of impoundment, fattened on roots and stems of cattails, duck potatoes, and bulrushes. Their little mound-shaped houses peppered the bottoms, sometimes so close together that a man could jump from one to another and another. But the boom didn't last. In 1975, my golden retriever, Toivo, and I reconnoitered one of the once-muskrat-rich areas on the bleak ice of winter. Toivo peed on the first house he came to and was able to mark every one of the sparse houses on a single bladderfull.

16

Diesel-Driven Towboats and Steel Barges

A Rebirth of Commercial River Transportation

There's a little bit of Huck Finn hiding out in most all of us. Nudged awake by the awesome splendor of a river sunset, we briefly nurture a wistful dream of renting or buying a houseboat and following the channel into the twilight and around the next bend. Curiously, most of the people who enjoy the Upper Mississippi River seldom, if ever, swim, boat, hunt, or fish it. They enjoy those pleasures vicariously through their car windows as they cruise the excellent highways that flank the river. They marvel at the rugged bluff-lands that contrast so sharply with wide expanses of river, especially when the hardwood forests are resplendent in their fall colors. And they like to see towboats, even if they don't understand them, wondering where they are going, what they are hauling, and maybe around suppertime, what the crew is having for dinner.

Towboats make the river environment different from ordinary inland aquatic environments. The river's a sterner, more rugged place, a place where work is done and business is conducted. A working river has more in common with the busy ocean shipping lanes than a placid lake where pampered power-boaters play on weekends. In the heart of the continent it is kin to the gritty Great Lakes ports, so much so that Wisconsin even boasts of the Mississippi River as its "third coast," the other two being Lake Superior and Lake Michigan.

The development of the Upper Mississippi as a commercial waterway has helped meet the need to move bulk raw materials and heavy, high-volume commodities over the wide geographical areas served by the river network. That water transport is an economically viable mode of transport is demonstrated by the fact that the routes of competing land transportation—railroads and highways—run parallel to the river taking advantage of the relatively gentle river valley terrain. Railroads, especially, hug the river, in some areas running right up the middle of the floodplain, as in Pool 6 between Trempealeau and Winona.

Barge shipping is more environmentally friendly and less intrusive to public welfare than trucks or trains. The towboat's horn is mellower and more pleasant than the train's, announcing the boat's presence rather than stridently warning of its approach. Most people don't have any direct contact with towboats. At night, a towboat's presence might go unnoticed unless it sounds its horn or shines the observer with its searchlight.

Towboats are less dangerous than trains or trucks; people are unlikely ever to collide with one or be killed by one. Towboats are fuel-efficient, and people don't have to breathe their exhaust fumes. If given a choice, the residents of most river towns would probably prefer to have one fifteen-barge tow slip silently past town than have two one-hundred-car freight trains clattering through town, blocking intersections, and shrieking at every unguarded crossing.

The Upper River's locks and dams are impressive structures, and most people enjoy viewing them. Observers also enjoy watching tows pass through the locks. The play of searchlights and the sound of amplified radio messages are dramatic and exciting. The public viewing stands at the locks are heavily patronized by visitors from most of the fifty states and many foreign countries.

Barge transport is slow but efficient because of economies of size. The barge industry pays a diesel-fuel tax of 24.4 cents per gallon that helps maintain the navigable channel and the locks and dams (but the money is used only for projects over eight million dollars).

The Towboat

In order to handle barges effectively, the steamboat had to evolve into the towboat. In the early days, a keelboat or barge was often towed behind a steamboat, permitting only one per boat and making for difficult control in currents or tight places. The technique of pushing barges ahead of the steamboat began in the 1840s and 1850s, and became standard after the Civil War. In order to push six large, wooden barges (the equivalent of one hundred rail cars), steamboats had to lose their gingerbread, unnecessary decks, and cargo spaces. They developed extraordinarily strong engines. Multiple rudders extended fore and aft of the paddlewheel for control going backward as well as forward (Tweet 1983). With the barges tightly connected to each other and to the towboat, the "tow" became a single unit with steering at the rear, providing increased maneuverability (just as an automobile can turn a tighter circle going backward than forward).

Technically, the towboat and its barges make up a *tow.* It is no more correct to call the entire unit a *barge* anymore than it is to call a railroad train a *boxcar.* The propeller or screw of a modern towboat is still referred to as a *wheel,* harkening back to the days of paddle wheelers.

A typical Upper-River towboat, the *Prairie State* is powered by two eighteen-cylinder diesel engines, each producing 2,150 horsepower. It has two nine-foot-diameter wheels (propellers), each working in a Kort nozzle that acts like a jet with only ¹⁄₁₆-inch clearance between the tips of the wheel blades and the Kort nozzle. A few Upper-River towboats generate as much as eight thousand horsepower. On the Lower River, which has been extremely channelized and straightened, towboats are generally more powerful, not to push bigger tows but to handle them in the swift currents.

The larger Lower-River towboats usually operate only as far north as St. Louis. The *Jean Gladders* is typical, 166 feet long, with two V-20, 3,600-horsepower engines. She uses about 1.1 gallons of diesel fuel per horsepower per day (almost eight thousand gallons), and also consumes eighty gallons of crankcase oil per day. The *Gladders* can take forty-six barges upstream, but only twenty-five downstream because she has less control then. The largest Lower-River towboats, like the *Argonaut,* are 10,500 horsepower.

Massive tows are not very agile. No movements are done suddenly. Their great momentum makes it virtually impossible for them to stop in less than a mile while going downstream. It takes the *Gladders* eighteen seconds to change the direction of rotation of her wheels, but smaller boats can change wheel direction in seven seconds.

Large towboats are equipped with sophisticated navigation gear: two powerful search lights, two radars (one is a spare), two sonars (port and starboard) at the bows of the lead barges, a gyro-operated swing meter, GPS (Global Positioning System), tow buoys on the port and starboard stern (not fancy, but very effective), and marine radio. The radar can clearly see shoreline, navigation markers, other boats, rain clouds, and even mayfly swarms.

Ideally, the tow would have raked barges—where the deck projects ahead of the waterline—at both ends, but if there aren't enough raked barges, box barges may be placed at the head of the tow. Box barges aren't preferred at the stern because wheel wash would hit the vertical surface while going in reverse. The most difficult steering is in reverse, and a box barge complicates it.

All modern barges are made of steel and are double hulled, with the outer hull protecting the inner cargo hull. The space between

A Monster Lower-River Tow

The immense power and hauling capacity of Lower-River towboats were dramatically illustrated to me during the spring flood of 1993. At Natchez, Mississippi, at 7 P.M. on April 24, I watched a towboat ponderously, and almost imperceptibly, pushing forty-two loaded barges (six wide and seven long) upriver against the strong current. Remarkably, it also had an idle towboat "on the hip" (lashed to its port side). The working towboat's three exhaust stacks belched black smoke, telling me that it was powered by three 3,500-horsepower diesel engines. At 2 P.M. the next day I watched the same tow pass Vicksburg, seventy-two miles upstream. It had made the journey in nineteen hours at an average speed of 3.8 miles per hour.

the hulls allows a lean inspector to check for leaks. If a leak is found, a gasoline-driven sump pump discharges the clean bilge water into the river automatically. Some tows may have several pumps in operation at any one time. Wooden shingles are still pounded into leaking seams, just as they were in days of steamboats and wooden barges. Because steel is such a good conductor of heat, barge decks may be fiercely cold in winter and hot enough to burn feet in summer. In winter, barges may drag bottom because of ice buildup on the bottoms of steel hulls (up to seven feet of ice has been reported).

The Crew

There are usually twelve crew members on a towboat: captain, pilot, two mates, four deck hands, two engineers, one oiler (striker), and one cook. They operate the tow twenty-four hours per day. On most union boats, each crewmember works two six-hour shifts per day for

Towboat Meals

The towboat crews of good boats are among the best-fed people on the planet. The tedium of work is relieved by excellent meals prepared by an experienced, hard-working cook, usually a woman. Fresh breads and pastries are baked daily. There is always a wide choice of fruits and vegetables. The galley is open for snacks between meals, and there is unlimited ice cream. Needless to say, becoming overweight is a common malady. Because many crew members are from the South and prefer southern cooking, some of the most interesting choices are unfamiliar to most Yankees—chicory coffee, eggs and brains, crackling corn bread, collards, black-eyed peas, biscuits and red-eye gravy, grits, and okra. The following was the hearty menu for a typical day when I was a passenger on the *Jean Gladders:*

Breakfast: orange and grapefruit juices, sausage, bacon, baking powder biscuits and gravy, eggs, milk, coffee.

Lunch: baked ham, mashed potatoes, gravy, relishes, sliced tomatoes, cucumbers in vinegar, lima beans and ham, peas, broccoli, corn on the cob, beets, pineapple-coconut desert.

Dinner: T-bone steak, baked potato, relishes, garlic bread, tossed salad, pears, cake, milk, tea, and lemonade.

Bedtime snack with the captain: chocolate cake and ice cream.

twenty-eight days and then has twenty-eight days off (with pay). The captain and pilot operate the boat, taking alternate six-hour shifts. The crew is not allowed to leave the ship except for emergencies. Alcohol and drugs are forbidden.

For modern tows, running the river is like skiing a slalom course over six hundred miles long, weaving between wing dams and through the narrow openings of railroad and highway bridges. A lock on one side of the river may be followed by a bridge opening on the opposite side. Going upstream is less difficult because the pilot has more control then, but going downstream is more dangerous because of the loss of steerage and the ability to stop quickly. Like an airplane pilot, the towboat pilot has "hours of boredom punctuated by moments of sheer terror." Running aground, especially on the Lower River, can be extremely dangerous at cruising speed because cables snap, sending shackles and ratchets flying. For the deck crew, being dismembered or crushed is always a danger. No one wants to fall between the tow and the lock wall.

Cargoes

The cargo capacity of barges is enormous. For example, a standard grain barge can carry fifteen hundred tons of soybeans, the harvest of about fifteen hundred acres. A fifteen-barge tow is equivalent to a 225-car freight train or nine hundred trucks. Standard barges are multipurpose. With cleaning, they can interchange gasoline, oil, and diesel fuel; coal, grain, and fertilizer; but not petroleum products and molasses. A standard barge is 195 feet long and 35 feet wide. New petroleum barges are 300 feet long and 52 feet wide. A standard lock is 600 feet long and 110 feet wide.

Typically, an Upper-River tow boat would take fifteen barges of corn or soybeans downstream just beyond Lock 27 at St. Louis, to be "turned" by a larger towboat that would assemble thirty-five such barges and guide them downriver to Baton Rouge or New Orleans where the grain would be pumped into oceangoing vessels for export. A barge load of corn typically changes ownership two to three times from St. Paul to New Orleans, but it may change twenty-five times. The river's commercial tows transport many cargoes, but their main function is to export America's surplus agricultural commodities—mainly corn, soybeans, wheat, oats, and rye. Scrap iron ranks second to grain in downbound tonnage. The Mississippi River handles 42 percent of the agricultural products exported from the United States. If it weren't for surpluses there wouldn't be a need for a Nine-Foot Channel.

The largest, single, downbound cargoes travel from the coal and agricultural fields of the central United States to New Orleans and Baton Rouge, which serve as major terminals for international waterborne commerce. New Orleans is the largest U.S. port; Baton Rouge is the fourth largest.

After their trip downstream, grain barges may transport coal or dry fertilizer on their return trip upstream. After being cleaned, they will transport grain again. Some specialized barges are usually for single use. Asphalt barges, for example, carry asphalt heated to 180°F downstream from Twin Cities refineries but return empty because they cannot be used to haul other products for fear of contamination.

Some barges are "high tech." Anhydrous ammonia barges, for example, are like giant thermos bottles whose pressurized contents are refrigerated. Barges that carry liquefied methane gas at a temperature of -258°F have foot-thick linings of specially treated balsa wood. At the other end of the spectrum, molten sulfur is transported at temperatures of 300°F to 350°F.

The largest bulk cargoes moved upstream on the river are petroleum products—gasoline, kerosene, fuel oil, lubricating oil—from the oil fields of Texas and Louisiana. Cheap, high-sulfur coal is also shipped upstream, mainly from Illinois and Kentucky. Many other products are carried in smaller volume, including iron and steel products, molten sulfur, liquefied methane gas, anhydrous ammonia and dry fertilizers, cement, aluminum ingots and plate, sugar, dehydrated molasses, and asphalt.

Because barges are so buoyant, great care must be taken when loading and unloading them. A loaded barge can break in two if unloaded in the middle. An empty barge can buckle if it is loaded in the middle.

During high river flows barges are often loaded to a depth of ten or ten and a half feet instead of the normal nine feet.

Hybrid Corn and Its Impact on the Mississippi River

Flowing toward the grain terminals of New Orleans and Baton Rouge, the Mississippi is truly a "river of grain"—especially hybrid corn.

Hybrid corn has revolutionized American farming since the 1920s when it was introduced. Farms began to be mechanized in the 1930s, leading to more widespread corn planting, which has, in turn, had immense impacts on the Mississippi River, mainly since the 1940s when the Nine-Foot Channel was completed. During the 1960s and '70s, some farmers opted for "continuous corn," disdaining the crop rotation lessons learned in the 1930s. Year after year, they planted corn on the same ground, for as long as twenty years, using a battery of fertilizers, insecticides, and herbicides. This monoculture devastated soils through erosion and nutrient depletion. Sediment and excess fertilizers washed into the river. In karst areas, nitrates polluted ground waters. Farming methods have improved in recent years, but the magnitude of the operation is apparent in the fall when, as corn is harvested and fall plowing is completed, the entire corn belt miraculously changes color from gold to black in a two- or three-week period. The dramatic color change would probably be visible from the moon.

Corn is genetically the most tinkered with plant in existence. Early on, Europeans adapted Indian beliefs and methods to their own ways of growing corn. They perfected types of corn by selecting cylindrical cobs that were the largest, with the healthiest seeds, the straightest rows, and the most uniform color. The best kernels from the cob were used as seed. In 1893 a farmer named James Reid introduced his prize-winning Reid's Yellow Dent corn. It was so good that it was planted across the United States, and other corns were forgotten.

Thousands of corn varieties, reverently preserved and kept distinct by the Indians, as well as those improved by American farmers, vanished from the face of the earth. Hybrid corns that were reliable, predictable, uniform, and capable of producing previously undreamed of yields began to become commercially available. One kernel of corn can multiply itself eight hundred times in four months. In 1900, 90 percent of North Americans lived on farms, producing food; now 97 percent live in towns and cities. Just over 3 percent of Americans now feed the entire population—and still produce a surplus of corn and soybeans shipped mainly by barge down the Mississippi for export overseas (Visser 1988).

The world now grows five hundred million metric tons of corn a year, nearly half it in the United States, and over 80 percent of that in the Upper Mississippi River watershed. The largest corn tonnage comes from Iowa and Illinois, with Iowa alone producing 1.8 billion bushels in 2000. With energy subsidies, mainly as fossil fuels like coal, oil, propane, and natural gas, the modern corn farmer attains corn yields unimaginable just fifty years ago. He uses machines powered by fossil fuels to prepare and fertilize his soil, plant the corn, control weeds, harvest the corn, usually to dry it, and haul it by truck to storage or shipment areas. His nitrogen fertilizers are manufactured from propane and natural gas.

By the time surplus corn reaches Baton Rouge or New Orleans by towboat for export, many more calories of fossil fuel will have been used in its production and delivery than there are in the corn itself.

In recent years it has been possible to grow corn with less use of insecticides. Genetically engineered corn, developed with recombinant DNA techniques, has been grown in the United States since the mid-1980s. In 2000, about one-fourth of the nation's corn crop was "Bt corn," containing the genes for the production of an insecticide produced by the soil organism *Bacillus thuringiensis* (Wheelwright 2001).

4

Ecological Relationships

17

The Corps Giveth, the Corps Taketh Away

Long-Term Impacts of the Nine-Foot Channel and Related Projects

The "Goods" and the "Bads"

The Nine-Foot Channel isn't simply a deep rut down the middle of the river. The creation and maintenance of the channel and the dams, pools, and backwaters resulted in changes in the river and the surrounding environment—spin-off effects—unanticipated by the project designers. Some of them are obvious—the acres of sand piled up by channel dredges are hard to avoid—but most are so subtle and happen so gradually that only the most astute river watchers even notice them.

An economist might think of these spin-offs as *externalities*—determining some to be *beneficial,* others *detrimental,* a determination much easier to achieve with a balance sheet than a complex riverine ecosystem.

Oftentimes that determination is relative, heavily influenced by the fact that the determiner is a human being. Many things judged as beneficial for humans are far from beneficent from the viewpoint of a crawdad, coot, or mosquito larva. For that matter, we don't have to venture outside our own species to stir up a heated argument on what is good or bad, with the argument on either side primarily reflecting "whose ox is being gored." The complex impact of channel maintenance dredging provides a good example of how on the river one action can generate a multitude of diverse reactions.

Regular dredging is needed to maintain a minimum nine-foot depth along approximately 20 percent of the river's navigable channel. Keeping the channel clear entails the disposal of several million cubic yards of sand and silt every year.

Until the late 1970s the Corps was free to dump the sand and silt sucked up from the channel bottom pretty much wherever it pleased. Although the technology of the day enabled dredgers to pump sediments for several miles to disposal sites as much as thirty feet above river level, they generally picked the handiest spots along the main channel to save time and money. These sand beaches were popular with local turtles as nesting areas and for local teens as party sites. Corps dredgers often remarked that if they made a new beach there would be a party there within twenty-four hours.

Channel Maintenance Dredging

Under authorization of the River and Harbor Act of 1930, the Corps of Engineers operates and maintains a nine-foot navigation channel for commercial navigation through the operation of a series of locks and dams and through annual maintenance dredging of the main channel. Average dredging volumes in the last seventy years have been almost 1.5 million cubic yards annually. Prior to 1980, placement of dredge spoil was confined primarily to lands and waters in the river floodplain, resulting in direct and indirect destruction and damage to the river's vital wetland habitats. The dredged material also decreased the flood storage capacity of the river floodplain. Fortuitously, the piles of dredged sand created and expanded the island system of the river in some reaches, forming popular beaches, and attracting thousands of recreational users. The sand piles are also valuable nesting habitat for turtles. In recent years, the Corps has preferred to use the euphemism "dredged material" instead of the traditional term "dredge spoil" that had been used for more than 150 years. "Spoil placement" is sometimes called "beach nourishment."

Turtles, campers, picnickers, sunbathers, swimmers, water skiers, and houseboaters doubtless find the sediment deposits a positive side effect of the channel project, but the sun-kissed beaches have a dark side as well.

These incidental recreation areas were often created when sediments were pumped directly into wetlands and floodplain forests, destroying them. The initial environmental destruction is aggravated when sediment is washed from the old spoil piles by boat wakes, wave action, and floods ending up in the backwaters, decreasing their life. Because the piles were so desertlike, the wind blew additional fine sediments from them into the backwaters. In time, slough openings were blocked, and spawning beds and food-producing areas were covered with sterile sand. The changes were continual, accumulative, and in most cases irreversible.

In the early 1960s, when professors and their students at several colleges along the river documented the adverse spin-offs from Corps activities, especially dredging, the Corps dismissed the complaints in a cavalier fashion by saying, "We made the reservoirs and if we fill them it's our business." When, in a public speech, I stated that the Corps' dredged sand was destroying backwaters and that increased sedimentation could ultimately cause increased flood crests, the commanding officer of the St. Paul District sent me a personal letter saying, "Frankly, Dr. Fremling, if you cannot prove your statements empirically you should not make any more speeches."

In the early 1970s there was an interesting turn of events. The St. Paul District got a new commanding officer, which they do

Mayflies by the Ton

Hexagenia mayfly nymphs thrive in areas where there is a silt bottom and well-oxygenated water. The navigation pools provide such habitat, and there is no doubt that *Hexagenia* mayfly populations increased because of the Nine-Foot Channel Project. Adult mayflies are a nuisance to most people, who probably view the increase as detrimental, but because the insects are excellent food for fish and diving ducks, the increase may also be viewed as beneficial. More recently, populations in some pools may be declining because silted areas are being blanketed with unproductive sand or are becoming anaerobic (without dissolved oxygen).

every three years or so, but this one was interested in the environment and receptive to new ideas. He invited me to come to St. Paul to make my case in an address to a combined meeting of the Corps of Engineers and the Society of Military Engineers. I felt like Daniel entering the lion's den as I walked into the large meeting room filled with river professionals, mainly engineers.

I presented a fifty-minute illustrated lecture documenting both the ecological "goods" and "bads" associated with the Nine-Foot Channel Project as well as earlier channelization projects. The lecture was well received, and I like to think that my presentation played a small part in fomenting the revolution in river management that ensued in the next three decades.

Some of the "Goods"

Increased Recreational Opportunities

Most people would agree that the huge expanses of water created in the Nine-Foot Channel impoundments improved the scenery of the area by opening new vistas to river residents, boaters, and motorists. To most, the sight of a modern towboat with a full complement of barges adds beauty, interest, and contrast to the naturalness of the river setting. Tow watching has become a popular pastime.

Recreational use and expenditures are highest in the reach from Minneapolis to Rock Island, where the river provides a rich mosaic of braided channels, islands, floodplain lakes and forests, and vegetated backwaters—almost all in public ownership. Recreational use and expenditures are low from Rock Island to Cairo where most of the broad, fertile floodplain has been separated from the river by levees and converted to agriculture, and where there are few backwaters and little public land.

Some of the "Bads"

Aging of the Reservoirs

The navigation pools and I are about the same age. We have become elderly together. Everybody knows that people age, but few realize that impoundments do too.

Opportunistic aquatic plants and animals multiply explosively in new reservoirs. Fish grow extremely fast, and populations may fluctuate wildly. For the first decade or so fishing is often phenomenal.

As a reservoir ages, populations stabilize, productivity drops, and fish grow more slowly. Eventually, the reservoir becomes senescent. Crowding, decreased food supplies, and competition with other species are among the many factors involved in the complex and poorly understood process of decline. The stable water levels created by the Nine-Foot Channel Project helped accelerate the decline. Since the water levels were strictly controlled, drying out, which is essential to maintaining healthy marsh habitats, never rejuvenated the backwaters created by the Nine-Foot Channel Project.

The biological productivity of the Nine-Foot Channel impoundments probably peaked out in the early 1960s. Since then, there has been a state of decline that is perceptible to most serious river users.

The first systematic studies of the Upper River impoundments followed the passage of the National Environmental Policy Act of 1969. This legislation required that governmental agencies address the environmental impacts of the operations and maintenance of all water-related projects—including the Nine-Foot Channel Project. Having insufficient trained personnel of its own, the Corps of Engineers employed contractors to conduct environmental impact studies on the Upper Mississippi. These studies elucidated some of the problems associated with the closure of the dams thirty years earlier. I was one of the contractors, responsible for Pools 4, 5, 5A, and 6.

The studies showed that the navigation pools were filling at an alarming rate with an attendant loss of water surface, fish and wildlife habitat, and boating areas. Wing dams, closing dams, and impoundment-producing dams had caused sand and silt to accumulate in the flood plain, effectively elevating it.

Loss of Productivity of Wing Dams and Closing Dams

The effectiveness of most wing dams as fish habitats and producers of fish food also decreased over the years. Not only has siltation taken a toll, but also many dams were buried by the Corps' dredge spoil before indiscriminant dumping was outlawed in 1973 (Fremling and Claflin 1984).

Most rock structures, like wing dams and closing dams not buried by sand, have slumped to varying degrees after construction but are now quite stable. Willow sticks still protrude from between rocks in most deep places, preserved by continual submersion in water. All of the rock surfaces that are exposed to river water, but not buried by sediments are dark, silt-covered, and teeming with living organisms. The buried portions of the rocks are generally light colored and show virtually no sign of biological growth. Below the first layer of rocks, almost all of the spaces are filled with sand or other sediments.

Increased Turbidity Caused by Increased River Traffic

Towboats scour the channel with their huge propellers (up to nine feet in diameter), increasing turbidity, eroding shorelines, and killing fish by entrainment or impingement through their Kort nozzles. The Nine-Foot Channel Project also enabled large, deep-draft pleasure boats to cruise the river, producing wakes that are usually larger and more damaging than those of towboats. Such craft are becoming increasingly numerous, and their operators are usually oblivious or uncaring about the shoreline erosion and collateral damage caused by their wakes.

Increased Danger of Spills of Hazardous Materials

Increased volume of commercial traffic on the river has also increased the chances for spills of oils and toxic materials.

Elevation of the Watertable

It is seldom noted that the creation of navigation pools raised the water table in adjacent bottomlands, facilitating the expansion of floodplain forests and a decrease in diversity of upland plant species.

Eutrophication

Many of the wetlands created by the Nine-Foot Channel Project are distant from the main channel, and water circulation through them is often poor during low-flow conditions. Entrapment and accumulation of organic debris such as tree leaves occur in these areas mainly during periods of flooding. This results in the accumulation of sediments and associated nutrients, and in the stimulation of the growth of aquatic plants. Collectively, these processes lead to accelerated rates of *eutrophication* (enrichment). The progression toward *hypertrophy* (overenrichment) has resulted in the reduction in diversity of invertebrate communities that occupy the river bottom.

Over the years, many channels of the river have been intercepted by dikes, closing dams, roads, railroads, and by barrier islands created by dredge spoil. Such isolated channels stagnate in the summer and the deeper ones stratify thermally. The rich, loose, organic ooze collecting on the bottom consumes oxygen from the lower stratum of water until it becomes a death zone. Most forms of life, clams included, cannot survive in these hostile areas.

Barriers to Fish Migration

Mississippi River dams hinder fish migration, and none of them have fish ladders to allow fish passage. Lock and Dam 19, the oldest navigation dam on the Upper Mississippi River, creates a formidable obstruction for migrating fish because it has a head of 38.2 feet, and water must flow either through the turbines of the associated power plant or over the top of the dam's regulatory gates. Ironically, the Keokuk dam is now serving as a filter, slowing the upstream spread of several undesirable species of exotic carp.

Navigation dams may also block the spawning movements of lake sturgeon, but the length of the sturgeon's immature life (eighteen to twenty years) and its susceptibility to commercial fishing nets and boat propellers have also been important factors in its decline. The same may be true for paddlefish, which frequently swim near the surface and are especially vulnerable to propellers. It's not rare to see a large lake sturgeon or paddlefish sliced like French bread (Fremling et al. 1989).

Recreational Overuse

On summer weekends and holidays, when a party atmosphere prevails, the sights and sounds of the river are overwhelmed by roaring swarms of pleasure boats and buzzing personal watercraft, too often piloted by intoxicated amateur skippers. In upper pools, especially from Lake Pepin to Prairie du Chien, recreational use is exceptionally high. Swimmers and campers flock to the public sand beaches that flank the main channel, and long past nightfall, the din of amplified music can be heard for miles, disrupting campers' sleep.

Most sportfishermen desert the main channel on weekends, unable to fish wing dams because of extreme boat wakes. Lockage times are annoyingly long, especially when they involve double lockages of towboats that take about one and three-quarter hours on average, nearly twice the time required for a single lockage (about fifty-five minutes).

The boaters and campers swarming the beaches and islands are contributing to the eventual demise of their government constructed playground—virtually all composed of dredged material pumped there more than thirty years ago by the Corps of Engineers as part of their routine channel maintenance. Most of the islands and shorelines are no longer being replenished with new spoil, and they are being eroded by currents during floodtime, by wind-driven waves, and by boat wakes. The effects of the latter are particularly obvious along the main channel where the shores are subject to the intense wakes of large, fast, deep-hulled pleasure boats that greatly outnumber towboats on upper pools. The sediments that wash into the main channel are carried along by the current, inexorably moving downstream until they are swept out into the backwaters where they will probably remain forever.

18

Vital Strands in the River's Web of Life

An ecologist is one who, instead of calling a spade a spade, calls a spade a geotome.

Author unknown

A Brief Primer

The ultimate energy source for virtually all of Earth's ecosystems is the sun. Powered by its radiation, green plants are able to make sugar (a carbohydrate) from carbon dioxide and water by the fundamental process called *photosynthesis* (using light to assemble). A by-product of this reaction is the oxygen that sustains us and makes Earth's atmosphere unique. Oxygen is vital to most of Earth's multicellular life forms. For most living things, photosynthesis is the most important of all chemical reactions. In essence, green plants trap the kinetic energy of the sun and store it as potential energy. From the sugar they make, the plants synthesize other organic compounds such as fat, starch, and cellulose. By adding nitrogen and other elements extracted from the soil or water they are able to make proteins and other vital compounds. Green plants are referred to as *producers* because, with few exceptions, only they can trap the sun's energy in new *biomass* (living stuff).

Animals that eat plants are called *primary consumers.* They, in turn, may be eaten by *secondary consumers.* When plants and animals die, they are disassembled into their elements by *decomposers* (mainly bacteria and fungi). As the biomass is trafficked through the river ecosystem, elements such as carbon, hydrogen, oxygen, nitrogen, phosphorous, and calcium may be *recycled* (used over and over again), but energy flows through the ecosystem and over 90 percent may be lost, as heat, at every step of a food chain. The shorter the food chain, the more efficient it is. The efficiency of transfer of energy is so low that most food chains do not exceed four or five steps. For example, it could take ten thousand pounds of plant biomass to produce one pound of a top predator like a northern pike or walleye at the end of a four-step food chain. Because the waters of the Upper Mississippi are so fertile, well-oxygenated, and green with *algae* and other plants, they may produce more than thirty-five times as many pounds of fish per acre as do the pristine, clear, trout lakes of the nutrient-poor Canadian Shield. Riverine habitats may produce over seven hundred pounds of fish, of many species, per acre per year.

Nutrients that would be recycled in the still waters of a lake or pond actually spiral downstream in the Mississippi toward the Gulf of Mexico, with the coils of the spiral being tighter in the quiet waters of impoundments and looser in areas of faster water. Biomass and nutrients from the watershed are continually being added to the spiral while others are being removed as they become entombed in the sediments of impoundments.

The Upper Mississippi is among the planet's most productive ecosystems. Its waters and soils are rich in nutrients, and because the river is flowing, its green plants are bathed in nutrient-rich water that is refreshed constantly. Unlike farm fields, and partly because of them, Mississippi River environments seldom suffer from nutrient depletion.

In the lakelike pool areas just above the river's dams and in Lake Pepin, *phytoplankton* (an assemblage of free-floating algae cells) is the main producer of new plant food. Indeed, phytoplankton can become so abundant during the heat of summer that river water may resemble pea soup. The tiny, free-floating animals that consume phytoplankton are collectively called *zooplankton,* which is consumed by many fish species, especially juveniles, but also by huge paddlefish and buffalofish. Small fish are eaten by carnivores like northern pike, walleyes, and bass. Some of the most vicious predators will eat almost any small thing that moves. The diets of northern pike and large-mouth bass may include ducklings, young blackbirds, snakes, young muskrats, and small turtles.

All of the river's interrelated food chains comprise the strands of a complex food web. Perhaps this chapter will provide some insights into the ecological relationships of a few of the diverse plants and animals (including humans) that comprise the links of the food chains.

Duck Potatoes and the Detritus Food Chain

Arrowhead, duck potato, and wapato are common names for the same valuable plant that thrives in lush, green stands in the river's shallowest waters, saturated soils, marshes, and swamps. Arrowhead is an excellent example of a *producer*—because it generates valuable new biomass both above and below the water surface. Arrowhead is a member of the genus *Sagittaria* (Latin *sagitta,* an arrow), which contains about twenty new world species, most of which have distinctive arrowhead-shaped leaves. A perennial, it grows in beds as high as three feet, but sometimes is almost completely submerged, especially in river reaches where water levels vary.

Sagittaria rootstocks produce starchy, potato-like tubers, favorite foods of ducks, swans, geese, and muskrats. Waterfowl swallow the tubers whole, but muskrats chew them up. Indians and early settlers collected the nutritious "spuds" for food. French explorers called them *pomme de terre* (fruit of the earth) and named many sloughs and marshes after them. Large numbers of duck potatoes can sometimes be harvested when they float on the water after being dislodged from the bottom by a wind storm. They may be cooked just like potatoes purchased at the grocers. I have eaten them many times in a variety of ways and notice a slightly bitter taste that is easily removed by discarding the first boiling water.

Sagittaria blooms July through September on the Upper Mississippi. Female flowers produce flattened nutlets packed into dense, round heads. The nutlets are also prime duck food.

Lush stands of *Sagittaria* look like they should provide food for lots of animals, yet hardly any of the leaves have bites taken out of them because the leaves are mainly cellulose, which is indigestible to most animals. In terrestrial environments, such plants are consumed by insects and ruminants like cows, bison, sheep, and deer, which have complex digestive systems that act like disassembly lines, employing bacteria and protozoans to do most of the actual disassembling. A ruminant harvests the plants with its teeth and tongue, chews the plant parts coarsely, adds saliva, and swallows them into a first stomach. Later, at its leisure, the animal regurgitates the wad of plants as the cud, chews it slowly a second time, and swallows it into a second stomach that serves as a fermentation vat, kept at a constant temperature of about 100°F. Here, microorganisms digest the cellulose into sugars that the animal can absorb. Herbivores generally have longer disassembly lines than carnivores of equal size. For example, the digestive tract of a muskrat is much longer than that of a mink of equal size.

Big aquatic herbivores are scarce in the river. The beaver and muskrat are notable exceptions, both employing microorganisms to digest cellulose.

Fish must be streamlined to swim, ruling out bulky, cellulose-digesting second stomachs. Also, fish temperatures are dictated by the temperature of the water they swim in, ruling out fermentation of cellulose most of the year. It should be noted that some river fishes such as suckers, catfishes, and carp supplement their animal matter diet with algae and other plants, making them *omnivores*.

Indirectly, *Sagittaria* is a vital food for invertebrate animals and fishes, but its greenery must first die. In early autumn, well before the first killing frost, *Sagittaria* withers and turns brown. By mid-October the lush, verdant beds have disappeared beneath the water surface. The water-soluble leaf components like sugars and tannins bleed out into the water where microscopic fungi and bacteria attack them, converting them into clumps of fungi and bacteria—biomass that filter-feeding invertebrates can harvest for food. (We all have seen water-soluble compounds converted into clumps of microorganisms in a cup of coffee or tea that has been forgotten for a few days.)

Dead *Sagittaria* leaves and stems, consisting mainly of cellulose, are set upon by bacteria and fungi as well as a host of invertebrates that chew and shred them into fragments. These particles become tiny open-faced sandwiches, slathered on both sides by microorganisms that can be digested and assimilated by filter-feeding animals. Mother nature has made vegetable soup! The leaf blades disappear first, but the fibrous, resistant stems remain for a long time. Every duck hunter knows this, having laboriously cut them from his fouled outboard motor propeller.

What Is a Detritus Food Chain?

Dead *Sagittaria*, like all other dead organic matter, is referred to as *detritus,* and there are many forms of detritus involved in a river system. In addition to water-based plants like *Sagittaria,* a major source of detritus is floodplain forests that deliver leaf litter at critical times, to be distributed by floods throughout the ecosystem. The fall rise in water levels is particularly important because it is concurrent with the loss of leaves from deciduous trees. Fall floods produce the best quality leaf litter, especially the leaves of silver maples, ash, willow, and cottonwood. The copious, nutritious, winged seeds of silver maples enter the detritus food chain in late May and early June. Nitrogen is added to detritus by the decomposers who absorb it from the surrounding water.

The Upper Mississippi receives lots of tree leaves and seeds because of the labyrinth of small channels that wind through floodplain forests in the tailwaters and midpool areas.

Such food chains, which originate with the biological energy contained within once-living matter, are called detritus food chains. They may be the planet's most important, but least understood, routes of energy transfer—not only in rivers and streams but also in oceans and terrestrial ecosystems. The river's detritus food chains are fueled by many species of emergent and submerged aquatic plants, tree leaves, animal remains, feces, and organic debris from the entire watershed, especially the floodplain. Insects and crustaceans that assimilate the detritus become high-energy fish food. These transformations are done most efficiently in water that is rich in dissolved oxygen. If dead plants sink into stagnant, *anaerobic* (oxygenless) areas, they don't enter the food chain. Entombed with silt, they may be preserved indefinitely, like peat.

Millions of years ago, organic matter was preserved anaerobically in swamps and ocean depths, ultimately to form coal, oil, and natural gas. Thus, the sun's energy was stored for eons, only to be released suddenly by humans in a few centuries, perhaps even warming our planet.

American Lotus: Fragrance and Food for the Soul

Like the duck potato, American lotus, *Nelumbo lutea,* is extraordinary. Its spectacular beds are usually adjacent to those of arrowhead, but in slightly deeper water. Found in Mississippi River backwaters as far north as Red Wing, Minnesota, it has thick, spreading rootstocks *(rhizomes)* and huge, bluish green, round leaves that may reach three feet in diameter. The remarkable leaves may be floating or raised above the water, often as high as four feet, and they are amazingly water repellant. If water is splashed on a leaf surface, it instantly shatters into silvery droplets that skitter around like mercury on a tabletop. The presence of only floating leaves is an indication that a colony is new or has survived some sort of catastrophe.

Lotus does best in quiet, nonturbulent water, two to six feet deep. In southern reaches it is often called the "yellow chinquapin lily" because its seeds resemble the acorns of the chinquapin oak. Where lotus exists in a pure stand, the rhizomes form a complex, extensive network whose total length may be forty-five miles per acre. The rate of colonization can be rapid, with a small patch extending itself radially over forty feet per growing season (Hall and Penfound 1942). The rhizomes produce starchy tubers that resemble hard bananas, lying six to twenty inches below the soil surface. The tubers were harvested and eaten by Native Americans, as were lotus seeds. The spherical, pea-sized seeds are difficult to crack; the meat is tasty but quite hard, and was usually ground into flour.

Flowering usually begins in late June, peaks in mid-July, and may continue until late August. The magnificent, fragrant, pale yellow flowers are truly spectacular, reaching a diameter of nine inches when fully opened. Unlike the flowers of water lilies, most do not float on the water, but rise upward eighteen inches or more on stiff

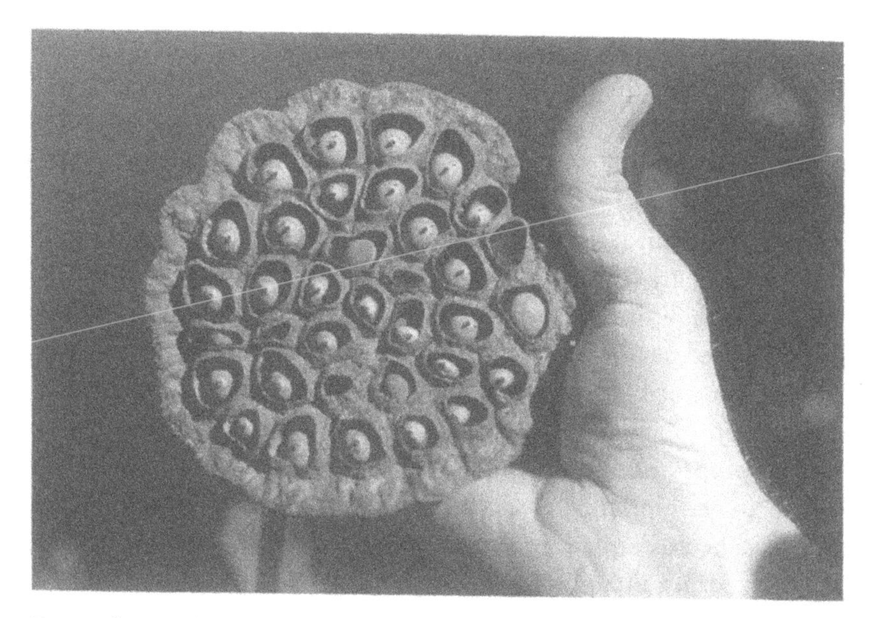

Buoyant lotus seedpods are remarkably adapted for dispersing seeds in river environments and are often sold in florist shops as "Hawaiian rattles."

stems. To me, lotus fragrance is almost overpowering, like cheap cologne, so potent that a big, flowering bed can be scented a half-mile downwind.

Lotus exhibits some fascinating physiological phenomena linked to dispersion of its seeds. The flower buds and flowers are constantly erect. Blossoms open in the morning and close at night, losing their petals in about two days. Within four days after petal drop, the seedpod, which resembles an astronaut's reentry capsule, suddenly bends downward to about a 45-degree angle and positions itself to face east. As the contained seeds mature, the seedpod returns to its erect position, turns brown, and then falls downward 180 degrees. At this time, the seeds are loose in their individual pockets, and the pod rattles when shaken. A few seeds may fall out at this stage.

Typically, the pod breaks off, lies on its side in the water, and floats away, wafted by the current or wind. On a windy fall day, a duck hunter may see hundreds float by. When the seedpod softens or tips upside down, the seeds fall out. Sometimes the pods bob along until they hit a snag, where they tip, releasing some of their seeds. The round seeds are "designed" to sink slowly to the bottom and to roll with the current until they settle in the quieter, silted areas where lotus thrives.

Lotus reproduces mainly via its tubers, which produce a flowering plant within one year, while seedlings require at least two years to flower and may take as long as four years. The seedlings are very delicate and their death rate in the wild is very high.

The seeds are remarkably adapted to ensure the existence of the species through droughts that may last for decades, perhaps centuries. The seed coat is impervious to water because it is composed of some of nature's best waterproofers—cellulose, lignin, chitin, cutin, and suberin (Shaw 1929). Germination will not occur until the seed coat is compromised somehow to allow water to enter. Most seeds will begin germinating in as little as twenty-four hours in water after they have been scarified by filing through the outer waterproof layer or by squeezing them in a vise until the seed coat is heard to crack. The simplest, but seemingly most drastic method is to presoak the seeds for thirty-six hours in concentrated sulfuric acid.

Nine years ago I put fifty seeds into a jar of water and kept them in my study where I could observe them regularly, adding and changing water when necessary. Over the years, seeds have occasionally germinated. A few have rotted. Twenty-one have yet to germinate, but I am confident that they would if I scarified them.

A popular belief is that in nature the seeds are swallowed by ducks or geese whose gizzards scarify the seeds. I doubt if this is so. I have presented a mixture of shelled corn and lotus seeds to ravenous wild mallards and Canada geese several times only to have them eat every

"Hawaiian Rattles"

Lotus seedpods are often used in winter bouquets, wreaths, and floral arrangements. When Arlayne and I had lived in Winona for only one year, and she was still unfamiliar with the Mississippi, I worked for a summer on the Rainy River that forms part of Minnesota's northern border with Canada. One day she and her buddy went on a shopping spree in Fort Francis, Ontario, returning happily with bouquets of "Hawaiian rattles" for which they had paid an outrageous price. I didn't have the heart to tell the ladies that their "rattles" were lotus seedpods, abundant in the Mississippi backwaters less than a mile from our front door.

kernel of corn but not a single lotus seed. In over sixty years of duck hunting, I have examined the crops and gizzards of many ducks and never found a lotus seed. Perhaps the seeds' spherical shape, dark color, and brittle exterior make them unrecognizable as food.

Lotus has little direct wildlife value. Muskrats eat some of the tubers. Dense stands provide brood cover for ducklings but may shade out valuable waterfowl food plants. A moth that lives in the flower, the American lotus borer, is the main consumer of the plant's greenery. It, in turn, is preyed upon by redwing blackbirds that often destroy flowers as they hunt for larvae. In winter, seedpods are shredded by nuthatches and chickadees hunting for overwintering moth pupae. As dead lotus plants decompose, they become part of the river's vital detritus food chain, feeding many filter-feeding invertebrates (Grubaugh et al. 1986).

Lotus provides food for the soul in summer and in winter. Canoeing through a massive, fragrant lotus bed on a beautiful, blue August day is unforgettable—but so are vivid winter scenes. In late November, on winter's first ice, I watched a strong wind break off seedpods, sending crowds of them scurrying like bands of elves across the slick ice surface for a mile. In calm winters, the seedpods may remain erect on their long stems until spring, frozen in the ice. During breakup, sheets of ice may float away, bearing intact thickets of long-stemmed lotus seedpods. Sadly, such observations of winter phenomena are reserved mainly for ice anglers, duck hunters, and trappers.

Silted areas of the Upper Mississippi serve as massive seed banks where myriad ungerminated lotus seeds accumulate, entombed in an anaerobic environment, waiting patiently for their day in the sun. They are insurance that this spectacular river dweller will thrive, as it has for millions of years, despite nature's fickleness.

Silver Maples Dominate the Floodplain

Floodplain forests through the entire Upper Mississippi River are increasingly lacking in diversity, trending toward forests dominated by silver maple (soft maple), a native tree that is marvelously adapted for life on the floodplain.

This "generalist species" has a wide tolerance to temperature extremes and is abundant along the entire Upper Mississippi River, all the way north to the river's source in northern Minnesota. Fast growing—it may attain a height of 120 feet—it is relatively shade tolerant and can withstand prolonged submersion during floods. In the floodplain, where soils are waterlogged and devoid of oxygen, its root system fans out, creating a dense mass of roots less than a foot thick. While this shallow root system enables it to flourish in wet soils, it also does well on drier sites (as in people's lawns).

Hardy and opportunistic, the silver maple is eager to seed and hard to kill. If cut by loggers or beavers, it sprouts readily from the stump, creating multiple trunks. If partially buried by sediment, it sprouts new roots from its trunk and primary roots. It blooms early in the spring, long before leaves appear, sometimes while there is still ice on the river. Blooming too early to be pollinated by many insects, it is wind pollinated.

Winged seeds mature in late spring, and flutter down like helicopters, spread by the wind. They are also disseminated by river currents during the usual "June rise." As river levels drop, the seeds may be stranded on freshly deposited sediment where they quickly germinate and, like a ring in the bathtub, show how high the water had been. They also germinate on the forest floor, where they may persist for years in dense carpets of stunted seedlings "as thick as hair on a dog's back," waiting in reserve for a sunlit opening to be created by the demise of a tree of the overstory. Once an opening is created, the stunted seedlings grow rapidly, powered by the sun. The loss of elms due to Dutch elm disease opened lots of new habitat for silver maples during the second half of the twentieth century.

Silver maple is a vital winter food source for beaver that stockpile green branches of several tree species in a submerged cache, swimming out from their house or den to nibble off tender bark during the winter. Some of the tree's energy and nutrients are routed through the beaver, whose feces fertilize the water. The cache itself provides habitat for aquatic invertebrates and fish.

Beaver: The River's Most Cost-Effective Engineers

Beaver play an important role in managing silver maples and, in turn, sediments. They cut or girdle older trees, which resprout, forming dense clones of saplings that slow flood flows, trap sediments, and raise the floodplain. Beaver cuttings from the upper, older end of an island often drift downstream to the lower, newer end where they wash ashore, take root, and form new tree growth.

Beavers are indeed busy seven days a week. Although they will work any shift, they prefer the night shift, industriously and unobtrusively cutting trees for food and construction materials as they build and maintain their dams and lodges, and make their winter food caches. In the process they manage the marsh and the floodplain forest. They are on call twenty-four hours per day, but receive no wages, holidays, vacations, fringe benefits, sick leaves, or coffee breaks. Thankfully, they aren't required to attend meetings, write grant proposals and progress reports, or plead to state and federal governments for funding. It's doubtful that they worry much about reciprocity between states. They never file environmental impact statements.

Beavers are North America's largest rodents, reaching a weight of fifty pounds. Like humans, they make their own environment, building dams of mud and sticks (and sometimes corn stalks) to create ponds. Their home may be a den dug into an earthen bank or a lodge built in the pond created by their dam. The massive lodge is made of sticks and mud and may be over six feet high. Beaver are mainly nocturnal; they are observed by most people in the early morning or late evening when a beaver often announces its presence by a loud, warning smack of its flat tail on the water.

The beaver's close relative, the muskrat, is far more numerous and usually weighs less than two and a half pounds. The muskrat prefers to dig its den in a bank, but if one isn't available, it builds a small, dome-shaped house in the marsh out of mud and soft aquatic vegetation. Its main foods are stems, leaves and tubers of aquatic plants, clams, and, more recently, zebra mussels.

Trapped to near extinction before the end of the nineteenth century, beaver were reestablished at various points in the Upper Mississippi River Wildlife and Fish Refuge during the late 1920s. One small colony established in 1929 had increased to about one hundred individuals four years later. Beavers are now abundant throughout the refuge. Their populations are managed by trappers who harvest them for their pelts.

Beavers are marvelously adapted for the busy life they lead. Each jaw has a pair of incisors adapted for gnawing. The front surface of each incisor is a thin, durable veneer of orange-colored enamel. Behind it lies softer dentin that makes up the bulk of the tooth. The beaver's teeth grow throughout life, with the dentin wearing down faster than the enamel that forms the cutting edge of a self-sharpening chisel. The beaver's molars are made of convoluted, vertical layers of enamel and dentin. As the beaver chews, the dentin wears down first, leaving ridges of enamel, creating a coarse grinding surface suitable for chewing tree bark, the beaver's main food. Beavers usually cut tree branches to build lodges and dams, but along agricultural tributaries they may use corn stalks, much to the chagrin of the farmer whose cornfield may be flooded by their dam.

Beavers are, without doubt, the most cost-effective habitat managers on the Upper Mississippi. An interesting beaver-managed area lies on the Minnesota–Iowa border where Winnebago Creek enters the upper end of Pool 9 from the west. Because Pool 9 is thirty-one miles long, the upper reach does not lie within the permanently impounded portion of the pool, and water levels fluctuate wildly, according to how many gates are open upstream on Lock and Dam 8.

The delta of Winnebago Creek is laced with distributaries that are dammed by beavers. Some of their interconnected low-head dams are over a half-mile long. Together they create about a square mile of rich, heavily vegetated, shallow ponds that are prime habitat for wood ducks, teal, mallards, widgeon, herons, great egrets, mink, muskrats, and raccoons.

When Lock and Dam 8 restricts river flow, most of the tailwater reach below the dam is reduced to an impassable mud flat, but the beaver ponds remain brimming with fertile water like green oases, filled with valuable duckweed and soft stem bulrush, surrounded by a sea of mud.

I wonder if the Four-and-One-Half- and Six-Foot Channel Projects could have been completed in the 1878 to 1912 period if beavers had not been trapped to near extinction. If populations had been at today's high levels, humans would have had serious competition for the willows necessary for the construction of wing dams, closing dams, and shoreline protection.

Floodplain Forest: A Hostile Summer Environment

The floodplain forest is a beautiful but hostile environment in the summer. Dappled sunlight filters through the dense, green forest canopy, but the landscape is foreboding, and visibility is limited to fifty yards or so. Yet, the gloom may be punctuated by patches of cardinal flowers that gleam like rubies in the somber environment. The understory of wood nettles and poison ivy is daunting, as are the hordes of mosquitoes and deer flies. The impenetrable stands of wood nettles are immediately painful, their microscopic stingers even penetrating heavy trousers. Those blundering into a big patch often panic and try to run out, but in so doing drive more stingers into their skin, increasing the agony. No one is immune to wood nettles. Democratically, they treat everyone the same. Fortunately, the intense pain subsides after five or ten minutes, and there are no after effects—just memories.

"Leaves three, quickly flee. Berries white, poisonous sight." All are not created equal when it comes to poison ivy. Some folks are extremely sensitive to its toxic oil; others (like me, so far) can wallow in poison ivy beds unharmed. For those who are sensitive, the pain, blisters, and itch come a day or so after the contact.

Floodplain poison ivy comes in two forms. The shrub form makes dense stands of erect stems that rise to eye level on a six-foot man. The liana form twines around tree trunks, climbing up to fifty feet in the air to get a glimpse of sun, and may be five inches in diameter at the butt. In autumn, poison ivy leaves turn bright red, shining like warning beacons amid the golden crowns of floodplain trees.

The popular sand beaches along the Upper River are usually ringed by poison ivy, lurking to attack unwary toileters. Anyone exposed to poison ivy is advised to shower thoroughly with lots of soap or detergent as soon as possible to wash away the toxic oil. Contaminated clothing should also be washed. The poison can also be contracted by petting a contaminated dog, by handling infected tools, or even standing in the smoke of burning poison ivy.

Frost grape and woodbine (Virginia creeper) are also abundant in the floodplain forest, but they cause no harm. Because they grow best on the edges of channels where they get sunlight, grapes are easy to pick from a boat. Frost grapes make excellent wine and jelly.

The floor of the floodplain is usually marked by the five-inch-high mud chimneys of crayfish (crawdads) whose burrows descend vertically to the water table. Crayfish burrows are constructed at night, and the chimneys are formed as the crayfish bring up mud pellets and deposit them around the burrow entrances. Crayfish emerge at night to scavenge on the forest floor, mainly for plant material. A person with a thin arm can gather crayfish by sticking his arm down the hole until he grasps the crayfish (or vice versa). Despite their smallness, crayfish are gourmet food for raccoons and adventurous humans.

Floodplain forests benefit the riverine ecosystem in many ways. They are vital corridors for migrating songbirds and serve as rich habitats for fish and wildlife during floods. They reduce soil erosion, improve water quality, and beautify and diversify the landscape. Leaves that fall from floodplain trees or wash in from tributaries are important energy sources that fuel complex food webs that culminate in organisms as diverse as mayflies, walleyes, and eagles.

Why Do Trees along the Edges of Side Channels Usually Topple Inland?

While the damage to the riverbank inflicted by boat-generated waves is easy to see along the main channel, less obvious is the insidious damage done by hunters, fishermen, and trappers in the backwaters. The damage they do often goes unseen because they are often out in their boats in the early morning or late evening when most observers are snug at home. The worst culprits may be duck hunters who penetrate the most remote locations in predawn darkness, competing at high speed for favorite hunting spots.

As they roar through narrow side channels with boats encumbered with bags of decoys and other heavy gear, their wakes wash sediment away from the shallow roots of the cottonwoods, silver maples, elms, and willows that line the channel banks. Still solidly rooted on the land side, the shallow roots act like hinges, set to swing toppling trunks and crowns onto the land. In time, a strong wind usually topples the trees inland, throwing up massive walls of roots about a foot thick. Thus, the shoreline retreats, islands become smaller, and the side channels become wider, shallower, and more monotonous. Interestingly, the sediment-free roots on the channel side are usually festooned with zebra mussels.

If storm winds are strong enough and from the right direction, some trees inevitably fall into the channel where the next flood will probably swing their crowns downstream, causing the trees, still anchored by their roots, to align themselves parallel to the shore. Scour holes develop among the branches, creating excellent summer habitat for bluegills, crappies, smallmouth bass, catfish, drum, redhorse, and many other species. The submerged tree branches furnish attachment for invertebrates that graze on the surfaces of the branches or filter food from the water that passes through them. The invertebrates, in turn, may be eaten by the fish.

Benthos: The "Yuck in the Muck"

Some folks just call them "the yuck in the muck"—leeches, sow bugs, crayfish, scuds, zebra mussels, and hundreds of other species of adult and immature insects, crustaceans, clams, snails, worms, and even sponges. The Upper Mississippi is incredibly rich in *macroinvertebrates* (animals without backbones that can be easily seen with the naked eye), both in numbers of species and in pounds per acre.

This diversity of critters inhabits all the nooks and crannies of the river bottom, including the water itself, sand, mud, and the surfaces of rocks, plants, and debris. Some, like caddisfly larvae, favor the submerged surfaces of man-made structures like locks and dams, bridges, navigation buoys and their anchoring chains, barges, towboats, and pleasure craft.

Macroinvertebrates that live on or within the river bottom are called *benthos.* They are commonly used as indicators of environmental quality because they are widely distributed and can exhibit dramatic community changes when exposed to water and sediment pollution. For decades, macroinvertebrates were absent or scarce in reaches where water quality was degraded by sewage. For example, the river downstream from the Twin Cities all the way into Lake Pepin suffered severe oxygen depletion caused by sewage. Pollution-sensitive organisms, such as burrowing mayflies, were missing.

Burrowing mayflies began recolonizing riverine reaches downstream from the Twin Cities in the early 1980s when dissolved oxygen concentrations increased in response to improved wastewater treatment (Fremling and Johnson 1990).

Fingernail clams and burrowing mayflies, target organisms for most studies, are important food for migrating diving ducks and coots, as well as many fish species (Wiener et al. 1998). Unfortunately, studies of macroinvertebrates are laborious and expensive because the animals are so difficult to sample, identify, count, and weigh.

Benthos communities that live on submerged hard surfaces such as rocks are called *epilithic*. In the unmodified river they would have been found on the rocks below falls and in rapids, and on cobble sediments in fast-water areas. Rock substrates in the untamed river were scarce, occurring mainly in the rapids at St. Anthony Falls, Rock Island, Keokuk, and St. Louis, but submerged, fallen trees and woody debris were abundant. They provided additional habitat. Before the river's immense clam populations were devastated by commercial exploitation and pollution, the shells of living and dead clams furnished hard substrate for epilithic fauna in a monotonous mud and sand environment.

Epilithic communities were enhanced by early channelization projects in the 1878 to 1912 period. These projects provided immense quantities of rock in the form of wing dams, closing dams, and shoreline protection.

Lock and Dam 19 at Keokuk, completed in 1913, and the Nine-Foot Channel dams, completed in the 1930s, provide great expanses of submerged concrete. Their tailwaters create a rapidslike environment, often full of huge stones placed there to prevent scouring. Navigation buoys and their anchoring chains, located at regular intervals along the edge of the navigation channel, serve as excellent substrate in the relatively fast current of the tailwaters and in the moderate current of impounded areas. They are especially important for net-spinning caddisflies, indicators of good water quality. Alarmingly, since about 1995, zebra mussels seem to be displacing most native epilithic fauna.

A number of benthos organisms are accomplished hitchhikers, stowing away on commercial and pleasure craft for a free ride to a new home. Towboats and their barges are especially important because their rough, rusted hulls are excellent habitat for many species. They transport sedentary species (like zebra mussels) upstream, enabling them to colonize new areas throughout the entire commercial waterway. Modern pleasure boats with smooth fiberglass hulls provide less surface for attachment, but the roughened metal of their propulsion units suffices for attachment by some species. Female caddisflies, for example, land on idle outboard motors and walk down the lower units into the water, laying patches of tiny adherent eggs as they go.

Adult insects are also transported by watercraft. Hordes of *Hexagenia* mayflies emerging at one locality may be transported more than one hundred miles on barges before they lay their eggs on the evening following their emergence from the water.

The construction of Lock and Dam 19 at Keokuk, Iowa, created an interesting combination of habitats for certain aquatic insects. Prior to impoundment, *hydropsychid* caddisfly larvae (net-spinning filter feeders that require swift water and hard substrate) must have thrived in the rocky Des Moines Rapids. *Hexagenia* mayflies that require a muddy substrate for construction of their burrows were probably fairly abundant above and below the rapids. When the dam was finished in 1913, creating Lake Cooper and its rich, muddy bottom, *Hexagenia* mayflies flourished. The rocky tailwaters of the dam, as well as the concrete and steel structure of the dam and powerhouse, provided increased habitat for the hydropsychid caddisfly larvae. Although other river cities have nuisance problems with mayflies, only Keokuk has severe nuisance problems with both mayflies and caddisflies (Fremling 1960a, 1960b).

Exotic Species: "Biological Pollutants"

Many people think the word *exotic* means sexy, strangely beautiful, or enticing. But in ecological circles it simply means animals and plants that are foreign, not native. With no natural predators to control them in their new environment, populations of invading exotic species can "explode," swiftly dominating habitats, and making exotics a major cause of the continuing loss of biological diversity throughout the world. In this sense, exotics can be thought of as "biological pollutants." The Mississippi and its tributaries already have their share of exotics, over 135 species. On the Upper River, the best-known exotics include curly-leaf pondweed, Eurasian water milfoil, purple loosestrife, common carp, grass carp, silver carp, goldfish, Asiatic clams, and zebra mussels.

Zebra Mussels: Invading Hordes from the Caspian Sea

Zebra mussels, native to the Caspian Sea region of Asia, entered Lake Michigan in the 1980s, probably in ballast water of oceangoing ships. They passed through Chicago's canals to the Illinois River and went downstream to the Upper Mississippi in 1991. The zebra mussel population exploded in the new environment, and by mid-1993 zebras were found throughout most of the Upper and Lower Mississippi Rivers. In the lower reach of the Illinois River, there were more than fifty thousand per square yard of river bottom. Subsequent high mortality reduced densities there to about four thousand per square yard by August 1994. By 2000, they virtually paved portions of the bottom of the Mississippi and moved up the St. Croix River. Control seemed impossible, but in the summer of 2001 a massive die-off occurred in the Upper Mississippi for reasons not yet understood. The die-off was not severe in Lake Pepin, which may function as a sanctuary for repopulating denuded downstream areas.

A single female zebra mussel can produce as many as a million eggs per year. They develop into microscopic, free-swimming larvae called *veligers* that quickly begin to develop shells. At about three weeks, the sand-grain-sized larvae settle out, attaching to hard surfaces using strong *byssal threads.* They feed by filtering plankton from the surrounding water, clarifying it.

Zebra mussels may attach to the shells of clams, where they interfere with the clams' feeding, reproduction, and movement. Consequently, the native clam fauna in the river is rapidly and severely declining. The mussels clog water-intake systems of power plants and water treatment facilities as well as the cooling systems of boat engines. They carpet the hulls of metal boats that are left in the water, increasing their weight and reducing their efficiency.

Zebras are unpopular with anglers because their sharp shells fray fishing lines and foul lures. Most anglers have had the experience of snagging a grapefruit-sized ball of zebras that has a single clam inside. Sometimes the snag contains nothing but zebras threaded to each other.

Zebra mussels affect some invertebrate species more than others. There is no question that they are altering the communities that inhabit the rocks of wing dams and other structures. There is no doubt that their filter feeding has increased the clarity of the river and reduced the food available to other organisms, but they are also affected by the river environment. Zebra mussels are not well suited to habitats with high water velocity or areas with no current. Zebra beds on top of most submerged wing dams are destroyed by winter's ice and cold, but they are repopulated the following summer.

Predators and environmental obstacles are also beginning to take a toll on zebra mussels in the Mississippi. Zebra mussels are eaten by some fish, like freshwater drum and carp. Muskrats like them. Diving ducks, like scaup, harvest them; so do mallards if they are within reach.

Mallards, which are puddle ducks, tip up to probe for mussels on the tops of wing dams during low-flow periods in late fall. Some have their crops so full of mussels, they make a crunching sound when retrieved by hunters. One observer watched a flock of mallards gorging on mussels covering a log that had washed up on shore, "just like they were eating corn on the cob."

Between 1999 and 2002, zebra mussel populations crashed throughout the Upper Mississippi River for unknown reasons. It seems likely that the invading zebras will forever be part of the river's fauna, but never in the incredible abundance observed during their initial population explosion.

Paddlefish: Gentle Giants from the Pre-Jurassic

A hundred-pound paddlefish has a mouth like a bucket, and it looks like it could bite your leg off, but this toothless river denizen is harmless.

Paddlefish occur only in the main stem Mississippi River drainage, including the Missouri River, and in the Yangtze Kiang River of China. The two paddlefish species were probably a single population eons ago when all continents were united in a single supercontinent,

These paddlefish were captured from the Mississippi River near Winona by commercial fishermen seining through the ice mainly for carp, buffalofish, and drum. Because it is illegal to harvest scarce paddlefish from the Mississippi, they were quickly returned to the river unharmed.

but continental drift over the past two hundred million years separated them. An individual of the species, which may weigh over one hundred pounds, exhibits features characteristic of its ancient pedigree. It has a cartilaginous "backbone" that extends into the upper lobe of its sharklike tailfin, and its paired pectoral fins are widely separated from its paired pelvic fins. The paddlefish collects plankton, small crustaceans, and insects as it cruises along, filter feeding with its huge, toothless mouth and sievelike gills. The paddlelike snout is not used for digging in the bottom for food as is popularly believed, but is a highly developed sense organ, apparently used to locate prey.

In the past, paddlefish were harvested commercially from the Mississippi. They were sold as boneless catfish or spoonbill catfish and commonly smoked. They are now fully protected on the Upper Mississippi but may be taken by sportfishing on the Missouri River.

Today, whether they are captured legally or illegally, paddlefish are treasured for their eggs. For centuries, the world's most expensive caviar (salted fish eggs) has been prepared from the beluga sturgeon of the Black Sea, but due to habitat deterioration that harvest has dropped steadily. In 2001, the Black Sea harvest was only 15 percent of what it was in 1999. Incredibly to most people, prime Russian Beluga caviar retailed in 2004 for about $125 per ounce while domestic caviar harvested from paddlefish and shovelnose sturgeon retailed for about $14 per ounce. Over the years, this price discrepancy has resulted in flagrant counterfeit labeling and sale of domestic caviar and other relatively inexpensive foreign caviars as more expensive imported Russian caviar.

In 2000, a $10.4 million federal fine and prison sentences were levied against U.S. Caviar & Caviar, Ltd. in connection with a U.S. Fish and Wildlife Service investigation of illegal caviar trade (U.S. Fish and Wildlife Service 2001).

King Carp: *Rex cyprinorum*

The carp is the Rodney Dangerfield of Mississippi River fish—it just gets no respect. A gourmet friend recommends serving red wine with red meat, white wine with white meat, and gray wine with carp. In America, the old "buglemouth bass," the "sewer inspector," gets no respect at all.

But elsewhere in the world things are different. Carp are among the world's most important freshwater fishes, furnishing high-protein food for millions of people in the Orient and central Europe.

Natives of Asia, carp were domesticated and reared in ponds in China and nearby countries for many centuries, and a number of strains or breeds were produced. Carp were brought into Germany and other European countries from Asia, probably in the eleventh or twelfth century in tanks of fresh water aboard sailing ships. Pond culture of carp became highly developed in central Europe, and carp were welcomed into England shortly before 1500. Izaak Walton called them the "queen of the rivers" (Eddy and Underhill 1974).

Carp have been in America so long that most people consider them naturalized citizens. They were introduced into the United States from Europe between 1870 and 1880, primarily to satisfy German immigrants yearning for fat carp. Germans call them "Karpfen" and most love them. They even pond raise "leather carp" that have no scales ("Lederkarpfen") and carp with only a few, big mirrorlike scales ("Spiegelkarpfen").

In 1877 the United States Fish Commission stocked ponds with carp in Washington, D.C., where the fish were considered to be so valuable that the ponds were guarded to prevent theft. Carp were raised in Iowa fish hatcheries beginning in 1880 and, at first, stocked into ponds constructed especially for their culture. Later they were distributed to all of the principal waters of the state. Minnesota stocked its first carp in 1883. Presently, carp are resident in all of the contiguous forty-eight states as well as in many localities in Canada and Mexico.

Carp are among the most adaptable of the world's fishes, flourishing in virtually every environment to which they were introduced. In recent years they have invaded the Red River of the North and are now found far up into Canada's Nelson River in areas formerly thought to be too cold for them.

An angler caught the Mississippi's first carp at Hannibal, Missouri, in 1880. By 1888 they were well established in bottomland lakes near Quincy, Illinois. Almost half a million pounds of carp were harvested at Lansing, Iowa, in 1894, and by 1899 the catch had increased sixfold. By this time, biologists had come to despise carp because they destroyed beds of aquatic plants vital to waterfowl and competed with native fishes. Yet, carp soon became one of the mainstays of the Mississippi's commercial fishery. Then, as now, the eastern states furnished the largest market for carp.

Carp are the champions of tackle busters, commonly weighing twenty-five to thirty pounds and sometimes over fifty pounds. What they lack in beauty they make up in sheer, brute strength. For their size, carp have small, sensitive mouths and are wary customers, easily spooked by noise or movement. Carp feed delicately and will drop bait immediately if they sense a hook or feel resistance from too heavy a line or from a sinker. The most successful anglers use small hooks and slip sinkers or no sinkers at all. Baits range from dough balls, whole-kernel corn, worms, and marshmallows to ripe mulberries.

Carp are opportunistic omnivores, eating tender roots and shoots of aquatic plants, but also insects and their larvae, aquatic earthworms, crustaceans, and small mollusks. Folks that see carp around sewer outlets usually assume that they are eating terrible things like feces, but such is not the case. Analyses of their stomach contents show that they are usually eating invertebrates that thrive in these enriched habitats. As for their scavenging in manure for corn, barnyard chickens and wild turkeys do it all the time. Turkeys are adept at flipping cow pies to harvest insects that feed on manure.

Because they eat relatively low in the food pyramid, carp are highly efficient at transferring food energy. They are at the top of short food chains, converting a variety of plant and animal matter into high-quality fish flesh.

My most vivid memory of carp fishing was on the Mississippi Headwaters at St. Cloud, Minnesota, in mid-June 1954 when I picked my way through a narrow, rocky strip of floodplain forest along the river's edge to reach the sewer outlet of a meat-packing company that discharged its slaughterhouse wastes into the river. The spring flood had passed, leaving the brush of the floodplain festooned with tampons, toilet tissue, and condoms, the usual legacy in those "good old days" of a falling, urban river. This did not shock me. In fact, I didn't even think it was unusual. It had been that way all my life. St. Cloud had no sewage treatment then, normal for most river cities.

The slaughterhouse sewer ran brown with the gut contents of cattle and pigs, and periodically red with their blood. Carp lazed around the sewer outlet, flashing gold as they rolled in the current. Excitedly, I baited with whole-kernel corn because I knew from previous experience that the carp were feasting on corn kernels that the livestock had eaten as their last meals. To a poor, unsophisticated fisherman this was analogous to "matching the hatch." I caught enough carp for smoking, being careful not to keep more than I could carry back to the car. When I cleaned them I found their stomachs gorged with corn—truly a gourmet diet! The carp were split down the back and the halves were brined for twenty-four hours in a chilled salt solution strong enough to float an egg. Smoking was done for about fourteen hours in an old icebox outfitted with a hot plate and a pan of maple chips. What could be more succulent, pink, and fragrant than corn-fed carp, maple-smoked and glazed with brown sugar?

I also have fond memories of free carp dinners at Jimmy's Bar in Keokuk, Iowa, in 1957. Jimmy served a delicious, "scored," batter-fried carp fillet, coleslaw, tartar sauce, potato chips, and a dill pickle on a paper plate to his patrons every Friday afternoon. For the uninitiated, scored carp was prepared by cutting a skinless fillet from each side of the fish and removing the dark meat along the lateral line with a sharp knife. (The dark meat was usually retained for later use as catfish bait.) The fillets were scored crosswise with a sharp knife, using strokes that went almost through the fillet, spaced one-eighth- to one-quarter-inch apart, cutting fine bones into small pieces that disappeared when the battered fillets were fried in lard hot enough to ignite a "farmer" match.

Because they are fatty fish, carp lend themselves to smoking, providing a delicacy thought by many equal or superior to any other fish. Smoked carp, which can be purchased in most river towns, is traditionally enjoyed with beer. The carps' oiliness, which makes them so smokable, also makes them susceptible to storing fat-soluble, persistent, organic chemicals like PCBs (polychlorinated biphenyls) and DDT that have been run up and concentrated through food chains. Check the fish consumption advisories for your section of the river. It makes sense to avoid eating large, old fish very often.

American Eels: Visitors from the Bermuda Triangle

For an angler on the northern reaches of the Upper Mississippi, catching a writhing, three-foot long eel is unusual (and unforgettable!), but the journey that eel made to get caught is truly extraordinary.

Among Mississippi River fishes the American eel is the champion traveler. Weighing over six pounds and more than a yard long, and invariably female, eels landed near the Mississippi Headwaters are but half way through an amazing round-trip odyssey of over four thousand miles.

The eel's incredible trip begins as an egg smaller than a grain of sand deposited in the Sargasso Sea, an area of the North Atlantic Ocean east of the Bahamas and southwest of Bermuda, familiar to

This female American eel was caught near Fountain City, Wisconsin, intercepted halfway through an incredible four-thousand-mile journey that normally begins and ends in the Sargasso Sea northeast of the West Indes. She was preserved as a university museum specimen, but could have been smoked and eaten. Because eel numbers are decreasing, anglers are advised to release them by cutting their fishing lines as close to the hook as possible. Removing the hook from the mouth of a slimy, writhing eel is an unforgettable experience!

most people as the mysterious Bermuda Triangle. The newly hatched larval eels do not resemble adult eels at all, but are transparent ribbonlike creatures that were long thought to be a separate species.

Much of the eels' oceanic existence is still a mystery. Unlike *anadromous* salmon, which ascend rivers to spawn in fresh water and descend to the ocean to mature in salt water, eels are said to be *catadromous* because they spawn in the ocean and ascend rivers to spend most of their lives in fresh water. They are found in many rivers along the eastern and Gulf coasts of North America.

After a year-long, perilous ocean journey of more than a thousand miles, larval eels may enter the mouth of the Mississippi River where they transform into miniature eels called elvers. The young females then work their way upstream, negotiating dams, and escaping predators. During their journey they feed voraciously on aquatic insects, crayfish, and small fish.

After growing for six to eight years in the Mississippi River, sexually mature females return to the Sargasso Sea to mate and lay as many as twenty million eggs at depths that may exceed one thousand feet. The much smaller males, most of whom never went upstream but loitered near the river's mouth awaiting the return of the females, accompany them on this journey.

In the Sargasso Sea, American eels may share spawning grounds with European eels, but are believed not to interbreed with them. The common breeding ground echoes the time when the continents were joined, and the two populations may have been a single species. After spawning, adult eels probably die because they are never seen again, but their larvae, guided by an unerring instinct, head for the continent (and perhaps the same river) from whence their parents came. No one knows how they find their way to a place they've never been.

Eels have been highly regarded as gourmet food for many centuries and command a high price in Europe and on eastern markets of the United States and Canada where they are baked, pickled, smoked, or jellied. It may be unwise to gorge on eels from the Mississippi River because their flesh is oily and presumably stores PCBs, which are thought to be toxic.

Dams have made the eels' journey more difficult, and theirs numbers are steadily declining on the Upper Mississippi River. Some eels make it through thirty dams, swimming as far north as St. Cloud in the Headwaters before returning to the Sargasso Sea. The odds of a newly hatched larval female eel ever completing its life cycle are incredibly poor, thus the need to produce millions of eggs.

Anglers: The River's Top Predators

It's not very difficult to fish an inland lake where water levels are usually quite constant from day to day and throughout the year. Landmarks like bays, points, islands, and bars don't move around much, and weed beds may look the same for decades. Predictably, lake fish spawn in certain areas, spend the summer in selected habitats, and overwinter in special areas. Well known to "old timers," these areas are "honey holes" that may have been productive for a lifetime. The main variables for the lake angler are season of the year, time of day, barometric pressure, water temperature, light intensity, dissolved oxygen concentration, structure, and the skill of the angler. Armed with outboard motorboat, electric trolling motor, hydrographic maps, graphite rods, monofilament lines, electronic fish locator, global positioning system, and a hundred assorted lures and baits, anglers usually succeed in catching fish but concentrate on a few select species. If they are unfamiliar with the lake, they can use the "seagull technique" (closely watching other anglers for extraneous movements and flashing landing nets).

It's much more of a challenge to fish the Mississippi, which has all of the lake variables plus many more—fluctuating water levels, variable currents, innumerable snags, wing dams, floating debris, varying turbidity, passing towboats, an infinite number of habitats, and many more species of fish, most of which have great seasonal movement patterns. Today's riverine landmarks like snags and bars may be replaced next week by new ones.

Current velocity in backwater channels changes by the season, day, and hour. Near the main channel, water flow may even change direction if a loaded fifteen-barge tow goes by. As the tow passes, river water rushes into the void created by the passing tow, sometimes dewatering shallow bays and side channels. In narrow main-channel reaches, the tow acts like a moving dam, pushing water ahead of it. A surge from a passing tow may overfill a marsh area, only to have the marsh run dry as the tow passes. Each surge stirs up loose bottom sediments, increasing turbidity.

Position is everything in river fishing. Because structure and current patterns are so variable, the difference of a few feet may spell success or failure.

Yet, the fisherman's rewards may be greater in the river than in the lake. The river is far more productive than most lakes because it is highly fertile, well oxygenated, and most areas do not thermally stratify in the summer. Anglers can fish all year round, often sliding boats over ice to fish a dam's tailwaters, and can use at least two lines in most areas. There are more species of fish to catch, with bait fishermen often catching nine or ten species in a day's fishing. Fishermen who learned to fish in lake environments are often unable to adapt to river fishing, leaving more room for those longtime river rats who encourage newcomers to play golf.

Surprising to many anglers is that river fish, especially sunfishes and yellow perch, have fewer parasites in their flesh than do fish from weedy inland lakes. This is because the parasites' life histories usually include stages that reproduce in certain snails that thrive in quiet waters with dense weed beds. Snail populations in the river are comparatively low, due mainly to predation by fish like freshwater drum but also because of currents and fewer dense weed beds.

Overwintering Panfish: Searching for a Snug Haven

Ice fishermen are experts at exploiting sunfish and crappie populations in overwintering habitats, some of which may be smaller than a tennis court. An army of prospectors sets out to find these sanctuaries in early winter when the ice is barely thick enough to support their weight. Most of them hike to get there, but others use outboard motorboats, airboats, picker boats, and hovercraft. When the ice gets a little thicker they turn to snowmobiles, all-terrain vehicles, cars, and trucks. Like gulls, they converge on the overwintering areas. Armed with sophisticated gear including powered ice augers, sonar, ultralight graphite rods, two-pound-test monofilament lines, pea-sized bobbers or spring bobbers, and tiny lures enhanced with insect larvae or special scents, they exploit the fish that bite aggressively in early winter. In their portable, darkened shelters, they can watch the fish bite if the water is shallow and clear enough. River water is surprisingly clear in winter when there is no runoff. At that time, the river is mainly the water table in motion (spring fed).

It is unlikely that many overwintering panfish habitats remain unknown to ice fishermen. Bluegills and crappies have done well for over a half-century in the lakelike environments created by the Nine-Foot Channel dams, but their overwintering habitats are deteriorating in ice-covered northern reaches of the Upper Mississippi.

There is ample habitat in summer. Bluegills and crappies can be caught on wing dams, in weedbeds, in snags, and around rock piles that lie along the edges of the main channel and side channels. However, in fall, as their metabolic rates drop in response to lowered water temperatures, they seek overwintering areas out of the current. And, like snowbirds, they are very fussy about where they'll spend the winter.

In the best areas, water depths may range from three feet to about twenty-five feet. Water temperatures must exceed 34°F, current velocities must be almost imperceptible, and dissolved oxygen concentrations must be above two parts per million. (At 32°F, water is saturated with dissolved oxygen at about twelve parts per million.) The fish also prefer areas of lessened light intensity. In deep areas at first ice, fish may congregate right on the bottom at depths of twenty-five feet or more. As the season progresses in the deepest areas, oxygen levels drop and light intensity decreases due to thicker ice and snow cover. These factors combine to produce a comfort zone that often occurs at a depth of about fifteen feet.

At first ice, some of the overwintering habitats may be less than four feet deep. Yet, water temperatures at the mud surface may be as high as 39°F. (Water is densest at that temperature and can be warmed slightly by rotting vegetation or inflow of groundwater.) However, the water temperature just under the ice will always be 32°F, the temperature at which water is least dense. By March, the ice may have thickened to three feet in northern impoundments, especially in winters with little insulating snow. The habitats seldom freeze to the bottom, but the space under the ice may be scarcely deeper than the overwintering fish are tall. Sunlight penetration decreases as the ice thickens and becomes snow covered. To make matters worse, the weight of heavy snows depresses the ice, causing water to ooze upward through cracks and at the shoreline, creating translucent slush that further decreases light penetration. This, in turn, lessens the ability of aquatic plants to produce oxygen by photosynthesis, causing dissolved oxygen levels to plummet, depressing the fish. Slowly, the water becomes colder than the critical 34°F. The fish become lethargic and refuse to bite, but they may still be curious enough to scrutinize lures. If conditions become bad enough, the fish may vacate their sanctuaries, often entering areas where increased current saps their energy reserves, further stressing them. If there is no easy escape route, a winter kill may occur.

In most fish, the production of disease-fighting antibodies falls off at winter temperatures. After a prolonged winter, stressed fish are doubly susceptible to bacterial infections. Their deaths usually go unnoticed because the spring ice is too rotten for observer access. Crows, eagles, ospreys, and gulls quickly clean up the mess, converting fish biomass to bird biomass. The birds' rich feces will stimulate plant growth that will trap the sun's energy, initiating myriad new food chains.

The navigation pools are aging and filling with sediment, reducing water depths and inexorably causing the irretrievable loss of panfish overwintering habitats. The long-term prognosis is not good.

The Strange Mystery of the Headless Gizzard Shad

Ice still covered the Mississippi backwaters and most of the main channel at Winona in late February 1999, but the turbulent tailwaters of Lock and Dam 5A were open. The sun shone brightly, and the windless day was balmy enough for fishing with spinning gear. We carefully slid a fourteen-foot aluminum boat over a few yards of ice into the open water, hopped in, motored up to the dam, dodged a few ice cakes, and settled back to drift with the current. As we soaked in the sun, we kept our lines vertical and our minnow-tipped jigs just off the bottom, sixteen to thirty-five feet below our boat. Fishing was good, and soon we had our limit of twelve nice saugers (sand pike).

As we fished, we watched a continuous, silvery parade of dead, hand-sized, young-of-the-year gizzard shad drifting downstream. Gizzard shad are silvery, herringlike fish that may weigh as much as one and a half pounds. Although they are excellent food for a variety of predators, the bony, soft-fleshed, plankton feeders are seldom caught or eaten by humans.

The winterkill didn't surprise us because gizzard shad are at about the northern limit of their range in the Winona area. Bacterial infections kill multitudes of stressed shad every winter when the icy water interferes with their ability to manufacture protective antibodies. But this time it was different. Virtually all of the shad were headless.

Shad are extremely prolific, and their rapid growth attests to the incredible productivity of the Upper Mississippi River. Those hatching early in the summer may be eight inches long by winter, while those hatching late may only grow three inches. During winterkills, the smallest ones die first, often in mid-January. Death comes slowly, and as they drift downstream they convulse and flutter erratically, furnishing irresistible meals for walleyes and saugers, and prompting winter-fishing doldrums. The stomach of one fourteen-inch sauger that we caught contained thirteen three-inch shad packed as neatly and compactly as canned sardines in a tin. The largest young-of-the-year shad die later, often in late February, as on the day of our fishing trip.

Puzzled, we theorized that bald eagles or crows were the decapitators. But why would they only eat the heads? It seemed unlikely that birds standing on ice would throw the fish back into the river after dining on the head. One fisherman suggested that the gill area decomposed first and the head simply fell off. Yet, the fish showed no sign of decay in the frigid water.

To our amazement, the decapitators turned out to be a flock of noisy, overwintering mallard ducks congregated below the dam, waiting for shad to drift down from the still-frozen pool above. Repeatedly, a vigilant duck would race up, grab a shad by the head with its bill, shake it so violently that its body flew off, and then scarf down the nutritious, oily head. They probably would have eaten the bodies too, but they were too big to swallow. Since then, we've seen mallards gorging on dead and dying three-inch shad, swallowing them whole.

Mallards are versatile feeders but typically eat plants. Only about 10 percent of the mallard diet is made up of animal matter, usually insects, crustaceans, and mollusks. These winter shad probably gave the ducks a welcome boost of protein and fat.

Such strange strands in the riverine food web are seen only by anglers crazy enough to fish the late-winter tailwaters. If we had stayed warmly snuggled up with our computers, we'd have missed the whole show.

The Seventy-two-Foot Hole

The Mississippi is usually portrayed as shallow and muddy, but it ain't necessarily so! On a bright blue day in the dead of winter (February 15, 2000) my buddy and I motored downriver for about two miles below Lock and Dam 4 at Alma, Wisconsin, picking our way through rafts of drifting ice to a sharp bend on the Minnesota side of the main channel, where we fished for saugers in seventy-two feet of water.

The water was as clear as weak iced tea, and when we lowered our minnow-tipped chartreuse jigs into the water, they disappeared from view at about nine feet. Some of the saugers we caught had their membranous stomachs turned inside out, protruding outward through their gills, symptoms of the bends and explosive decompression. Several small fish that we released bobbed at the surface, buoyed up by escaped gases from their ruptured air bladders, but they were quickly snatched by bald eagles that flew down from the cottonwoods about fifty yards away. Seven immature eagles chirped and trilled, pleading for fish. We fished this "hot spot" reluctantly because we knew that the mortality rate of released fish would be high.

Autopsy of our catch showed that the fishes' swim bladders had perforated, releasing contained gases into their body cavities, pushing their stomachs outward, and inflating them inside out. The stomachs couldn't come out the fishes' mouths because their "one-way" teeth prevented it, but their blunt gill rakers allowed the thin-walled stomachs to squeeze outward through the gills. Most fishermen erroneously think that it is the air bladder that protrudes through the gill opening, but this is not possible in saugers because the air bladder is firmly attached to the body wall along the vertebral column for its entire length.

We enjoyed the fish fry that evening, but with remorse for the probable deaths of the small fish that we released. We rationalized that the fish would not be wasted but would furnish fuel and body-building nutrients for bald eagles and other river denizens. As they decomposed, dead fish would produce detritus that would nourish filter-feeding invertebrates. Released nutrient elements would stimulate the growth of green plants, the river's primary producers of new food.

19

"Mayflies! What the Hell Good Are They?"

Indicator Species

Hanging in my office is a large, framed display of beautiful butterflies that I collected along roadsides in the mid-1950s near Keokuk, Iowa. In those days the fields, roadsides, and fencerows still supported a rich variety of plant life, weeds if you will, that provided a veritable summer smorgasbord to caterpillars of many species.

Caterpillars are voracious feeders but often very picky eaters, their diet confined to a single plant species. For example, while the adult monarch butterfly sips nectar from a variety of flowers, as a caterpillar the monarch's diet is strictly milkweed leaves.

Today, it would be a real challenge to duplicate my collection. Over the past half-century Iowa's diverse prairies have been converted into a great corn-soybean-hog-cow *biome* (a huge ecosystem), kept that way by intensive cultivation, wetland drainage, batteries of insecticides and herbicides, and genetically engineered crops.

After World War II, roadside ditches and rights of way were often bombarded with the herbicides 2,4-D and 2,4,5-T, killing most broad-leaved plants. As a result, most weed patches and the butterfly-supporting plants they contained are gone, and the butterflies with them.

The phenomenon of butterfly disappearance isn't peculiar to Iowa. It is nationwide. Butterflies, by their relative abundance and their diversity of species, are excellent indicators of environmental quality and diversity.

The Mississippi River has its own, but usually less obvious, cast of indicator species, including invertebrates, fishes, mammals, birds, and plants. In this chapter I delve into the ecology of *Hexagenia* mayflies, true big-river denizens, excellent indicator species, and vital links in hundreds of food chains.

I've studied the biology of these aquatic insects for almost fifty years. Friends have often suggested that I had tunnel vision, studying so few species for so long, but my specialty has been somewhat like being in a shack with a tiny hole in the roof on a clear night. From

Like Canaries in a Coal Mine

Canaries were used for many years in coalmines to monitor air quality because of their extreme sensitivity to poisonous, odorless gases such as methane. If the canaries passed out, miners abandoned the shaft because they knew they'd be next, either through asphyxiation or explosion of the flammable methane. Similarly, *Hexagenia* mayflies are good monitors of water quality.

While chemical tests are expensive and describe water quality only in terms of specific toxins and times, *Hexagenia* distribution inexpensively indicates combined effects of many toxins, anoxia (no oxygen), and other stresses throughout the year.

Burrowing *Hexagenia* nymphs live in sediments where toxins accumulate. They are unable to tolerate anaerobic conditions, and they cannot swim long distances to escape environmental stress. Since they have a relatively long life cycle, the condition of the *Hexagenia* population is a good indicator of water quality over a period of time.

The presence of mayflies does not mean that the river is unpolluted, but it does indicate that it has been of *Hexagenia* quality every day for up to a year. Sampling mayflies can be compared to taking a blood sample and using it as a measure of the well being of the entire body.

Hexagenia mayfly nymphs live in burrows in the river bottom that provide seclusion and serve as respiratory tubes. Nymphs pump oxygenated water through them with undulatory movements of the feathery gills that arise from their abdomens. Elephant-like tusks are used for digging, as are mole-like front legs. Three pairs of legs bear combs and brushes for straining food particles from the water and mud. The hind pair of legs has special brushes for cleaning the gills. Gleaned food particles are collected and concentrated from the hind legs by the second and first pairs of legs, and then eaten. Nymphs are a prime, high-energy, year-round food source for hungry fish and diving ducks.

where I sit, nothing can be seen through the hole, but as I stand on a stool and put my eye to the hole and peer through it I can view a star-spangled, infinitely large universe on the other side. My specialty has been like that, enabling me to view Mississippi River ecosystems through the eyes of mayflies, some of the river's most important life forms (Fremling 1960a, 1968, 1989; Fremling and Johnson 1990).

Fascinating Life Histories

Delicate and ephemeral, mayflies live out their adult lives in a single day, albeit in the company of uncountable numbers of their kind.

Despite the individual fragility of the gossamer-winged adults, the durability, adaptability, and tenacity of the species is evidenced by their long existence on earth. Predating the dinosaurs, mayflies have been around for so many millions of years they hitched a ride on the drifting continents, and the resulting worldwide distribution of various species provides excellent biological support for the theory of continental drift.

Having survived the cataclysm that wiped out the dinosaurs, as well as other mass extinctions that destroyed millions of species, the greatest threat they have faced is water pollution caused by humans.

Mayflies live virtually everywhere in the world. In North America alone, biologists have listed about 625 species living north of the Rio Grande.

Mayflies spend most of their lives as aquatic nymphs (the British call them larvae). They live in a wide variety of standing and running freshwater habitats and then develop into winged stages that mate, lay eggs, and soon die. Most have an adult life of only a day or two.

When the winged adult emerges from its nymphal "skin" at the water's surface it is called a *subimago* (subadult), which later molts into a final winged stage, or *imago* (adult). Mayflies are the only insects that molt as winged adults.

Those Big Black Bastards

Many species of mayflies inhabit the Mississippi River. Most are inconspicuous and cause no problems, but two species *(Hexagenia bilineata* and *Hexagenia limbata)* are notorious for the nuisance problems they cause. They are the largest mayflies in North America. *H. bilineata* are generally more abundant than *H. limbata*, but *H. limbata* become increasingly abundant northward and make up the first hatches of the summer when water temperatures rise to 53°F.

H. limbata are easy to recognize because of their yellowish color. Towboat personnel call them "forerunners" because they portend the big hatches of somber-colored *H. bilineata* that will come later when river temperatures rise to 66°F.

Crews of riverboats also have special names for *H. bilineata*; probably the most colorful, as well as descriptive, one is "those big black bastards." River residents use other names such as fishfly, willowbug, Junebug, and shadfly. In Savanna, Illinois, the annual summer celebration is called "Shadfly Days."

Masses of *Hexagenia bilineata* mayflies were attracted to car headlights on a Mississippi River bridge at Winona, Minnesota, on the night of July 8, 1966. They were virtually all females, intercepted during their upstream flight to lay eggs. The height of the pile tapered off with distance but was still an inch deep thirty feet away. The insects were still flying in undiminished numbers when the car was shoveled free of the wet, greasy mass at 12:35 A.M. Collections made by ship captains, lock personnel, and other cooperators revealed that the awesome "hatch" encompassed the great expanse of river from Wabasha, Minnesota, to Alton, Illinois, a distance of 550 river miles. Another hatch of almost the same magnitude occurred on the nights of July 15–17 over much of the Upper Mississippi River.

Nymphs of the genus *Hexagenia* are burrowers that spend up to a year as filter-feeders in U-shaped burrows in the muddy bottoms of lakes and rivers. Once mature, they swim to the river's surface where they shed their skins, mainly at night, and emerge en masse as winged, somber-colored subimagoes (or duns). This mass emergence is usually referred to as a "hatch." The newly emerged adults fly, usually with a breeze, to the windward shore where they may cluster in incredible numbers, weighing down tree branches and slathering houses and cars. There they remain motionless during the following day, moving only when disturbed or to keep themselves shaded. Birds feast on the mayflies, but are confronted with more than they can eat. The adult molt begins in the afternoon, peaking about four o'clock. The resultant imagoes are noticeably more delicate and more brightly colored than the subimagoes, and they are much better fliers.

Swarms form along the riverside in the evening when the males begin their aerial mating dance. Females flying through the swarms are intercepted by males; copulation takes place in flight. Exhausted males remain near the mating area and die within hours, but females fly upstream to lay their eggs, apparently flying until they

Talking to a Mayfly

A little mayfly biology is necessary to tell this tale. In order for a gravid *Hexagenia* mayfly to expel her egg masses into the river, she must alight on the water and spread her wings so that each adheres to the surface film. Thus bracing herself, she spasmodically muscles about eight thousand tiny eggs out into the water, getting an assist by a push from her air bladder.

I walked into my favorite tavern one muggy July evening in the midst of a *Hexagenia* hatch. My old friend the proprietor knew I'd been studying mayflies for about thirty years. "Hey, Doc, have you learned anything new about mayflies?" he smirked.

"Kermie," I said, "would you believe that I've finally learned to communicate with them? Treat me to a beer and I'll prove it."

I stepped outside and gently plucked a motionless, gravid female imago *Hexagenia* from the screen door. Returning my prize to the bar, I declared that I was going to command her to lay her eggs at my discretion. I requested a glass of water and dropped her in, being careful to make sure she landed on her side. A crowd of beery onlookers peered intently as she floundered. Someone chanted, "Lay eggs, lay eggs," but nothing happened. Then, predictably, she righted herself and pressed all four of her gossamer wings firmly onto the water surface! I ordered, "*Hexagenia,* please lay eight thousand eggs." Quivering, almost ecstatically, she spasmodically released a magnificent cloud of eggs that sifted downward to the bottom of the glass, amid a thunderous ovation.

Kermie stole my mayfly trick and performed it for years, sometimes embellishing it by substituting a glass of beer for the glass of water.

have exhausted their energy reserves. They may travel several miles, depending on air temperature and wind, and then crash-land on the water where each female lays about eight thousand tiny eggs. The eggs separate from each other and drift slowly to the silted riverbed where they hatch in about twelve days, depending on water temperature. The eggs are remarkably adapted to behave like silt particles, remaining suspended in the current until they settle in the silted habitats so vital for nymphal life. The minute, newly hatched nymphs are just visible to the naked eye. They slowly increase in size through a series of many molts during the ensuing warm months of the year.

All feeding is done during nymphal life. Adult mayflies have no functional mouths or digestive tracts. The adult digestive tract is modified into an air bladder that keeps the female afloat when she returns to the river to lay her eggs. Folks who have walked or driven over masses of adult mayflies have heard the air bladders snapping.

Nuisance Problems

Hexagenia swarms are often so large and dense that they can be observed on a towboat's radar screen. Mayflies are exceedingly unpopular with towboat crews. Adults of both *Hexagenia* species are attracted to the boats' powerful carbon-arc and mercury-vapor searchlights. The incredible swarms of insects can virtually block the lights, making it difficult for the pilot and crew to spot unlighted channel markers. Crushed insects render decks, ladders, and equipment of the boats slippery, smelly, and dangerous. Towboats must be completely hosed off with water after every major mayfly encounter.

Along with their regular cargoes, tows carry hitchhiking mayflies up and down the river. During periods of heavy *Hexagenia* emergence, most towboats carry huge cargoes of newly emerged subimagoes. A towboat with a full complement of fifteen barges is about 2.5 acres in size, and when the open-top barges are empty, their surface area is much greater. The boats travel twenty-four hours per day, averaging about ten miles per hour when downbound and about eight miles per hour upbound. Thus, during a cool-weather emergence, millions of *Hexagenia* may be transported more than 350 miles downstream or 250 miles upstream before they transform to imagoes and lay their eggs.

A sly towboat captain, with a mayfly-covered tow, has been known to turn out his lights while locking, slipping out unlighted and leaving his burden of mayflies with the lockmen at the well-lighted lock.

Hexagenia subimagoes create nuisance problems by their sheer numbers, but the most severe problems are caused by female imagoes being diverted to light sources as they fly upstream to lay their eggs. (They apparently fly upstream to compensate for the inevitable downstream drift of eggs and nymphs.) Mayfly eggs have a high fat content, and they stick to whatever they touch. Drifts of decaying mayflies and their eggs smell like rotting fish, hence the name "fish-fly." Attracted by the bright lights of the city, drifts of the insects form under streetlights, often blanketing cars parked beneath them. Traffic is impeded, shoppers desert the streets, and in extreme cases snow plows are called out to reopen highway bridges that have become impassable.

On July 14, 1958, the *Des Moines Register* News Service carried the following dispatch:

Welcome to Minnesota

An interesting mayfly-caused accident occurred some years ago in Winona. About 10 P.M., a man roared up the ramp of the interstate highway bridge on a motorcycle, approaching from Wisconsin at high speed. Unknown to him, a slight breeze from the north had concentrated the upstream mayfly flight around the lights at the summit of the bridge where a thick, sodden carpet of crushed mayflies had formed. Blinded by swarms of mayflies, the driver lost control of his bike, skidded, capsized, and slid down the Minnesota approach, luckily not encountering any traffic. The bridge deck was well lubricated with squashed mayflies, sparing the cyclist a bad case of road rash, but he and the cycle were caked with the quivering mass—in his hair, in his mouths, down his neck. I'd heard that you could tell a happy motorcyclist by the bugs stuck in his teeth, but this was too much. I doubt if he's forgotten the stench of mayflies incinerating on his cycle's exhaust manifold.

> Dubuque, Iowa—Fishflies controlled the Julian Dubuque Bridge here for forty minutes, then surrendered with heavy losses under an armored counterattack by highway commission scraper-trucks. Traffic was stopped on both sides of the Mississippi River Bridge by multitudes of fishflies, starting about 9 P.M. Sunday. The battered bugs caused slipperiness on the bridge until highway commission trucks plowed a path through and sanded the surface.

A similar situation occurred on July 9, 1966. The annual Steamboat Days celebration was in full swing in downtown Winona, Minnesota, when *Hexagenia* crashed the party. By 11 P.M. insects were six inches deep on the floor of the carousel. At 11:45 P.M. the carnival had to be shut down because mayflies had clogged the radiators of the diesel-powered generators, causing them to overheat.

Problems caused by the upstream egg-laying flight have been lessened in recent years by the use of sodium vapor lights whose longer wave lengths (yellow) are less attractive to insects. In some areas, bridge lights are simply turned off. Grated bridge decks allow smashed mayflies to drop through.

Vital Links in Many Food Chains

Nymphs are primarily filter feeders, consuming organic *detritus* (particulate matter), algae, bacteria, and small organisms that they strain from the water and gather from the sediment with elaborate systems of combs and brushes on their legs. Growth rate is mainly a function of temperature; life cycles are completed in as little as twelve weeks in the lab but require as long as two years in northern lakes. In the Upper Mississippi, the life cycle is one year or less. In prime habitats, there may be over eight hundred nymphs of various sizes per square yard of river bottom. In Pool 19, the June 1962 *Hexagenia* population was estimated at 23.6 billion (Carlander et al. 1967).

It's hard to overstate the importance of *Hexagenia* to the ecology of the Mississippi River. In a short, efficient food chain, nymphs convert organic detritus into high-quality food for fish and waterfowl. Their long life cycles make them available during all seasons, and because nymphs pass through as many as twenty to thirty molts, they are sized to suit a wide variety of fish species.

Vigorous *Hexagenia* populations increase the river's turbidity. Their burrowing and feeding activities churn the muddy bottom and may prolong the life of impoundments by reducing the organic content of the sediments and by resuspending sediments to be swept down river.

Unfortunately, nymphs concentrate persistent toxins like chlorinated hydrocarbons (e.g., DDT, PCBs) and heavy metals (e.g., mercury, lead) in their tissues and pass them up the food chain where they are further concentrated at every step. Their digging activities may resuspend toxins that otherwise would have remained entombed in the sediment.

By creating silted impoundments, the navigation dams of the Nine-Foot Channel Project undoubtedly increased the ability of the Upper Mississippi River to produce *Hexagenia* mayflies. Pollution made it impossible for *Hexagenia* to exist in some river reaches, but improved sewage treatment during the past forty years has enabled *Hexagenia* to thrive in formerly desecrated areas. Yet, the long-term prognosis is not good. In some pools, "sterile" sand is smothering highly productive silt habitats.

Natural ecological succession dictates that lakelike navigation pools and backwaters will be ultimately replaced by dry land. In the name of "environmental management," and at great expense, federal and state agencies are replacing *Hexagenia* habitats with islands constructed of dredged sand and/or rock. In some pools, the halcyon days of spectacular *Hexagenia* emergences may be over.

Irresistible Fish Food

A *Hexagenia* nymph is an irresistible tidbit for a hungry fish. Fish gorge on *Hexagenia* nymphs as they rise from the bottom to emerge as adults. Commercial fishermen catch fat "cats" whose stomachs are so packed with nymphs that their swollen bellies are "tight as a drum." Fish also feast on adult mayflies during the hatch and when females return to the water to lay their eggs. Fishing is usually poor during and immediately after these feeding frenzies. Nymphs are also eaten by fish, winter and summer, when they abandon their burrows in response to lowered levels of dissolved oxygen, crowding, or for unknown reasons.

If *Hexagenia* habitat becomes deficient in dissolved oxygen, nymphs abandon their burrows, swimming until they find better habitat or become exhausted. Ice fishermen often observe them when fishing shallow backwaters in late winter. Nymphs also congregate in muskrat and beaver runs. If nymphs do not find oxygenated water, they become lethargic and can go into "suspended animation" for twenty-four hours or more in icy water, perhaps being lucky enough to drift with a current to an oxygenated environment. Swimming and drifting nymphs are easy marks for hungry fish, as well as over wintering mallard ducks.

Carp and other foragers dislodge nymphs as they root about for food or thrash in the shallows during spawning. Some fish actually take bites out of the river bottom, forcefully expelling clouds of muddy water from their mouths, and then gobbling nymphs that swim from the clouds.

Unlike a baitfish, which swims with a side-to-side motion of its abdomen and tail, the *Hexagenia* nymph swims with an up-and-down motion of its abdomen, much like a swimmer doing the butterfly stroke. Its waving, feathery gills and tail are additional enticement. A nymph dislodged from its burrow will try continually to reenter the bottom, bouncing along with head down, with its undulating abdomen and tail pointing upward. A feathered, hair, or twistertail jig

Using SCUBA to Feed *Hexagenia* Nymphs to Bass

Using SCUBA and a diving light, I have explored the bottoms of many Headwaters lakes at night. I've never observed a free-swimming *Hexagenia* nymph, but I can easily dislodge nymphs from their muddy burrows by digging with my hands. Smallmouth bass and rock bass are quick to take advantage of the opportunity for an easy meal. We swim together, shoulder to shoulder, and as I dig in the bottom they close in, so eager that they nudge my hands. Dislodged nymphs are inhaled before they have swum a foot. If the bass work this closely with a diver, it's easy to see how they might work similarly with bottom-disturbing fish like carp. Some expert walleye fishermen use an electric trolling motor to follow grazing carp in shallow water, adeptly employing jigs that resemble *Hexagenia* nymphs to entice walleyes that are hunting nymphs dislodged by the carp.

closely simulates a nymph and can be deadly in the hands of an expert angler.

Hexagenia nymphs are popular bait for ice fishing in Wisconsin and Michigan, but they've never really caught on in Minnesota. Collecting nymphs through the ice in winter is hard work, but some families have done it commercially for two or more generations.

Trout fishermen have long been the most avid amateur students of mayflies, skillfully tying artificial flies to "match the hatch" and keeping detailed records of when "hatches" occur on the streams they fish.

Survival Strategies

In the eons of their existence, *Hexagenia* mayflies have evolved effective survival strategies. Adult activity takes place mainly at night when most insect-eating birds are inactive. By synchronizing their

Good News? The Mayflies Are Back!

Devastating pollution of the Mississippi resulted in the disappearance of *Hexagenia* mayflies from St. Paul to Hastings, Minnesota, for more than a half century. In 1927, the U.S. Bureau of Fisheries reported that during August between St. Paul and Hastings the river lacked sufficient dissolved oxygen to support fish life. Beginning in the 1970s, growing public awareness of environmental issues and fear of contaminated fish resulted in increased monitoring, restrictions, and cleanup of domestic and industrial sewage from the Twin Cities area. *Hexagenia* responded and by 1984 they were appearing again in the river below St. Paul and in the upper end of Lake Pepin, even attaining nuisance levels in some places—causing some people to shiver and others to rejoice.

The mayflies' return was due to improved water quality. Like the mayflies, and for the same reason, people are also returning to the river in the Metro area.

Mayflies have also returned to Lake Erie and Lake Michigan's Green Bay, areas famous for their spectacular *Hexagenia* hatches until pollution almost annihilated the insects in the 1950s. Happily for most people, *Hexagenia* made a dramatic reappearance in both areas in the 1990s due to improved water quality. In the summer of 1999 on Lake Erie, clouds of *Hexagenia* fourteen miles long and over four miles wide were observed on Doppler radar (Masteller and Obert 2000) hatches, mayflies satiate predators by suddenly providing more food than the predators can eat, thus enabling the bulk of the population to survive and reproduce. The nonconformist fly that emerges early or late is more likely to be eaten, and less likely to find a mate. The species' abandonment of adult mouths and digestive tracts guarantees that there will be no stragglers.

Under the best of conditions, the chances of a newly hatched nymph ever reaching adulthood and reproducing are extremely poor. That's why each female must produce so many eggs.

People often lump mayflies with other noisome animals like leeches, ticks, and mosquitoes, indignantly asking, "What the hell good are they?" I have always been tempted to respond by asking the questioner, "First, tell me what the hell good are you?"

20

Flooding

"The River Is a Strong Brown God—Sullen, Untamed and Intractable"

I do not know much about gods; but I think that the river
Is a strong brown god—sullen, untamed and intractable,
Patient to some degree, at first recognised as a frontier;
Useful, untrustworthy, as a conveyor of commerce;
Then only a problem confronting the builder of bridges.
The problem once solved, the brown god is almost forgotten
By the dwellers in cities—ever, however, implacable,
Keeping his seasons and rages, destroyer, reminder
Of what men choose to forget. Unhonored, unpropitiated
By worshippers of the machine, but waiting, watching and waiting.*

Indians and frontiersmen were well aware that the Upper Mississippi could be counted on to flood every spring, but the U.S. Army underestimated the river when it built Fort Crawford at Prairie du Chien in 1816. Twelve years later the river went on a rampage. Water ran four feet deep in the barracks at the fort, and the entire settlement was no more than an island.

For years the flood of 1828 set the standard for high water on the Upper Mississippi, although, with no white men around to measure them, who can tell what floods came before?

Notable floods swept through the region in the springs of 1844, 1851, 1859, 1862, 1870, 1880, 1888, and 1892—in 1881 high water didn't hit until October during the fall rise, a most unusual occurrence (Petersen 1965).

Nineteenth-century floods could be devastating, but so could low water caused by drought. During the summer of 1864, river levels

*Excerpt from "The Dry Salvages" in *Four Quarters,* copyright 1941 by T.S. Eliot and renewed 1969 by Esme Valerie Eliot, reprinted by permission of Harcourt, Inc.

were the lowest ever recorded. In early May 1864, after a winter with little snow, steamboat captains began reporting low water conditions all the way north to St. Anthony Falls at Minneapolis. The usual June rise didn't happen, and low water persisted for most of the navigation season. Lack of water in the backwaters and tributaries, as well as the main channel, meant the usual fleets of logs and lumber failed to arrive downstream, dealing a heavy blow to the economy of river towns. With most of the channel above Lake Pepin less than three feet deep, Hastings became the head of navigation. Water was so low that steamboats frequently had to blow their whistles to drive cattle out of the channel. Huge sandbars formed along the waterfronts of Dubuque and Muscatine, threatening the river towns with the prospect of becoming inland cities. Large boats coming upstream couldn't negotiate the Keokuk Rapids and had to portage their cargoes around the rapids to smaller boats at Montrose (Petersen 1965).

Floods and Low Waters Are Natural Events

A flood can be defined as a flow so great that it overtops the natural or man-made banks of the river channel. If a floodplain exists, any flow that spreads out over the floodplain constitutes a flood. As a rule, the volume of floodwater is not large. Generally, flood flows account for only about 5 percent of the total annual discharge of the basin (Leopold 1994).

Floods are natural occurrences along the Mississippi. The main flood is usually in the spring, caused by snowmelt and rains. There is usually a second, less severe high-water period in the fall after trees of the forests have lost their leaves, resulting in reduced loss of groundwater to the atmosphere via transpiration. Typically there is also a June rise, but severe midsummer floods have been unusual.

During periods of high water the river recharges the groundwater, raising the water table to river level. After the flood crest has passed, the water table is higher than river level. Groundwater then slowly seeps into the river, keeping levels relatively high.

The riverside land, which is periodically inundated by floodwaters, is called the floodplain. This low, sodden, buggy corridor is generally unappreciated, yet it is vital to the river's ecology. In their natural state, floodplains reduce flood crests by temporarily storing floodwaters; they improve water quality, provide vital habitat for wildlife, and provide opportunities for recreation.

The floodplain forest increases the floodplain's roughness, reducing the velocity of flood flows. Trees and other plants anchor the river's banks, reducing erosion and providing shade that reduces water temperatures. Leaves and organic debris from floodplain plants serve as foundations for detritus food chains. Floodplain trees provide important roosting and nesting habitat not only for songbirds but also for larger birds like herons, egrets, cormorants, wood ducks, hawks, ospreys, and bald eagles.

For rivers and their inhabitants, floods are natural, inevitable events to which fish and wildlife have adapted over the millennia. For scores of river fish species, the annual "flood pulse" acts as a reproductive cue, stimulating them to migrate out of the river to spawn in the floodplain and to feast in new, untapped pantries.

"The river used to 'get up' twice every year. In the spring it would rise and 'put water and fish in the backwaters,' and in the fall it would rise and 'come back to get them,'" an old-timer said of the annual Missouri River floods.

What Do "River Stages" Mean?

Extreme fluctuations in river level have been observed for many years on the Mississippi. Zero points with which to compare subsequent river stages were established during the low water of 1864. In the spring of 1880, at Winona, the flood crest rose sixteen feet, eight inches above the zero point, but during the severe drought of 1932, the river level dropped forty-two inches below the zero reference level. Confounding the issue, the impoundment created by Lock and Dam 6 (downstream at Trempealeau) extends upstream past Winona, permanently raising the water level such that "normal" river level is now five and one-quarter feet. It's interesting to note that cities located in tailwaters areas or along the upper reaches of navigation pools have low water levels today that approximate the 1864 benchmark level, while those in pool areas have higher levels. Consequently, casual scanning of river stages at successive cities along the upper river can be confusing because it appears that the river often flows uphill! To add to the confusion, water levels are increasingly being expressed as "feet above mean sea level."

Most important of all, the flood provides a "reset mechanism" for the floodplain environment that renews and stops plant succession, which, if left unchecked, would destroy the diversity necessary to maintain cover and nesting habitat for many native fish and wildlife. Flooding is as important to the maintenance of the river ecosystem as the sun is to photosynthesis (MICRA 1993).

While flooding is a natural phenomenon, flood damage is a result of human use of the floodplain. The natural flood regimen changed dramatically when Caucasian invaders removed the land's protective vegetative cover, tilled the soils, drained wetlands, tiled croplands, waterproofed portions of the watershed with asphalt and concrete, and sped the rate of runoff with storm sewers and ditches.

Until about three hundred years ago, North America's rain and snowmelt were soaked up by thick layers of vegetation, humus-rich soils, and reserved in lush depressional areas called wetlands. During periods of high water, wetlands served as natural sponges, storing and slowly releasing floodwaters and trapping sediments. Millions of beaver dams created successive impoundments along the courses of smaller streams. Much of the precipitation that fell to earth was returned to the atmosphere through evaporation or was trapped and retained in soils and underlying rock formations as groundwater.

Swelled by spring snowmelt, streams spread out across the wetlands. When they finally reached main stem rivers, the waters benignly spread out across wide floodplains that slowed and stored them until they evaporated, infiltrated the soil, or withdrew back into the channel to continue on their inevitable journey toward the sea.

These conditions didn't last long after white men began arriving in North America. Fur traders indelibly altered the landscape as the beaver, whose dams had exerted direct control over streams and small rivers, were trapped to near extinction. By the late 1600s, European fashion had already drastically altered the Mississippi watershed's hydrologic cycle. As the old beaver dams washed out they were not replaced; later, many were deliberately removed to promote drainage.

By the 1850s, farming was in full swing in the Upper Mississippi watershed, and little prairie or forestland was spared from the plow.

As the virgin prairie soils were turned over and their humus content was reduced by erosion and oxidation, their water-holding capacity was diminished.

Outlet ditches were constructed and tile drains installed to extend agricultural development into the myriad wetlands that were viewed as disease-ridden impediments to economic activity. The 1848 Swamp Land Acts transferred one hundred thousand acres of Mississippi floodplain to the states for conversion to farmland by drainage and levee construction.

The runoff storage capacity of wetlands in the Upper Mississippi watershed has decreased markedly since settlement, raising flood crests downstream. By the 1980s, Minnesota had lost 42 percent of its wetlands to drainage and development; Wisconsin had lost 46 percent, Iowa 89 percent, Illinois 85 percent, and Missouri 87 percent (Dahl 1990).

As cities grew and sprawled across the landscape, elaborate systems of storm sewers were built to dispatch urban runoff into local streams, which were often dredged and straightened to handle increased flows.

When all of these factors were combined, water was rapidly redistributed from the upper to the lower watershed, resulting in more frequent floods and floods of greater magnitude (Hey and Philippi 1997).

Upper Mississippi River floods don't necessarily produce floods on the Lower Mississippi. Once the floodwaters of the Upper Mississippi have squeezed through the half-mile-wide constriction at Thebes Gap south of Cape Girardeau, Missouri, which acts like the stem of an inverted funnel, they flatten out in the much wider space between the huge levees of the Lower Mississippi Valley. Floods on the Lower Mississippi are a product of Upper Mississippi flows, but also of the combined flows of the Ohio River, the Missouri River, and many other tributaries. The monstrous, mainline southern levees can accommodate crests in excess of fifty feet. Historically this has meant that major flooding in the Lower Mississippi Valley was often caused when Mississippi flood levels were higher than the water levels in its tributary streams. The Mississippi then acted as a hydraulic dam, causing tributaries in flood stage to back up around their less substantial levees, flooding the Mississippi bottomlands from the rear.

Flood control on the Lower River began when the French constructed the first levees to protect New Orleans in 1727. By 1844, levees were continuous along the river's west bank as far north as the Arkansas River. On the east bank they ran all the way to Baton Rouge.

In response to worsening Mississippi delta flooding, Congress appropriated fifty thousand dollars in 1852 for two studies. Charles Ellet Jr., an engineer who understood the hydraulics of the river and the hydrology of the watershed, published first. He concluded that the more frequent and extensive delta floods were attributable mainly to (1) extension of cultivation throughout the Mississippi Valley (2) extension of levees that prevented the flooding river from spreading out onto the floodplain (3) shortening the river with cutoffs, thus increasing its slope and velocity, and (4) the gradual progress of the delta into the sea, lengthening the river, decreasing its slope and velocity, causing flood waters to back up over the land. Ellet, however, was far ahead of his time (Ellet 1852; Hey and Philippi 1997).

The second study, "Report upon the physics and hydraulics of the Mississippi River" by Captain Andrew A. Humphreys and Lieutenant Henry L. Abbot, endorsed levees as the only appropriate technique for preventing flood damage. They held that by concentrating the river's flow between levees, energy of the flood volume would scour and enlarge the channel to carry away the increased flow. The gospel according to Humphreys and Abbot became the foundation for Mississippi River flood-control strategy for the next 140 years (Arnold 1988; Hey and Philippi 1997).

The Mississippi River Commission followed a "levees only" policy along the Lower Mississippi. Levees up to four stories high were constructed to limit the high-water width of the river and concentrate the flood discharge. The river was sealed off from its floodplain, natural reservoirs, and outlets. The river's volume and speed increased, but the increased current did not scour the bottom enough to compensate, and the capacity of the channel shrank rather than increased. As a result, the flood of 1912 devastated the

Lower Mississippi region although it carried far less water than the great flood of 1882 (Barry 1997a).

Today, the river is flanked by levees or natural loessial bluffs (as at Natchez and Vicksburg) all the way north to Dubuque, Iowa (Fremling et al. 1989). From St. Louis southward the river is also flanked by agricultural levee districts within which parcels of fertile, often waterlogged, bottomland have been "reclaimed" by ringing them with flood levees. The wettest parcels are usually ditched to conduct excess water to sumps where it is collected and pumped over the levees and into the river.

Flood stages have increased between St. Louis and Cairo, due mainly to constriction of the high-water channel by wing dams and the loss of floodplain capacity due to leveeing and development. The Corps of Engineers has squeezed the river's natural flow into a tighter channel. This causes water to rise higher within levees, prevents it from fanning out, accelerates the river's velocity, and transfers upstream flooding problems downstream.

Ironically, present-day river elevations during low flows are lower than they were in the premodification days, due mainly to scouring of the low-water channel by wing dams. River stage fluctuates as much as fifty feet annually, effectively dewatering some secondary channels during low flow.

The Classic 1965 Spring Flood

For almost a quarter-century following 1892, the Upper Mississippi caused little damage, but it returned with a vengeance with the

The 1922 Rat Harvest

As with previous floods, the rising April waters of 1922 produced a highly undesirable harvest at Dubuque, Iowa, where hordes of rats were driven from submerged dumps to seek food and shelter in town (Petersen 1965).

On April 22, the *Dubuque Times-Journal* reported, "A total of 232 rats have been reported killed in the rat crusade being conducted by the Boy Scouts of the city. . . . Walter Kemp, of Scout Troop 8, broke all previous individual records, when he turned in 89 rat tails at Scout headquarters a few days ago."

floods of 1916, 1920, 1922, 1938, 1942, 1951, 1952, 1965, 1969, 1973, 1996, 1997, and 2001.

The 1965 "hundred-year flood" stands out as a classic, a textbook example of how nature and human development can combine to bring devastation to the river valley.

Fall rains had charged the ground with moisture, raising the water table in most areas of the basin. The winter of 1964–65 was severe but with little early snow, giving the ground plenty of time to freeze deeper than usual and making it impervious to the infiltration of spring snowmelt. In late winter heavy snows fell relentlessly in the watersheds of the Mississippi Headwaters, St. Croix River, Minnesota River, Chippewa River, Wisconsin River, and dozens of smaller tributaries. In a normal winter St. Paul may get fifty inches of snow. By the spring of 1965 there were seventy-three inches on the ground.

As winter moved into spring, it became apparent that a potential severe flood was contained in the snow cover.

Prolonged frigid temperatures kept the snowpack from melting during the winter. Instead the snow cover condensed and packed, increasing its water content. By mid-March, daytime temperatures remained cold, preventing gradual melting. At the end of March the rains came, but the snow pack absorbed them with little or no runoff. Then, in early April, the weather warmed suddenly, accompanied in some areas by rain. The ground was still frozen, sending the runoff straight into the river, pushing flood crests to record highs, creating havoc downstream.

One river city, Winona, Minnesota, was billed as "the city that saved itself." City leaders gave heed to the ample warning offered by flood forecasters and took action. For almost a month, the city was like a war zone, with trucks, heavy machinery, and hundreds of people working twenty-four hours a day. Volunteers and contactors saved the city, which lay almost entirely within the floodplain, by building a temporary levee over twenty feet high and more than nine miles long around the city.

Hailed as a "hundred-year flood," the great spring flood of 1965 was the most devastating in Upper Mississippi Valley history. It was followed, in 1969, by a second "hundred-year flood." The floods spurred the building of flood-control facilities in many river cities. The dike systems were designed to not only protect the cities but also their sewage treatment systems, preventing the discharge of raw municipal sewage that had previously accompanied even minor floods.

Downstream at St. Louis, the 1965 flood was a virtual nonevent, not even making the list of the top ten floods by volume or stage. This was because flood conditions were limited to the Upper Mississippi. The Missouri, Illinois, and other area tributaries had remained within their banks.

The Great Midsummer Flood of 1993

Localized floods are not uncommon during the summer. A severe thunderstorm may send a tributary stream roaring over its banks, devastating floodplain developments. But the Great Midsummer Flood of 1993 was unusual because it was so expansive, and it occurred during the growing season when there was no melting snowpack. It was a slow-moving disaster that established itself as one of the most destructive floods in U.S. history and one of the costliest natural disasters of the twentieth century.

Great floods occur when a drainage basin is incapable of absorbing additional water, and all rainfall or snowmelt must run off the surface (Leopold 1994). In the spring and summer of 1993, the jet stream that flows eastward across the United States detoured southward across the Midwest where it met a high-pressure system, which pumped warm moist air northward from the Gulf of Mexico, causing unusually persistent, heavy local rains to fall, month after month, on land that was still saturated from spring runoff. Much of the land was agricultural, planted mainly to corn and soybeans.

On July 22/23, a large area of southern Iowa and northern Missouri received over six inches of rain in about thirty-six hours. Floodwaters poured into the Mississippi from the Wapsipinicon, the Maquoketa, the Turkey, the Volga, and countless smaller tributaries to form a crest that moved south at an inexorable eight miles per hour. In an unusual orchestration, the crest was joined at St. Louis by high water coming down the Missouri from the west.

Flooding was particularly extensive along the Lower Missouri, Des Moines, and Middle Mississippi Rivers. At Cairo, the Mississippi had a discharge three times greater than the levees squeezing the river were meant to contain. The Great Flood of 1993 that swamped the Farm Belt resulted from unusual weather conditions, but most flood damage was caused by the failure of the same levees, dams, and other control works that allowed the development of the floodplain for urban and agricultural use.

The Great Flood was reported in the media as a 500-year flood. Actually, the greatest flood in history at St. Louis was in 1844 when the river's flow was about 1.3 million cubic feet per second, and the crest (stage) was estimated at 41.32 feet. In 1993 the peak flow was only about 1.03 million cubic feet per second, but the crest was 49.47 feet; also in 1993, there was less water but a higher crest. If today's levees had been in place during the 1844 flood, the river would have crested at 52.0 feet.

The Corps of Engineers maintains that the 1844 flood wasn't measured, only estimated, and that it cannot be proven that river engineering has made flooding worse. The Mississippi is extremely complex, and its actions are hard to predict with physical models with concrete banks. The real river is not pure; it is full of dirt and debris; forests and man's structures roughen its floodplain. Because of these factors and many other variables, computer models aren't perfect either.

Before settlement, the Mississippi eroded its bottom and banks during flood peaks, increasing the storage capacity of the channel. Additionally, the river used its broad floodplain as a natural flood reservoir. By the time the Great Flood of 1993 arrived, the Middle Mississippi's channel had lost about a third of its volume, and the floodplain had been largely cut off from the river by levees.

Levees can only provide a limited level of protection. No levee is flood proof. But levees create a false sense of security that encourages floodplain development. That complacency is enhanced by federal policies that dictate that homeowners protected by a hundred-year-flood levee are no longer required to purchase flood insurance. This uninsured floodplain development is ripe for an economic catastrophe when a big flood does occur.

The Great Flood of 1993 was a "battle of levees." It pitted landowner against landowner and city against city as they tried frantically to raise their levees high enough to beat the rising river. The levees may not have failed, but they contributed to flood height, which may have caused other levees upstream to fail. During the Great Flood of 1993, all kinds of levees failed, including twenty levees constructed by the Corps of Engineers (Faber 1995).

People who made their living on the river were hurt nearly as badly by the flood as the people who lived along side it. The Great Flood was a nightmare to the towing companies that had to suspend operations for fear of damaging waterlogged levees with their wakes. The barge lines estimated that they lost at least one million dollars a day due to the flood.

Ominously, the 1993 flood was followed by repeat flooding in 1996, 1997, and 2001.

Ecological Impacts of Levee Building

Below St. Louis, levees have isolated the river and its fisheries from its floodplain, reduced flood storage capacity, and led to higher flood crests (Simons et al. 1975). Levees have encouraged development, and, as a result, fisheries habitats behind levees have been drained and filled. Flood control works have greatly decreased the amount of floodplain available as nursery, spawning, and feeding habitat. In addition, many floodplain lakes have been isolated from river overflow and no longer serve as habitat for river fishes.

The Nine-Foot Channel dams and the wing dams and closing dams of earlier channelization projects have caused sediment to accumulate, which in the natural river would have been exported downstream. Eventually, most of it would have ended up in the Louisiana deltas or in the Gulf of Mexico. The accumulated sediment has raised the bed of the Upper Mississippi. It seems an inescapable conclusion that as navigation pools fill with sediments, the rise in the river bottom must be attended by a commensurate rise in flood crests, thereby creating a need for more levees in previously unflooded areas and along the lower reaches of tributaries.

The Human Tragedy

> Unlike a human enemy, the river has no weakness, makes no mistakes, is perfect; unlike a human enemy, it will find and exploit any weakness.
>
> Barry 1997, 156

Because the federal government has built and repaired levees and paid disaster relief, local officials have had no reason to direct new development away from the floodplain. On the contrary, for river communities floodplain development brought benefits—including more tax revenue—but few costs. Ironically, rather than reducing flood losses, levees have increased flood losses by encouraging a false sense of security, which in turn encourages floodplain development. The promise of "no more need for flood insurance" has been the "carrot" for building dikes around parts of cities. Levees have reduced the frequency of floods but increased the costs of floods that inevitably overtop levees. In years past, such floods mainly impacted cropland, but levee failures today result in catastrophic losses of life and infrastructure (Faber 1995).

Contrary to popular belief, a "hundred-year" levee does not protect an area from floods for one hundred years. Rather, a person living behind such a levee has one chance in a hundred of being flooded in any given year. Survivors of a hundred-year flood are not exempted from flooding for the next ninety-nine years. They still have a one-in-a-hundred chance of being flooded the next year—and the next. There is a hundred-year flood somewhere in the United States every year. Moreover, the frequency of hundred-year floods is predicated on the climate of the past two centuries. No one yet knows how global warming or natural climate change will affect Mississippi River floods.

Because we've been keeping records for only about a century and a half, we don't know what a big flood is. A hundred-year flood may only be a nice medium-sized flood. Today's "hundred-year floodplain" may become the future's "fifty-year floodplain" as a result of our alterations of the watershed.

Federal policies still encourage floodplain development. Congress pays at least two-thirds of the cost of flood control projects and until recently paid the full cost. Congress has provided no-strings-attached disaster relief over the past quarter-century. Federal flood insurance has provided subsidized rates to homes and businesses. Many flood-prone homes are repaired repeatedly at federal expense or with flood insurance premiums. In many instances, flood insurance payments have cumulatively exceeded a home's value.

After a flood, victims face a pressing economic question: to return or not return to the floodplain. Most realize that it's going to happen again, that moving back is like standing in line to be robbed. If people choose to live in the floodplain, the federal government shouldn't insure them.

Fortunately, flood protection at the beginning of the twenty-first century involves more than just building more and higher levees. Today, up to 15 percent of federal funds provided to communities for disaster relief can be used for voluntary relocations and buyouts. Hundreds of homes and businesses in Davenport, Iowa, for example, were relocated to higher ground after the Great Flood of 1993. In subsequent floods, damage to those properties has been eliminated entirely. Forty-one percent of St. Charles County, Missouri, was inundated in 1993, flooding thousands of homes and businesses, but flood damage was reduced by 99 percent in recent floods of the same magnitude through voluntary buyouts.

Leveeing cities is probably justifiable, but leveeing farmland is not. Levees actually *increase* flood crests, transferring flooding problems upstream and downstream. The levees that line the Upper Mississippi squeeze the river's natural flow into ever-tighter channels. The Corps of Engineers concedes that channnelization levees raise flood crests, but it maintains that upstream reservoirs on tributaries compensate sufficiently to make it a "wash."

The main determiner of a flood's threat is the height of its crest, but the slower the crest moves, the more dangerous it becomes because it exerts pressure on levees for a longer time, saturating them to the height of the crest. Then, the water-saturated levee is essentially a water column, and as the flood recedes, the dike's water flows riverward, taking the levee with it.

Likewise, the homeowner who surrounds his home with sandbags and pumps water from his basement may be courting disaster. As the pressure of water increases outside his basement it may cause the basement walls to collapse inward, dropping the house into the basement. To prevent this, it may advisable to fill the basement with clean water before the flood.

Lessons Taught by Pool 19

Almost continuous daily records of water surface elevation at Burlington, Iowa, and river discharge at Keokuk, Iowa, have been kept since 1878, documenting the impact hydroelectric and navigation dams have had on the river over the past century and a quarter.

The records show that sedimentation has caused a gradual loss of water storage capacity and river area in forty-six-mile-long Pool 19 since the Keokuk Dam was completed in 1913. This trend is predicted to continue until the mid-twenty-first century when equilibrium will be reached and river width should be similar to predam conditions. In the postdam era, up until 1988, recurrence of annual floods that surpassed both flood stage and major flood stage steadily increased, and major floods occurred four times as often.

Loss of river volume is not confined to Pool 19, but is occurring in all twenty-seven navigation pools. As the pools lose their reservoir capacity, increased flood heights and number of flood days will be more common (Grubaugh and Anderson 1988).

What Should Be Done?

Levees, dams, and dikes cannot fully protect people from flooding. The only certain protection is relocation of vulnerable homes, businesses, and farms out of the floodplain to higher ground. Agricultural lands that flood frequently should be converted to wetlands.

Flood assistance programs, other than aid to households and individuals, should be redirected to accomplish economically and environmentally sustainable land use. Sustainable use for frequently flooded areas includes grasslands, wetlands, and undeveloped recreational lands. Sustainable land use for occasionally flooded areas includes pastures, rotational cropland with emphasis on hay production, and wildlife habitat. Sustainable land use for infrequently flooded areas may include the uses listed above, plus row crop agriculture, and buildings and developments designed or planned with realistic expectations of flood frequency and costs.

Wherever we can, we should reunite the Mississippi with its floodplain. "A river-floodplain ecosystem not only requires a flood, it also requires a floodplain. An analysis should be done of the costs and benefits of buying out levee districts and breaching the levees. Aside from the benefits to fish and wildlife from restoring access to the floodplain, flood damage would be reduced and sedimentation rates might be lessened. Flood heights and damage would be reduced because the broad river-floodplain system would have a greater flood conveyance capacity than the present constricted channel and there would be no risk of catastrophic levee failure during a record flood because there would be no levee" (Sparks 1992).

In the early days, wetlands were seen as wastelands. Today, there is a growing recognition that wetlands can reduce flooding by absorbing excess rainfall and snowmelt while filtering pollutants, trapping silt, and providing refuge for wildlife. Federal and local governments should be restoring floodplains and wetlands to reduce flooding, while protecting residents and businesses by helping them move out of harm's way (Faber 1995).

Direct flood damage is only one of the problems caused by present land uses and management practices of the Upper Mississippi River Basin. Flooding exacerbates the surface-water quality problems of high turbidity, excess nutrients, and toxic substances that extend all the way into the Gulf of Mexico.

Floods are "acts of God," but flood damages are caused by acts of man. It has become obvious that floodplain development raises flood crests. Levees generally increase flooding upstream and water velocity downstream. Once engineering is started it cannot be discontinued. An engineered river must be maintained and repaired forever. Ultimately, water always wins.

Old Man River has been telling us for over 150 years, "I am trying to get out of this straitjacket and one way or another I'll do it."

Postscript: A Flood of Development at St. Louis

The disastrous Flood of 1993 covered seventeen thousand square miles in nine states, forced the evacuation of about fifty-four thousand people and caused damages estimated at twelve to twenty billion dollars, making it the nation's costliest flood. Missouri, at the nexus of two of the nation's most powerful rivers, was hardest hit, suffering direct damages of at least three billion dollars. Yet today, less than ten years post-flood, more than $2.2 billion in new office space, shopping centers, and roads now stand on 4,200 acres of land in the St. Louis area that was under water during the 1993 Flood. Incredibly, projects planned or underway call for an additional 14,000 acres in the Missouri River–Mississippi River agricultural floodplain to be converted to commercial and residential developments. Projects include massive new levees and the upgrading of existing levees to withstand 500-year and 1,000-year floods. Working with developers, city officials and landowners have taken advantage of liberal regulations, public subsidies, and tax-increment financing to develop the floodplain. The U.S. Army Corps of Engineers still pays for up to 65 percent of new levee construction and 80 percent of levee repair after a flood. Subsidies have enabled developers to transfer a large portion of their risk to the nation's taxpayers, thus putting short-term economic gains of individuals ahead of the long-term safety and environmental quality interests of the public (Shipley et al. 2003).

21

Pollution

Has a Defiled River Been Reborn?

A Witch's Brew

The word *pollution* means different things to different people. To some, it brings to mind sickness—water-borne diseases like typhoid fever, cholera, diphtheria, dysentery, and hepatitis. Others think of a river flowing with foam, foul odors, and litter, choked with weeds and home to tainted fish swimming through water that looks like pea soup. A river can be overwhelmed by putrescible pollutants as diverse as animal manure, slaughterhouse wastes, human sewage, ground garbage, sugar, milk, tree leaves, grass clippings, and other plant and animal materials that have a high biochemical oxygen demand (BOD). In decomposing them, the agents of decay rob the river of dissolved oxygen, vital to fish and other aquatic animals. (Most aquarium owners know the consequences of too much fish food.)

A river overloaded with such pollutants may go *anaerobic* (without oxygen), generating flammable, but odorless, methane gas that makes it bubble and belch. Another gas, hydrogen sulfide, adds the characteristic smell of rotten eggs. These, plus other foul gases, produce the noxious vapor historically called *miasma*. Until the 1870s when Louis Pasteur, Robert Kock, and Joseph Lister finally convinced the scientific community that bacteria caused typhoid fever and cholera, people thought that they were contracted by inhaling miasma, especially in combination with moist night air.

Pollution includes all of this, and much more. In fact, the most insidious pollutants may be invisible and odorless. For example, *heavy metals* and *chlorinated hydrocarbons* have accumulated in the river corridor, especially in backwater areas that are deemed most valuable as fish and wildlife habitats. There is no question that their presence has stigmatized the river, harmed wildlife, and curtailed human consumption of river fish.

Sedimentation

Sediment is the ultimate pollutant. While the river may ultimately flush itself of many pollutants, sediment is usually permanent, displacing the very water itself. Today, sedimentation is the Upper Mississippi's major cause of habitat degradation. The main sediment

---

Dead Zones in the Gulf of Mexico

Nitrogen and phosphorus loading of the Mississippi River is largely to blame for the development of large, oxygen-depleted *hypoxia* areas, also known as *dead zones* in the Gulf of Mexico, responsible for massive kills of fish and invertebrates. The farm fields of the Upper Mississippi and Ohio River Basins are the Gulf's major sources of nitrogen and phosphorus during most of the year.

In the Upper Mississippi River Basin, more than 60 percent of the land area is devoted to cropland or pasture, and the major sources of nitrogen to most river waters are commercial fertilizers, manure, organic soils, and plant debris. The basin, excluding the Missouri River watershed, accounted for 31 percent of the total nitrogen delivered from the Mississippi River to the Gulf of Mexico between 1985 and 1988. Nutrient loading has contributed to the development of a seven-thousand-square-mile zone (about the size of New Jersey) of reduced dissolved oxygen in the Gulf of Mexico stretching across the Louisiana coast and onto the upper Texas coast near Galveston (Alexander et al. 1995).

---

source is upland erosion from farmland, but sediment is also furnished by construction activities, forest exploitation, bank erosion, and virtually all of man's outdoor activities in the watershed. Sediments include sand, silt, and clay, as well as innumerable organic substances and precipitated marl (calcium carbonate). Unfortunately, the river's sediment deposits are not even noticed by most people until they rise above the river surface or interfere with navigation.

Large amounts of topsoil and subsoil were eroded from the land during days before soil conservation. Contrary to popular belief, the soils didn't all wash into the Gulf of Mexico. They are still haunting us. Large volumes are stored in the lower reaches of tributaries where they have raised valley floors. They stand poised, waiting to be resuspended by bank erosion and transported to the Mississippi during heavy rains and floods.

Fertilizers

The nutrient elements nitrogen and phosphorus, so vital for agricultural production and lush lawns, are major water pollutants. In small amounts, they can increase the productivity of the river, but they can also wreak havoc in lakes and impoundments by causing *eutrophication* (overenrichment), resulting in proliferation of nuisance algae and aquatic plants, speeding the impoundment's rate of

Fecal Coliform Bacteria: Guilt by Association

It's virtually impossible to sample river water for the presence of all *pathogenic* (disease-causing) organisms that could be present as a result of pollution by the excrement of humans and other animals. Instead, a "guilt by association" test is usually run.

Fecal coliform bacteria live in the colons (large intestines) of animals, including man. They are normal, beneficial components of the intestinal flora—and they are easy to test for. Most are not harmful, but some strains can cause severe illness and even death. If they are present in the water, we know that feces are present (human or otherwise), and that therefore deadly water-borne disease organisms like cholera, typhoid fever, and hepatitis could also be present. The coliform bacteria test is the standard for swimming pools, public beaches, municipal water supplies, and industries like dairies and breweries.

aging. The nutrients get into the river from many sources—mainly animal manure, chemical fertilizers, and sewage treatment plant effluent—but also from the decomposition of virtually all naturally occurring organic substances.

Nitrates are a serious problem in karst areas where fissures and sinkholes conduct animal manure and chemical fertilizers downward into the groundwater. Nitrate contamination of drinking water can cause *methemoglobinemia* in infants and newborn pigs, reducing the ability of their blood's hemoglobin to transport oxygen (in effect, much like carbon monoxide poisoning).

Heavy Metals

Heavy metals have been known to be toxic for centuries. Some of the most notorious are arsenic, lead, mercury, copper, chromium, cadmium, and zinc. They entered the river in the untreated wastes of many industries.

In years past, mercury was flushed directly into the nation's waterways, primarily from paper mills that used it as a fungicide. These flows have essentially been stopped. Today, the primary sources of mercury in the Mississippi River are from rain and snow that wash mercury vapor out of the atmosphere.

It's hard to imagine that a heavy, liquid metal like mercury—over thirteen times heavier than water—could become airborne, but it can. Mercury will evaporate at room temperature. It was widely used for centuries to extract gold from pulverized ore. Mixed with the ore, it formed an amalgam with the gold and was later boiled away, leaving the pure gold behind but thus causing the deaths of thousands who inhaled the vapors. Mercury, found naturally in coal, also enters the atmosphere from coal-burning power plants as well as from incinerators. It was also used for years as a mold inhibitor in latex house paint, from which it gradually evaporated as the paint aged.

Because the Upper Mississippi is a hard-water stream, heavy metals like lead, cadmium, and mercury tend to combine with carbonates to form insoluble compounds like lead carbonate that settle to the bottom to become entombed in sediments. If resuspended, they may be flushed downstream. Lead concentrations have decreased primarily due to the elimination of tetraethyl lead as an antiknock compound in gasoline.

In a study of cadmium and mercury concentrations of sediments and *Hexagenia* mayfly nymphs in a 354-mile reach of the Upper Mississippi extending from Pool 2 through Pool 16, the highest concentrations of both contaminants in sediments and nymphs were in Pools 2, 3, and 4, just downstream from Minneapolis–St. Paul, Minnesota (Beauvais et al. 1995).

Chlorinated Hydrocarbons

In the United States, bioaccumulation of DDT, the famous chlorinated hydrocarbon insecticide of World War II vintage, led to eggshell thinning in fish-eating birds such as bald eagles. Their fragile eggshells were easily crushed during incubation. As a result, eagle populations plummeted, causing near extermination in the lower forty-eight states.

Beginning in the 1960s, the use of persistent chlorinated hydrocarbon insecticides such as DDT, heptachlor, dieldrin, endrin, and chlordane was greatly curtailed, and concentrations in fish and wildlife declined (Schmitt and Bunck 1995).

Much like DDT, *polychorinated biphenyls* (PCBs) are waxy, manmade compounds that are persistent, fat-soluble, bioaccumulatable, and toxic. Various PCBs—more than two hundred different compounds have been formulated—were used since 1929 as fire retardants, hydraulic fluids, lubricants, as heat transfer agents in electrical equipment, and as a component of carbonless copy papers. Long thought to be relatively innocuous, large quantities of PCBs were discharged directly into the Mississippi River and were also present in landfills and urban runoff.

PCBs were often dispersed via transformer oil. Most of the large, familiar canisters on power poles are oil-immersion transformers. Oil is an excellent insulator but flammable, so PCBs were added to the oil as fire retardants. When a transformer failed, the valuable metal components, especially copper, were salvaged, but the oil was usually given away, often to people who spread it on dirt roads to keep the dust down. From the roads the PCBs washed into tributaries of the Mississippi.

The presence of PCBs in the river is attributed mainly to industrial sources. In sediments and mayflies sampled in 1988, concentrations were highest from the Twin Cities through Lake Pepin. Downstream from Lake Pepin, concentrations were much lower. Farther downstream, concentrations were greatest in pools with cities, especially in the Quad Cities area (Rock Island, Moline, Davenport, Bettendorf).

Although PCBs were banned in the United States in the 1970s, countless tons of them still move about the planet because they are very stable, gluing themselves into food webs. Like DDT and other persistent organic pollutants, PCBs are fat-loving and are concentrated in animals at the top of the food chains because they store most of the combined mass of PCBs that every species below them on the food chain has eaten and stored, a process called *bioaccumulation.*

Along with DDT and other chlorinated insecticides, PCBs were run up through long riverine food chains. Together they limited reproduction in fish-eating birds like bald eagles, double-crested cormorants, pelicans, and ospreys. The compounds also impacted mink and otter, and nearly caused the extinction of peregrine falcons. By 1980, the direct discharge of PCBs to the Upper Mississippi had been greatly restricted. Today the main source is via atmospheric transport—a worldwide problem.

It would be unethical to feed PCB-laden food to humans to see if they develop health problems, but there are hundreds of research papers showing that PCBs adversely affect rats and mice.

Prior to the mid-1970s, the Upper Mississippi had elevated levels of PCBs and heavy metals like mercury, cadmium, and lead downstream from every major industrial city. Even though DDT use in the United States was banned in 1972, it is still stored in riverine sediments and can be detected in some fish. Since 1972, inputs of most toxic organic industrial pollutants have been decreased by 99 percent.

In general, concentrations of persistent contaminants that accumulate in fish and wildlife are lower now than at any time for which accurate data exist (Schmitt and Bunck 1995).

Because the Mississippi has the stigma of being polluted, people are often reluctant to eat river fish, turning instead to fish from the oceans, thinking that they must be pure. But they should know that the oceans' top carnivores like shark, tuna, mackerel, and swordfish have even higher levels of PCBs, mercury, and other pollutants than do fish from the cleanest reaches of the Upper Mississippi. Various agencies have recently issued consumption advisories warning of high mercury levels in albacore tuna.

The Pristine River

Early explorers were impressed with the quality of the Mississippi's waters. Above the St. Croix, Stephen Long noted that Mississippi water was "entirely colorless and free from everything that would render it impure, either to the sight or taste." Both Long and Zebulon Pike described the waters of the Mississippi below its confluence with the St. Croix as reddish in color in the shallows. In deep water, Pike said it was as "black as ink." Long incorrectly interpreted the reddish color as being due to the color of sand on the bottom. We now know that the waters of the St. Croix were naturally tannin-stained, due to their origin in northern bogs, and that they colored the waters of the Mississippi at their confluence. Contrary to popular belief, before white settlers broke the sod and felled the woodlands along its banks and the banks of its tributaries the Upper Mississippi was seldom muddy.

Defiling the River

The first pollution complaints on the Upper Mississippi concerned sawmill refuse, not because of aesthetics or fear of toxicity, but because it had become a navigation hazard. By the late 1870s, so much sawdust had been poured into the river from the mills along its banks that steamboat pilots reported bars composed of sawdust obstructing navigation above Lake Pepin and as far south as Winona. They also complained that the sawdust-choked river water was permeated with pine resins that caused foaming in steamboat boilers.

The sawmill waste problem solved itself when the lumbering era petered out early in the twentieth century, but the problem of urban wastes continued to plague the river. Many citizens thought that treating these wastes was unnecessary, theorizing that the river would purify any material dumped into it. Most people assumed that rivers would forever be the common sewers and dumping grounds for everybody.

At the dawn of the twentieth century, the River and Harbors Act of 1899 was the most broad and effective water pollution legislation in existence. It outlawed casting refuse into navigable waters and also stipulated that refuse could not be dumped on the banks of tributaries if it was liable to wash into navigable waters. But by the end of the nineteenth century, the river was more important as a sewer than a navigation channel. So long as what went into the river didn't impede the progress of passing watercraft, the floodgates were open. Out of sight, out of mind was the word of the day.

As a result, for about sixty miles, through Metropolitan Minneapolis and St. Paul and downstream to Lake Pepin, the river ran for years as a huge open sewer (Wiebe 1927). By the late 1880s, Minneapolis was dumping about five hundred tons of raw garbage into the

The Lake Pepin Story

Lake Pepin has been severely impacted by pollutants from the Twin Cities and from the Minnesota River Basin for about a century and a half. With an average depth of about fifteen feet and an average water-retention time of nineteen days, Lake Pepin acts as a natural settling basin, greatly enhancing water quality of the river downstream. Recent sedimentation rates in Lake Pepin range from 1.2 inches per year or greater in upstream reaches to about 0.25 inches per year in downstream reaches. About 21 percent of the lake's volume was lost between 1897 and 1986. Although the sediment-trapping ability of Lake Pepin substantially reduces contamination downstream, this ability is diminishing as the lake fills and its volume decreases. Today, the main sediment input to Lake Pepin is in the form of organic sediments, the result of enrichment of the river by urban and agricultural pollutants. The good news is that many contaminants that settled out in the lake in the past 150 years have been buried and aren't being mobilized.

Improved waste treatment facilities in the Twin Cities area have caused marked improvement in general water quality, resulting in recurrence of desirable aquatic insects like *Hexagenia* mayflies and increased fish diversity (Wiener et al. 1998).

Mississippi River below St. Anthony Falls each day in addition to uncalculated amounts of untreated domestic sewage and industrial wastes. St. Paul added an even greater amount of garbage, plus slaughterhouse wastes that included blood, gut contents, grease, and wash water. All cities on the Upper Mississippi were doing similar things—or worse.

The demands of navigation only made the problem worse. To assure a stable navigation channel, locks and dams were built on the Upper Mississippi. Although navigation dams did not cause the pollution problem, they exacerbated the situation and focused attention on deteriorating water quality.

Lock and Dam 1, completed just below St. Paul in 1917, collected most of the raw sewage of Minneapolis and St. Paul in its impoundment. This fetid, festering accumulation led the U.S. Bureau of Fisheries to report that during August of 1927, forty-five miles of the river below St. Paul lacked sufficient oxygen to sustain fish life of any kind.

The urban cesspool extended further downstream when Lock and Dam 2 was completed in 1930 at Hastings and accumulated the remainder of the Metro sewage and that of growing suburbs, and southside packinghouses and stockyards.

This growing public disgust led to the first sewage treatment facilities being built in 1938. The plants significantly improved water quality, and most fish species could again live in the reach below St. Paul (Scarpino 1985).

Even so, until the 1980s, the Mississippi often ran anaerobic and foul-smelling below St. Paul—defiled by domestic sewage, industrial wastes, and slaughterhouse wastes that had a high BOD.

This persistent befouling of the river was in part due to the infrastructure of the cities along its banks. Instead of building separate sewers for sanitary sewage and storm water, some cities like St. Louis and the Twin Cities of Minneapolis and St. Paul built combined sewers and ran their combined flows directly into the river.

When treatment plants were built, no effort was made to separate the city's storm and sanitary sewers. As a result, along with municipal and industrial sewage, the combined flows directed into the wastewater treatment plants included tree leaves, grass clippings, animal feces, and litter, challenging the plant capacities in the best of times. Sand, used in ice abatement on city streets, posed a major problem by regularly overloading the grit-removal facilities at the sewage treatment plants. Following a major storm or during a rapid snowmelt, treatment plants couldn't handle the volume of the combined flows, so the valves were opened and the untreated wastes of the entire city again flowed directly into the river. Not only did the river get a slug of raw sewage every time it rained, the rushing waters scoured out the sludge that had accumulated in the sewers since the last downpour.

Eventually, toughened water quality standards and stiffer penalties forced the Twin Cities to separate their sanitary and storm sewers, ending the dumping of raw sewage that, for decades, gave the Pig's Eye wastewater treatment plant a bad name downstream.

It was a totally unintended benefit to communities downstream that the navigation pools of the Nine-Foot Channel Project served as massive sewage treatment lagoons. With a settling out at each subsequent impoundment and the aeration of the tailwaters passing through the dam, the putrescible portion of the metropolitan pollutant load was decreased. Although downstream cities added their own pollutants, the additions were very small compared to those of Minneapolis and St. Paul; the river was bigger downstream, allowing for greater dilution. The large tributary rivers like the

"I'll Race You to the Sewer!"

No Iowa city along the Mississippi had municipal sewage treatment in the late 1950s when I was doing research for my doctorate at Keokuk. Human sewage entered the river raw. It was there that I first heard fisheries biologists facetiously refer to the "white trout index" (the number of condoms per gill-net haul) as a measure of municipal river pollution.

I occasionally studied the river as a guest aboard the Keokuk-based U.S. Coast Guard buoy tender *Lantana*, collecting caddisfly larvae as the crew serviced navigation buoys. At the end of an unforgettable, hot, stifling July day in 1958, the *Lantana* tied up for the night along the Burlington, Iowa, waterfront, upstream from a huge municipal sewer. The sweltering crew needed to cool off, and so did I. Oblivious to water quality, we refreshed ourselves by high diving off the *Lantana*'s bridge into the cool river. I'll never forget the challenge, "I'll race you to the sewer!"

St. Croix, Chippewa, and Wisconsin added relatively clean water to the Mississippi, increasing its ability to assimilate its pollutant load. As a result, biologically at least, the Upper Mississippi was comparatively clean from Wabasha, Minnesota, downstream to the Quad Cities.

Status at the Close of the Twentieth Century

To most observers, water quality in the Upper Mississippi has improved in recent decades. Gross pollution by domestic sewage has been reduced since passage of the *Federal Water Pollution Control Act of 1972,* which mandated secondary treatment of sewage effluent. The purposeful dumping of raw industrial sewage or municipal wastes has been virtually eliminated on the Upper River in the last thirty years. But the river still receives an array of contaminants from agricultural, industrial, municipal, and residential sources. The impacts of these contaminants on river biota are still largely unknown.

Incredibly, many knowledgeable, conscientious people believed—as late as the early 1970s—that the United States could not afford sewage treatment facilities in every city and still finance its Cold War with the Soviet Union. They earnestly believed that the Mississippi should have been written off as a national sewer—Uncle Sam's colon—a *cloaca maxima.*

The major statutory influence on the Mississippi's water quality was the *Federal Clean Water Act of 1972.* The act had ambitious goals: (1) to restore and maintain the chemical, physical, and biological integrity of the nation's waters (2) to eliminate the discharge of pollutants into navigable waters by 1985, and (3) to make all waters swimmable and fishable.

A "Charmin" House

The Mississippi's caddisfly larvae build small, tubular homes in which they live. Most of these aquatic insects build their abodes of sand grains or pieces of vegetation, but in the 1950s one species at Keokuk preferred toilet paper. Charmin was a popular brand, hence the riddle, "What did one Keokuk caddisfly larva say to the other Keokuk caddisfly larva?" Answer: "My, what a Charmin house you have!"

Needless to say, implementing the Clean Water Act has been an incredibly complex, massive, politically charged task. To most river rats, it has been an incomprehensible, bureaucratic mess. Perhaps, one has to be old enough to remember the "bad old days" to appreciate the phenomenal cleanup of the Upper Mississippi.

Much of the power of the Clean Water Act lies in a single short phrase. Section 301 states: "the discharge of any pollutant by any person shall be unlawful."

Point sources of pollution were the easiest to address. Much of the progress in pollution control was due to the requirement that all point sources of pollution had to obtain National Pollutant Discharge Elimination System (NPDES) permits from the U.S. Environmental Protection Agency (USEPA). This provision placed the burden of proof on the polluter, who now had to explain why the discharge could not be eliminated—or face penalties under law. The government no longer had to prove harm to justify action (Robinson and Marks 1994).

The USEPA went farther, defining effluent guidelines for the permits. For example, municipal sewage dischargers had to provide at least secondary treatment using bacteria in an aerated tank *(activated sludge treatment)* to further break down organic matter. Industrial

"It's Asparagus Season on the Upper River."

Prior to the 1970s, most American cities ran their sewage into a river (or the ocean) untreated or partially treated. Typically, each river city obtained its drinking water upstream from the city but voided its sewage downstream. Inevitably, river water was thus consumed and voided repeatedly by successive downstream river dwellers. Chlorination of city water, probably the single greatest public health measure of all time, prevented epidemics of waterborne diseases like typhoid fever and cholera.

In 1958 I made an April visit to a laboratory near the mouth of the Ohio River. A technician took a drink from the water fountain and dramatically exclaimed , "Aha! It's asparagus season on the upper river!" On cue, I asked him how he knew, and he responded that he could detect the odor of aspartic acid (asparagine) in the city water. Aspartic acid, excreted in the urine of asparagus eaters, gives their urine its characteristic, unpleasant odor.

and commercial dischargers were required to pretreat their effluents to control toxic pollutants.

Implementing mechanisms included federal money for sewage treatment plant construction, waste treatment management planning, and increased research in pollution-control technology. States were required to develop water quality standards for in-state and interstate waters, to identify all waters not meeting the standards, to calculate the additional pollution reductions needed to achieve the standards, and to incorporate these requirements into permits. If a state failed to do any of these things, the USEPA would do it for them. The act was hardly perfect, and massive violations occurred repeatedly and flagrantly. Lack of monitoring was one problem, but even with monitoring, there were high levels of noncompliance (Robinson and Marks 1994).

Point source discharges from municipal runoff, sewage treatment plants, and industries were common until the mid-1970s, but nonpoint sources like agricultural runoff and runoff from urban areas are now more significant sources of pollution.

Though many of the tools for controlling polluted runoff have been included in previous Clean Water Acts, they have lacked commitment and funding.

In 1987, Congress created section 319 of the Clean Water Act to pull together and strengthen nonpoint runoff controls dispersed throughout the Act, requiring nonpoint source reduction to the maximum extent practicable. But the 319 programs were plagued by insufficient direction and oversight by the USEPA, lack of adequate implementing mechanisms, and inadequate funding.

Although nonpoint pollution dwarfed the sewage treatment challenge of the early 1970s, the average annual appropriation for section 319 was about fifty million dollars in the late 1980s—a fraction of program needs. In contrast, in the eighteen years between the passage of the Clean Water Act and 1990, fifty billion dollars had been invested in sewage treatment alone. "Even if the legislation had been fully funded and fully implemented, water pollution problems would persist because the legislation relies almost entirely on a piecemeal approach, failing to require pollution prevention programs be administered on a watershed basis" (Robinson and Marks 1994, 19).

Because the Mississippi's drainage extends over such a large area, runoff-abatement laws enacted by the states that border the river have little impact on the greater volume of the water flowing into the river. The Mississippi is the sink for the pollution that runs off about 40 percent of the land in the United States. To significantly reduce the burden of polluted runoff to the Mississippi and to the Gulf of Mexico, guidelines must apply to vast areas within the Mississippi drainage. We now have fifty individual state runoff assessment and management programs that vary in terms of comprehensiveness, stringency, degree of public participation, accountability, funding, and effectiveness.

As a result, toxic polluted runoff from urban areas and farms continues to be a persistent problem in the Mississippi River watershed. A wide range of chemicals can be traced to runoff, including animal manure, chemical fertilizers, dioxins, heavy metals, and millions of pounds of herbicides like atrazine and alachlor. The use of PCBs, and insecticides like DDT, chlordane, and endrin has been banned, but they still show up in fish tissues, prompting fish consumption advisories, and they still persist in the environment. Newer organophosphate insecticides and carbamates do not bioaccumulate, but they are extremely toxic, especially to aquatic invertebrates and fish.

Today, the Upper Mississippi receives a complex mixture of agricultural chemicals, primarily herbicides and their degradation products, from the surrounding rich agricultural land that is intensively cultivated for corn and soybeans. The Minnesota and Des Moines Rivers, for example, are the primary contributors of alachlor, cyanazine, and metachlor.

The highest human population densities in the Upper Mississippi River watershed are in cities along its rivers. Urban development has increased the rate of runoff to the river because of the conversion of spongelike soils to waterproof concrete, asphalt, and rooftops. Urban storm runoff contains trash, tree leaves, grass clippings, pet feces, crankcase oil and grease, lawn fertilizers, road salt, antifreeze, herbicides, insecticides, detergents, and whatever unwanted substances, like paint and solvents, that people throw down storm drains. While municipal and industrial pollution have been controlled to a great extent in most municipalities, most urban runoff still enters the river untreated.

Soaring Bald Eagle Populations: An Environmental Success Story

When the bald eagle became our national symbol in 1782, it soared over the entire Mississippi River watershed, but by the mid-1970s there were only about a thousand breeding pairs surviving in all of the lower forty-eight states. Because of the population crash, due mainly to egg shell thinning caused by DDT, the magnificent bird was declared an endangered species.

Since the 1970s, decreases in the Mississippi's content of DDT, PCBs, and other chlorinated hydrocarbons have enabled bald eagles to return to the river en masse. Most of the river's overwintering eagles migrate southward yearly from Alaska or Canada, and the river's open-water areas provide good fishing for them. It's not unusual to count more than a hundred in late winter during a nine-mile boat trip between Lock and Dam 5A and Lock and Dam 5.

Several river cities have erected eagle watch facilities that attract hundreds of eagle watchers. Alma, Wisconsin, is strategically located right at the site of Lock and Dam 4 and a power plant that pours warm water into the river, and this creates a large expanse of ice-free water. Shoppers strolling down the main street of Alma have grown accustomed to seeing eagles flying at treetop height. In spring, active eagle nests can be seen in many areas from highways that flank the river.

In concert with the return of eagles, there have been increases in populations of double-crested cormorants, gulls, and peregrine falcons.

5

The River Today and Tomorrow

22

Who's in Charge?

A centipede was happy, quite,
Until a toad in fun
Said, "Pray, which leg moves after which?"
Which raised her doubts to such a pitch,
She fell exhausted in the ditch,
Not knowing how to run.

Metcalf and Flint 1939, 79

Federal actions have historically had dramatic effects on the river, as exemplified by construction of the lock and dam system of the 1930s and the creation of national fish and wildlife refuges beginning in 1924. The states' role in river management began with fisheries work in the late 1800s but has expanded significantly since then—especially since the early 1970s. Today, states assume most responsibility for floodplain management, fisheries management, drinking water regulation, regulation of pollutant discharges, and enforcement of water quality standards.

This chapter is an attempt to unravel and simplify the complex story of how the Upper Mississippi River is managed as a multiple-use resource, supposedly to wisely provide the greatest good for the greatest number of humans, and also to protect, maintain and restore the riverine environment. To an outsider, and perhaps to some insiders as well, the multilayered bureaucratic web of management agencies has become bewildering and incomprehensible (as will be illustrated in the flow chart on page 341). Because so many federal and state agencies are involved, acronyms are used liberally to save space. A list is included at the front of the book. I think that I have approached my task factually and objectively, having been associated with the staffs of federal and state agencies as well as river-oriented academicians for over forty years. But because I've tried to "tell it like it is," portions of the chapter will be upsetting to some.

River Managers and River Rats

Throughout this chapter I frequently use the terms "manager" and "management" when referring to how the river and its resources are exploited, conserved, rehabilitated, and "improved." These actions are relative and subjective, of course, as seen from a human perspective. The term "improve" is especially subjective. As viewed by canoeists that seek solitude, for example, the construction of a big marina is hardly an improvement.

River managers are federal, state, county, or city decision makers who interpret data, formulate policy, propose legislation, devise projects, recommend regulations, enforce laws, and implement a host of other actions that have an impact on the river and those who use it. Their deliberations are theoretically based on data collected by scientists, economists, and technicians, but are influenced by political pressures and public opinion. Traditionally, technicians, usually trained in the sciences, collected and analyzed data. In recent years, however, technicians have assumed greater roles in management decisions. Managers ultimately determine how the river is used for things like commercial transportation, flood control, sewage disposal, water supply, water power, lumbering, recreational boating, hunting, sport fishing, commercial fishing, trapping, bird-watching, camping, aesthetic enjoyment, and the training of hunting dogs.

The term "river rats" is not meant to be derogatory. Rather, it describes a heterogeneous group of people who "have the river in their blood" and are impacted by the decisions and actions of the managers. They include a knowledgeable but not necessarily scientific group of bird-watchers, sport fishermen, commercial fishermen, hunters, trappers, hikers, boaters, and river-watchers who would rather live along the Mississippi River than anywhere else. Some owe their livelihood to the river, but most use it for recreation. Many just like to watch it. Bird-watching is the fastest-growing pastime nationally, and in Minnesota the amount of money spent on wildlife-watching presently exceeds the amount spent on hunting. Nonconsumptive river users spend large sums on things like bird seed, bird feeders, photography, binoculars, telescopes, videos, books, travel, and outdoor clothing.

River rats are increasingly becoming sophisticated in the ways of the river, its wildlife, its management, and its politics. The Internet affords them incredible amounts of information.

Most of today's river rats are conservation oriented because they realize that river resources are in peril, but their numbers still include game hogs who boast of shooting a hundred ducks a year and anglers who routinely take over their legal limits of fish.

"Lots of Them New-Age River Managers Ain't River Rats"

Older river rats tend to be suspicious of those who manage the Mississippi's natural resources from 8 A.M. to 5 P.M., Monday through Friday, but who don't hunt, fish, trap, or boat, and can't carry on intelligent conversations with those who do. Today's river managers and technicians may include state or federal employees who have had no long-term love affair with the river, but who happen to work on it because they have skills needed in projects sponsored by the agency that hired them. They may have had little or no direct experience with the river. Fewer and fewer river workers actually come in physical contact with the Mississippi or its environs, either as part of their job or as recreation. Some became administrators and river managers by default via bureaucratic policies or interagency transfers.

Critics charge that new threats to the environment (real or perceived) mean jobs, funds, and credibility for bureaucratic behemoths that must nourish themselves and justify their existence. Agencies must, therefore, "sell" each threat, convincing the public and the media, but especially legislators, that the threat is real. If they are successful, there may be a "funding frenzy" where tax dollars are poured into river programs that sprout like mushrooms. Once created, the programs are difficult to kill, even if the threat should pass.

Landmark Federal Mandates that Determine How the River Is Managed

The *Northwest Ordinance of 1787* was the first and perhaps the most far-reaching document to impact the Upper Mississippi River. It dealt with settlement of United States territory northwest of the Ohio River that was bounded on the west by the Upper Mississippi. The ordinance stated that all navigable waters leading into the Mississippi and St. Lawrence Rivers shall be common highways and "forever free" to U.S. citizens, without any tax, impost, or duty. To this day, the ordinance plays a major role in legal matters regarding interstate river navigation. For example, bridges are still considered obstructions to navigation and are maintained by their owners at no expense to the navigation industry for any towboat/bridge collisions.

Direct user fees to defray channel maintenance costs were prohibited by the language of the ordinance for over 150 years, but since 1978 a "user fee" has existed in the form of an excise tax on diesel fuel used by commercial vessels. The tax was Congress's way around the old ordinance. The fuel tax, 24.4 cents per gallon in 2004, accumulates in a Waterways Trust Fund to be used on a cost-share basis to fund navigation improvements such as new locks (Davis 1982; Rasmussen 1994).

The Trust Fund has never accumulated enough revenue to contribute significantly to the cost of operating and maintaining the Upper Mississippi River navigation system. Yet, the barge operations continue to use this as an example of how they are "paying their fair share."

The *Upper Mississippi Refuge Act of 1924* created the Upper Mississippi River Wildlife and Fish Refuge, preceding the Nine-Foot Channel Project by about six years.

The *National Environmental Policy Act of 1969* (NEPA) required the filing of an Environmental Impact Statement (EIS) for every future major federal action and federal project that may produce significant impacts on the environment. This included the Mississippi's Nine-Foot Channel Project of the 1930s because it was ongoing.

The *Clean Water Act of 1972* (CWA) prohibited discharge of any pollutant (including dredged material) by anyone, including the Corps. The act explicitly preserved the states' right to impose more stringent standards. The U.S. Environmental Protection Agency (USEPA) and the U.S. Army Corps of Engineers (USACE or, simply, the Corps) under the National Pollution Discharge Elimination System (NPDES) administer the program.

The *Endangered Species Act of 1973* (ESA) directed the U.S. Fish and Wildlife Service (USFWS) to identify and develop protection plans for threatened and endangered species as well as evaluate the potential impacts of specific proposed actions on such species.

The *Water Resources Development Act of 1986* (WRDA) authorized creation of the Upper Mississippi River System Environmental Management Program (EMP) and the Long Term Resource Monitoring Program (LTRMP), to be funded by the Corps and implemented by the USFWS and the five states that border the Upper Mississippi River and Illinois Rivers.

The *WRDA of 1987* recognized that states have the primary responsibility to implement federal water quality policies. It required each state to develop and submit to the EPA a comprehensive management plan to address nonpoint pollution problems.

A Bureaucratic Maze

Flowing as an interstate river, the Mississippi belongs to everyone, and it belongs to no one.

Flowing amid five states and, as a "working river" subject to a variety of demands and multiplicity of uses, state and local authorities historically found it convenient to ignore the river and pass off management of the waterway and its resources as the responsibility of "someone else"—at least so long as "someone's" interests didn't conflict with their own.

This attitude and practice is reflected in the often-contentious hodgepodge of federal and state agencies managing the river today. Because of the river's interjurisdictional character, the federal government has played the major role in its management. That central role was first formalized on June 28, 1879, when Congress created the

Mississippi River Commission (MRC), a mix of Army and civilian engineers, to "control" the entire Mississippi. In addition to federal authorities, there are many quasi-governmental commissions and associations, as well as city and county governments, that have a stake in river management. Examples of city or county management would include no-wake zones for boats, commercial harbors, barge fleeting areas, marinas, river patrols, zoning ordinances, and port authorities.

A diverse spectrum of individuals and citizens' pressure groups also work on river-related issues, seeking to further conservation, public health and safety, recreation, tourism, and economic development. At times these pressure groups have exerted extraordinary influence on developing policy. For example, the action group Citizens for a Clean Mississippi—mobilizing public support in downriver communities with the slogan, "We can't all live upstream!"—was instrumental during the 1970s, in concert with a lawsuit by the State of Wisconsin, in forcing the Twin Cities of Minneapolis and St. Paul to cease dumping raw sewage into the Mississippi.

At this point, a brief summary of some of the major policy players up and down the river will illustrate how administrative management and decision making are decentralized and dispersed among various federal and state agencies, providing an extreme example of multilevel bureaucracy, shown as a flow chart on page 341.

Federal Agencies

The Mississippi River Commission (MRC) is comprised of three Corps officers (one of whom serves as president), an admiral of the National Oceanic and Atmospheric Administration (NOAA), and three civilian members (two of whom must be civil engineers). Authorizing legislation of 1879 gave the MRC responsibility for development and improvement of the entire Mississippi, but since the 1928 authorization of the Mississippi River and Tributaries Project for the Lower Mississippi, development of that project has been the primary focus of the commission. Technically, the MRC's mission includes the river from Cairo, Illinois, downstream to the Head of Passes at the river's mouth. There never has been an MRC member appointed from the Upper Mississippi region.

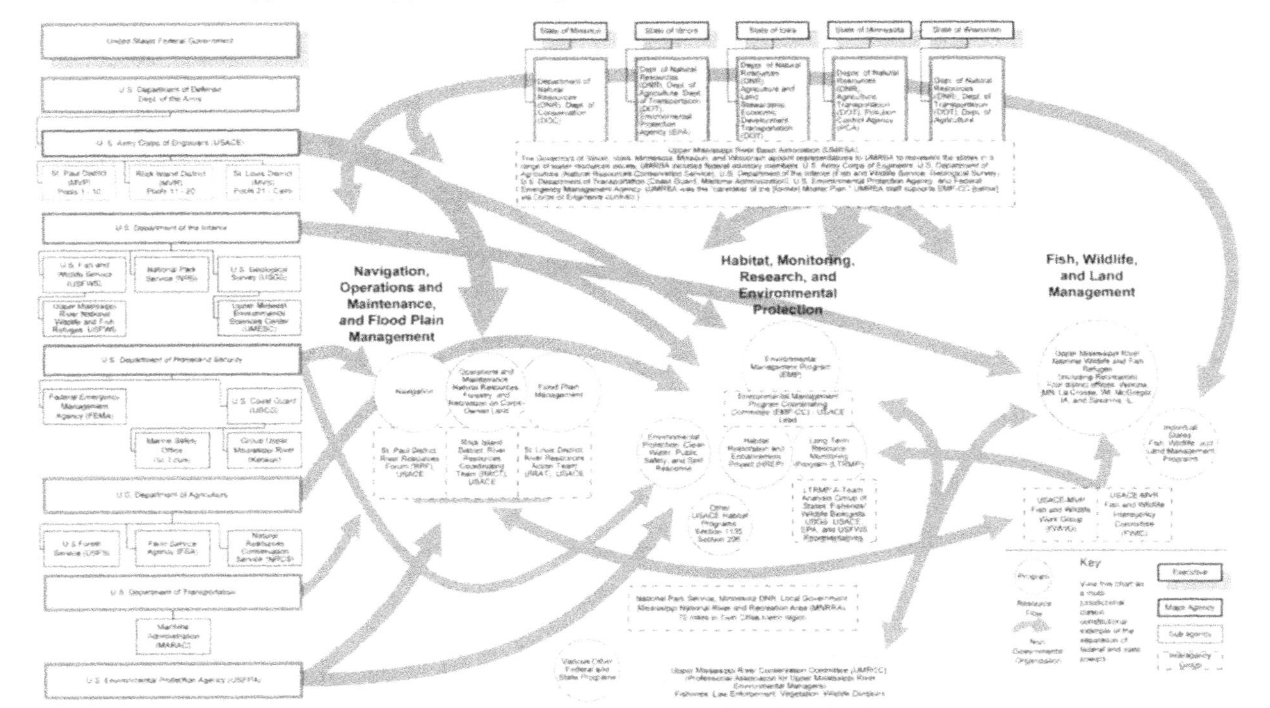

This mind-boggling flow chart shows jurisdictions and routes of funding for federal and state agencies managing the Upper Mississippi River (graphic by Upper Mississippi River Basin Stakeholder Network, St. Mary's University of Minnesota; C. R. Fremling, major contributor).

The U.S. Fish and Wildlife Service (USFWS) of the Department of the Interior is charged with protecting and conserving fish and wildlife and their habitats. Although the agency is charged with managing the river for both fish and wildlife, fish were historically shortchanged because of strong pressure from organized hunter groups to manage for waterfowl. These same groups also ensured that authorizing legislation for new refuges favored wetlands and waterfowl management. The USFWS "culture" and staff are still weighted in favor of waterfowl and migratory bird management over all else.

In spite of the river being a unique "dual purpose" river (i.e., navigation and wildlife refuges), its refuges are seldom actively managed by the USFWS. Historically, refuge managers "watched" the refuges; conducted surveys; maintained signage, kiosks, parking lots, and boat ramps; and enforced laws. With perennial budgetary limitations and lack of staff there was little else that they could do. Presently, virtually all active Mississippi River management is done, not with USFWS funds, but with external funds appropriated by Congress and doled out by the Corps. The present "drawdown" experiments (to dry out sediments and encourage growth of marsh plants) are the first significant attempts to manage the river for something other than barges. The USFWS is currently restructuring its existing programs and policies to better support an ecosystem approach to resource management.

The U.S. Environmental Protection Agency (USEPA) is responsible for protecting environmental quality, including surface and groundwater quality and wetlands. It ensures that state water quality programs meet the minimum federal standards under the Clean Water Act and is the primary federal responder for oil and hazardous materials spills from land-based sources.

The U.S. Geological Survey (USGS) is responsible for providing reliable survey data, including topography, biological monitoring, and contaminant and sediment loading. The Long Term Resource Monitoring Program (LTRMP) has recently been transferred to the USGS after it had previously been transferred from the USFWS to the National Biological Survey (NBS).

The U.S. Coast Guard (USCG) is responsible for river-traffic safety and is the primary federal responder for hazardous spills from vessels and river transfer facilities. It regularly services the river's navigation aids, conducts search and rescue, sponsors boating safety classes, and licenses vessel operators. Keokuk, Iowa, is the headquarters for the Upper Mississippi River Group. More recently, the USCG has assumed a greater role for national security on the river and now operates under the jurisdiction of the Department of Homeland Security, which was created in 2001. All towboat pilots are examined and licensed by the USCG. One percent of the USCG budget prior to September 11, 2001, was dedicated to port security, but after the World Trade Center attacks 57 percent of the budget was allocated for those efforts, causing other functions to suffer due to increased work load without additional resources.

The National Park Service (NPS) administers the National Wild and Scenic Rivers Program, which includes part of the St. Croix River. In cooperation with local governments, the NPS manages the seventy-two-mile Mississippi National River and Recreation Area (MNRRA) that runs through the Twin Cities Metro corridor.

The U.S. Department of Agriculture (USDA) has the single greatest responsibility in influencing the watershed and the river due to its immense influence on land-use practices throughout the vast Mississippi River watershed. They, in turn, have major impacts on water quality, sedimentation, and flooding. USDA subsidy programs usually determine how much agricultural land is laid bare by agriculture, and how much land is protected by cover crops in "soil bank," "set aside," and other conservation programs.

Legislation, like the *Farm Bill of 2002,* that provides high subsidies for row crops, mainly corn and soybeans, promotes "fence to fence" tillage and commensurate soil erosion and flooding. USDA programs are also ultimately responsible for how much of the river floodplain is leveed and farmed. However, the Corps, the Federal Emergency Management Agency (FEMA), the Small Business Administration (SBA), and others also play roles in levee repair and maintenance, especially during flooding disasters. FEMA, which became part of the Department of Homeland Security in 2003, coordinates federal emergency response operations and administers the National Flood Insurance Program.

The U.S. Army Corps of Engineers (USACE), usually referred to as the "Corps," is more powerful and influential than any other agency operating on the Mississippi River. The Corps is part of the Department of Defense, but its civil works program has a civilian construction mission, receiving much of its funding from Congress on a project-by-project basis. Its deep-seated bias for construction projects indulges congressional interest in delivering federal dollars to members' districts. "The Corps [is] one of the largest and most unusual agencies in the federal government. It is an executive branch bureaucracy that takes marching orders from Congress, a military-run organization with an overwhelmingly civilian work force, an environmental regulator despised by environmentalists" (Grunwald, September 10, 2000, A01).

The Corps has provided navigation and flood control facilities on the Mississippi for over 150 years, and since 1977 has administered Section 404 of the Clean Water Act to protect wetlands and other aquatic systems on the Mississippi. Historically, the Corps' main goal was to manage the river for navigation, often at the expense of water quality, fish and wildlife, recreation, and tourism.

The Corps has divided the Upper Mississippi into St. Paul, Rock Island, and St. Louis Districts, designated as MVP, MVR, and MVS in the flow chart on page 341. The commanding officer of each district is usually a full colonel whose job is temporary (about three years). He is assisted by other military personnel. In practice, the Corps' full-time civilian personnel make policy and maintain continuity. With over two hundred years of experience, the Corps is the master administrator of contracts.

The Corps prides itself in its long-term, "keep busy" mind set. While opponents of Corps projects think tactically in terms of years and decades, the Corps seems to think strategically in terms of centuries, often outlasting its adversaries. The Corps typically responds to opposition by putting "out of favor" studies and projects (e.g., Twelve-Foot Channel) "on the shelf" until the time is right to resurrect them. If the Twelve-Foot Channel Project doesn't "fly" this year, it may be shelved for a decade or so, then dusted off, updated, and reactivated.

To know where power lies, one needs only follow the money trail (shown in the flow chart). The Corps receives appropriations from

Congress and then functions as a contractor to state agencies (such as state Departments of Natural Resources) and other federal agencies (such as the USFWS), and is often accused of placating them by funding diversionary projects. The Corps' largess is compensatory for the environmental damages its other projects do. It mollifies environmentalists and improves the Corps' image. The Corps also barters—"If you permit seven new 1,200-foot locks on the Mississippi and Illinois Rivers we'll give you another EMP."

Because the Corps controls the federal funding available for the river's major resource management projects, it also controls how the money is apportioned and what resource questions are answered. Consequently, natural resource managers usually operate in a "data-poor, reactionary mode" rather than in the relative "data-rich, proactive" perspective enjoyed by Corps engineers and developers—the bottom line being that information is power (Rasmussen et al. 2004).

On more than one occasion, the Corps has exaggerated the economic benefits of projects it seeks to build. Three cases serve as examples. Within Minneapolis, locks and dams were completed below and above St. Anthony Falls in 1956 and 1963 to extend the Nine-Foot Channel above the falls and into the city's Upper Harbor in anticipation of Minneapolis becoming a major river port. That hasn't happened. Commercial barge traffic has never come close to meeting the Corps' original projections. Yet, federal tax payers pay more than three million dollars yearly to operate and maintain the locks and dams (Faber 2003a).

A half century ago, the Corps built six major dams on the Missouri River to provide water levels sufficient to float barges, forecasting that the Missouri would become a vital link in the nation's inland water transportation system. Tonnage was projected to reach eight million tons a year, but topped out at three million tons in 1977, and today is less than 1.5 million tons. Most grain exported from Nebraska and western Iowa now travels by railroad to Pacific ports, with increasing amounts consumed locally by cattle and ethanol plants (Nygren 2003).

In 1982, the Corps predicted that 123.2 million tons of traffic would flow through new Locks and Dam 26 near St. Louis, but in 1998, traffic flow was only 73.7 million tons (Faber 2003b).

The Corps often justifies its projects with flawed analyses, conveniently miscalculating the true economic and environmental costs of projects. For years, members of Congress have blindly accepted these analyses and thrown unquestioning support behind projects in their colleagues' districts, expecting reciprocal help when their own "pork-barrel" projects come up for consideration.

The passage of the *National Environmental Policy Act,* coupled with the litigation it spawned nationally, forced the Corps to revisit almost all of its basic policies regarding environmental management. Recently the Corps has been trying to recast itself in a more ecosensitive light.

State Agencies

All of the five states bordering the Upper Mississippi have Departments of Natural Resources (DNR) or equivalent departments concerned with hunting, trapping, fishing, forestry, agriculture, mining, economic development, water quality, air quality, transportation, and many other things that impact the river. They also do a variety of river research and inventory projects, serve as advisors to the federal agencies, and enforce laws pertaining to the river's natural resources. Funding for river studies often depends upon doles from the Corps.

The Minnesota-Wisconsin Boundary Area Commission (MWBAC), now defunct, served as an ombudsman, hosting informational public meetings, conducting boating studies, and creating a public forum for issue resolution. Both states recently withdrew their support for the MWBAC, illustrating a lack of state commitment regarding river resources.

Federal-State Cooperative Associations

The Upper Mississippi River Conservation Committee (UMRCC) has a longer track record than any other coordinating agency. Founded in 1943, the UMRCC brought resource managers from the five states along the Upper Mississippi and the federal government together to discuss common problems and coordinate efforts to protect and manage Mississippi River fish, wildlife, and recreational resources. The U.S. Fish and Wildlife Service provides a part-time coordinator for the UMRCC whose membership includes virtually all of the wildlife, fisheries, recreation, and enforcement managers working on the Upper Mississippi River System (which includes the Illinois River).

Located in Rock Island, Illinois, its objectives are to promote the preservation and wise use of the natural and recreational resources of the Upper Mississippi River, and to formulate policies, plans, and programs for conducting cooperative studies. The UMRCC provides biologists with a platform for organized action and a level of communication previously unattainable. Although most members are agency personnel, academicians have always been welcomed and invited to participate in annual meetings.

The UMRCC initiated the river's first long-term resource monitoring effort in the late 1940s and early 1950s, and to this day maintains one of the best and most complete sets of commercial fisheries data in the world (Rasmussen 1994). The UMRCC's major emphasis continues to be fisheries oriented.

The Great River Environmental Action Team (GREAT) was formed in 1974, under the leadership of the Corps and the USFWS. The interagency team (federal and state) was organized to identify and assess the problems associated with multipurpose use of the river and to develop recommendations for improved management of the river. The Upper Mississippi River was divided into three study reaches (GREAT I, II, III, respectively, in the Corps' St. Paul, Rock Island, and St. Louis Districts). Each was covered by its own study team, under the organizational umbrella of the Upper Mississippi River Basin Commission (UMRBC). GREAT I, the first one deployed, was responsible for the reach from the head of navigation in Minneapolis–St. Paul to Guttenberg, Iowa.

From 1974 through 1980, GREAT I proved that federal agencies, the states, and the public could work together and approach problem-solving in the best interests of total river-resource management. GREAT I was a model for interagency cooperation. Agencies that in the past had often been adversaries, like the Corps and the USFWS, found that they could work together.

Among GREAT recommendations were to develop uniform floodplain management standards for the affected states; develop a comprehensive management plan for the Upper Mississippi River

National Wildlife and Fish Refuge; rehabilitate backwaters; evaluate dredging and island creation; facilitate the maintenance and development of the waterway system; develop a comprehensive recreation management plan; develop a more environmentally and economically sound channel maintenance program coupled with a reduction in sediment yields from tributary rivers; select specific sites for placement of all material to be dredged until 2025; carry out an extensive cultural resource inventory; develop an effective program for keeping soil on the land; and support pollution-control programs. As would be expected, among GREAT's prime recommendations were pleas for federal funding to implement these ambitious recommendations.

Subsequent to the completion of the GREAT studies, two other interagency task forces were established to continue interagency coordination of resource management activities on the Upper Mississippi River. They were the River Resource Forum (RRF) in the St. Paul Corps District, the River Resources Coordinating Team (RRCT) in the Rock Island Corps District, and the River Resources Action Team (RRAT) in the St. Louis Corps District.

The Upper Mississippi River Basin Association (UMRBA) is the single largest voice for states to leverage their influence on river management and policy locally and in Congress. It serves as a forum to assist Wisconsin, Minnesota, Iowa, Illinois, and Missouri in the coordinated management of their water resources. A gubernatorial appointee represents each member state, usually from the agency with the most direct responsibility for water resource management. The UMRBA also has federal nonvoting participants, including the USEPA, the Corps, USDA (Natural Resources Conservation Service), and the Departments of Interior (USFWS, USGS) and Homeland Security (USCG). The UMRBA was formed in 1981 as successor to the former Upper Mississippi River Basin Commission, which was eliminated by Presidential Executive Order after completion of its master plan for the river in 1981. The UMRBA serves as a partner along with the USFWS and the Corps on the Upper Mississippi River System Environmental Management Program (EMP) and is the "overseer" of the EMP. The Upper Mississippi River Basin Association prepared the 1989 Upper Mississippi River Basin Charter that primarily addressed interbasin water diversions.

The EMP and LTRMP are programs of the Corps and the USFWS (and more recently the USGS). They are intended to provide coordination between the agencies on research and monitoring issues. The EMP began in 1987 through the Corps with the USFWS and States cooperating. The LTRMP was authorized under the WRDA 1986 as an element of the EMP implemented by the USFWS and funded by the Corps. Funds were to be used primarily for habitat restoration projects, mainly to counteract habitat loss due to sedimentation and navigation development (Rasmussen et al. 2004). "The impetus for the creation of EMP and the LTRMP was not a generous and altruistic action of the Corps but as a mandate driven by the Federal Court system and their findings in the Lock and Dam 26 litigation. As one of the witnesses for the plaintiff in that litigation, I vividly recall how the Corps was literally forced into a corner by Judge Ritchie. The irony of the situation lies in the fact that the Corps was put in charge and they immediately began using the court mandate as a bartering tool" (Thomas Claflin, personal communiqué September 2002).

The mission of the LTRMP is to provide decision makers with the information needed to maintain the Upper Mississippi River System as a viable multiple-use large river ecosystem. Its long-term goals are to understand the system, determine resource trends and impacts, develop management alternatives, manage information, and develop useful products. Monitoring from the six remote state-operated field stations includes water levels and quality, sedimentation, fish, vegetation, and invertebrates, as well as land cover/use.

The balance and effectiveness of the EMP has been open to question. Over the years, "significant amounts of EMP money have been spent on research and remedial projects to patch up local habitat problems, but the major influence on the river in terms of money spent and impacts to the system, has been the Corps' work to advance the single purpose of navigation. . . . Despite all this effort and spending, little is happening in the way of systematic, on-the-ground watershed planning or restoration" (Robinson and Marks 1994, 29).

As a result of the 1986 WRDA, Congress appropriates money to the Corps to fund the EMP. One-third of the funds are allotted annually to the Geological Survey's LTRMP, which in turn dispenses 50 percent to the USGS Upper Mississippi Environmental Science Center at La Crosse and 50 percent to six USGS field stations that gather and process data. The six LTRMP field stations are located at Lake City, Minnesota; Onalaska, Wisconsin; Bellvue, Iowa; Cape Girardeau, Missouri; Alton, Illinois; and Havana, Illinois (on the Illinois River). The field stations are operated by the DNRs of the various states. In Illinois, they are operated in cooperation with the Illinois Natural History Survey.

The Corps funnels the other two-thirds of the WRDA funds to Habitat Rehabilitation and Enhancement Projects (HREP) within each state. A team made up of the Corps, USFWS, and state DNRs manages each project, deciding which and where projects should be done. The Corps contracts the actual construction. In general, HREP funds have been divided about equally among the five states to accomplish projects devised mainly by fisheries biologists. The purpose of projects upstream from Clinton, Iowa, has mainly been preservation and enhancement, while below Clinton it has been restoration.

Money is lost at every level of this multilayered bureaucracy. Of the original congressional appropriation, about 30 to 40 percent is lost as overhead at the Federal (LTRMP) level. Another 25 to 30 percent is lost as overhead at the state (DNR) level. Only an estimated 30 to 55 percent of the original appropriation ever "touches the water" as bonafide habitat improvement. Peer-reviewed publication of research results has been scanty.

Great numbers of interagency meetings are held annually at sites along the river, attended by many salaried managers. But by the time decisions are made about which construction projects to fund, there may not be sufficient funds for work implementation.

A disproportionate amount of EMP funds have been allotted to the La Crosse area and Pool 8. The Upper Midwest Environmental Sciences Center (UMESC) of the USGS is located in La Crosse. It includes two major building complexes, usually referred to as the East and West Campuses. Together, until severe budget cuts in 2002, they had a staff of about two hundred and an annual budget of twelve million dollars.

The Mississippi River Interstate Cooperative Resource Agreement (MICRA) is an organization of all twenty-eight state fishery resource management agencies in the Mississippi River Basin. The USFWS provides staff for MICRA and works with the group to manage and preserve the basin's fishery resources more effectively. MICRA publishes an informative, influential, bimonthly newsletter, *River Crossings*.

Who's on First?

A system as complex as the Mississippi cannot be managed successfully in bits and pieces, yet that is how it is managed, and probably always will be. Holistic management seems impossible because there are too many agencies, with new ones being formed continually. For example, there is a new ad hoc organization of Minnesota state and county employees, the Basin Alliance for the Lower Mississippi in Minnesota (BALMM). To create a catchy acronym, members moved the Lower Mississippi upstream from Cairo, Illinois, to southern Minnesota.

About twenty federal agencies, forty-five state agencies, and hundreds of local units of government have management and regulatory responsibilities for the Upper Mississippi River environment. Usually, each agency focuses on only a few elements of any river issue. None has exclusive jurisdiction over the river or the exclusive responsibility to protect its resources. Because management objectives are often vague and contradictory, progress is difficult to measure. Managers are seldom interested in critical appraisals that might question the continuation of current programs, even if they do not seem to be working. The Upper Mississippi's continuing deterioration is evidence that piecemeal management does not work (McKnight Foundation 1996).

Management of the Upper Mississippi and its tributaries is often disorganized, contradictory, or ineffective. To preserve and restore the river, a unified vision must emerge, but the task falls mainly to public agencies charged with resource management. However, their objectives for river management are often at odds because they serve vastly different constituencies that have different demands for the river. For example, the Corps has traditionally served shipping interests and worked at flood control. The USFWS has served hunters, trappers, anglers, and bird-watchers. The clients of the USDA include farmers and foresters.

Another complication is that most of the land within the watershed is privately owned and managed. Most people recognize that farmers manage ecosystems but may not realize that the millions of owners of city lots are minimanagers who also tend crops (grass, vegetables, and ornamentals), spread fertilizers and biocides, grow trees, manage livestock (dogs and cats), and generate wastes. Sadly, many of the property owners are either oblivious to environmental matters, too impoverished or stressed to manage properly, or simply don't care.

Because there are no concrete management plans for the Upper Mississippi, agencies and property owners tend to pursue actions without a clear understanding of their impact on the overall river system.

Because there is yet no overarching program that requires all agencies to work together in the interests of the river, agencies charged with managing the river each address their own piece of the issue, neglecting the big picture (Robinson and Marks 1994).

There is a degree of consensus that river management should be on a watershed basis, and that perhaps a new superorganization is necessary to implement the concept. Yet, many fear that such a superprogram could be disastrous if dominated by development agencies (e.g., the Corps), which seems probable. Also, most river managers and river users are extremely provincial. Like the birds and mammals they manage, they are themselves strongly territorial and will defend their turfs and resources.

Even some of the talented, dedicated people who have spent their professional careers trying to promote river issues have become disenchanted because their dream professions have become mind-numbing nightmares of budgetary crises, report writing, reorganizations, empire building, funding frenzies, and meetings.

An Historic Perspective

Half a century ago, duck hunters, fishermen, trappers, and towboat captains were about the only people paying serious attention to the river. Even at the University of Minnesota, with the great river flowing as an open sewer through its Minneapolis Campus, serious academic study of the Mississippi was ignored. Rather, university professors usually concentrated their aquatic investigations in the Mississippi Headwaters of northern Minnesota where the lakes and river were pristine and esthetically pleasing. In a "publish or perish" atmosphere, working on an open sewer was tantamount to committing academic suicide. Likewise, at the University of Wisconsin, pioneering aquatic biologists looked almost exclusively at lakes, ponds, and small streams, as did their successors. Limnology, a relatively new science, was taught at most prestigious schools but was mainly concerned with standing waters.

The ecology of the Upper Mississippi River was hardly studied until the 1960s, with the exception of pioneering work done by Upper Mississippi River Wildlife and Fish Refuge Biologist Dr. William E. Green beginning in the early 1940s. Remarkably, he conducted his extensive field studies by canoe, producing highly accurate, colored maps of river vegetation in the critical years right after the big impoundments were created.

The Upper Mississippi was virtually a forgotten resource until the 1960s. All bordering states were equally guilty of ignoring the river and using it as a convenience for sewage disposal and transportation. Unfortunately, this was reflective of our national attitude toward large rivers.

During the 1960 to 1975 era, most basic studies of the ecology of the Upper Mississippi were conducted by professors and their students from small colleges scattered along the river. Their research was supported by meager college funds and grants from the National Science Foundation (NSF), the Federal Water Pollution Control Administration (FWPCA), and the USEPA. In 1967 there had been only four researchers gathered around a single table at an organizational meeting in Winona. In 1968 they formed an organization called the Upper Mississippi River Research Consortium (UMRRC) so that they and their students could meet annually to exchange information and present their research results before their peers. Annual meetings were held sequentially at colleges along the river.

Federal funding for river research dried up in the late 1960s, but passage of the *Clean Water Act of 1972* put new money and, predictably, new interest into river studies. The Act directed the Corps to do environmental assessments of the impacts of its ongoing Nine-Foot Channel Project. Since the Corps and the USFWS lacked sufficient qualified personnel, they awarded most assessment contracts to the riverside colleges. The contracts were bargains for the federal agencies because the indirect costs charged by the colleges were very low, and the college investigators worked long hours and on weekends with esprit de corps. The research provided excellent work experience for advanced students of biology, chemistry, and geology. The research faculty members were highly trained, virtually all with advanced academic degrees and years of experience. The colleges had good laboratory facilities, boats, and field equipment.

For a brief time, Mississippi River research flourished as a result of multi-investigator, multi-institutional projects that involved academic researchers doing exciting science together. This was a heady period because such multi-investigator projects were rare. Regrettably, thirty years later they are rarer still.

"Everybody's Getting into the Act"

Upon graduation, the students who participated in these early investigations provided federal agencies with cadres of river-knowledgeable personnel. Some augmented the growth of the great bureaucracies and adapted culturally. As agency empires grew, and federal money and personnel poured into the Mississippi, federal agencies did more and more of their river research themselves.

In 1986, mainly as mitigation for authorization of the highly controversial "super Locks and Dam 26," at Alton, Illinois, the U.S. Congress officially designated the Upper Mississippi River System as a nationally significant ecosystem and a nationally significant commercial navigation system. It authorized a $124.6-million, ten-year HREP for the Upper Mississippi as part of a larger $190 million EMP for the Upper Mississippi and selected navigable tributaries, to be implemented through an interagency (state and federal) effort.

LTRMP developers made deliberate decisions to not use existing university research capabilities. Instead, a new, redundant federal and state research and monitoring infrastructure was created at a cost of over five million dollars per year.

Today, most research and habitat management being done on the Upper Mississippi is being done with federal funds by technicians of the USFWS, the Corps, the USGS, and the DNRs of the five states. Habitat improvement projects are usually done in cavalier fashion without consulting experienced academicians, even if the projects are done in the colleges' "back yards." The prevailing attitude among technicians seems to be, "We know what's wrong with the river and we know how to fix it." To learn about projects in their study areas, academicians must read about them in the newspaper or attend public meetings.

The cost effectiveness and advisability of the work done by state and federal agencies may be questioned, but they provide continuity over the long haul that may not have been attainable by university personnel.

Interest in big rivers, especially the Mississippi, has increased dramatically in recent years. The Upper Mississippi River Research Consortium, founded and nurtured by academicians, has included increasing numbers of agency personnel, and because membership now includes workers from the Lower Mississippi as well as other rivers within the basin, the "Upper" has been dropped from the name. In 2001, there were 327 people on the MRRC mailing list, with 130 participating in the annual meeting. The MRRC is today's primary professional organization devoted to Mississippi River Research.

The Big Picture

> The Upper Mississippi River is more than just a river; it is a unique resource and the best example of a multipurpose river in the United States. Through congressional designation, it is the only inland river in the Nation serving as a Federal inland waterway for commercial shipping and a national wildlife and fish system.
>
> GREAT I executive summary 1980

The Mississippi was the only river in the United States designated for two major federal purposes—commercial navigation and wildlife refuges—purposes more often than not perceived as being at odds.

Conflicts between these two authorizations and project purposes peaked in the 1970s.

At various times, the Corps promoted a twelve-foot navigation channel, year-round navigation (fifty-two weeks to Burlington, Iowa, and forty weeks to Cassville, Wisconsin), new Locks and Dam 26 at Alton, Illinois, and bigger locks at the busiest sites upstream.

Simultaneously, growing public support for environmental protection and management led to lawsuits over the environmental consequences of the Corps' channel operation, maintenance (mainly dredging), and expansion of the Nine-Foot Channel Project. Lawsuits challenged the Corps practices and sought an even playing field in environmental matters. The resolution of the conflict resulted in major interagency studies on the interplay between habitat and channel management practices.

A number of habitat management and rehabilitation projects followed in the wake of these studies. They included backwater dredging, wing dam and levee construction, island creation, bank stabilization, side-channel openings and closures, wing- and closing-dam modifications, aeration and water-control systems, waterfowl-nesting cover, acquisition of wildlife lands, and forest management. Critics contended that the projects were merely "Band-Aids" to treat something that shouldn't have been done in the first place.

Under *Section 404 of the Clean Water Act,* the discharge of dredged material into all waterways was regulated through permits issued by the Corps under EPA guidelines. The 1977 reauthorization of the CWA gave explicit powers to the Corps for Section 404 jurisdiction over waters used in interstate navigation, including adjacent wetlands (Robinson and Marks 1994).

The CWA required that ordinary citizens and agencies proposing projects within the floodplain had to file a 404 Assessment and perhaps an EIS through the Corps' regulatory branch in cooperation with the USFWS and state agencies like the DNR. However, evaluations for Corps' own projects were done "in-house" without going through the same regulatory branch. Concurrently, the Corps, in the name of habitat improvement, dumped millions of cubic yards of dredge spoil and rock into productive backwater areas to create artificial islands and other structures, mainly in the name of creating

These two artificial islands are typical of federally financed Habitat Rehabilitation and Enhancement Projects along the Upper Mississippi River. Completed in 2003 in Pool 5A just upstream from Lock and Dam 5A, they were designed to restore habitat diversity and to reduce the effects of wind and wave action within Polander Lake. Their main component is sand dredged from the river's main channel, stored in the Corps' Wild's Bend containment site and pumped over a mile to the construction site. Dolomite rock, quarried from the bluffs near Fountain City, was used to build protective groins to retard erosion during the fifty-year life expectancy of the project. The islands were capped with fine sediment dredged from the surrounding river bottom, then planted with a variety of native trees, shrubs, grasses, and forbes. The construction cost of the project was $1,467,808.

Such projects are not without negative impacts. This seven-acre project displaced *Hexagenia* mayfly nymphs, midge larvae, annelid worms, clams, and other invertebrate animals that are important food for many species of fish and waterfowl. The area was also widely known as paddlefish habitat (U.S. Fish and Wildlife Service).

or enhancing wildlife habitat. Some massive island-building projects seem to have been done primarily to find a place within the floodway to dispose of the Corps' unwanted dredged material, the product of channel maintenance. Some of these projects destroyed valuable fisheries and invertebrate animal habitat. Preoperational hearings were held, and public comments were solicited but usually ignored.

The last three decades of the twentieth Century were contentious. In 1973, the Organization of Petroleum Exporting Countries (OPEC) imposed an oil embargo on the United States; subsequent gas lines and energy shortages precipitated an increased environmental awareness throughout the nation. In reaction to the crisis, "Earth Day" almost became a national holiday; people began to recycle aluminum cans, glass, and paper; cars became smaller and more fuel-efficient (due mainly to foreign competition); home insulation increased; thermostats were turned down; people installed third panes on their windows; outdoor Christmas lighting was discouraged; solar heating and earth homes became fashionable; and folks littered less. "Woodsie Owl" reminded children, "Give a hoot. Don't pollute."

As oil imports increased and environmental enthusiasm began to wane nationwide, new terms like "tree hugger," "environmental activist," and "econut" entered the public discourse. In time, just breathing the words "environmental impact statement," "wetland," "mitigation," "endangered species," or "*Higginsi* clam" made hearts beat faster and blood pressures rise.

The new environmental awareness had been accompanied by new federal and state laws that generated additional federal and state enforcement agencies, which, in turn, were accused of arrogance in their enforcement practices, as well as foot dragging in granting permits to industries and municipalities for construction projects. At Winona, Minnesota, the permitting process for a project to deepen the city's urban lake by dredging and using the dredged sand to build an industrial park inside the city's dike system dragged on for more than eighteen years, during which a new state regulatory agency, the Board of Water and Soil Resources (BOWSR) was created, with authority overlapping and even superseding that of the Corps. Developers complained about harassment that forced them to "jump through increasing numbers of hoops" that were getting smaller and being held higher.

There was continued frustration with regulatory processes, showing a need for reform so that developers could have "one-stop shopping" when seeking necessary permits, rather than going to several agencies.

For most Americans, Earth Day was only a memory by the mid-1990s. Big cars and sport utility vehicles were in vogue with the "Me Generation." Recycling had lost much of its popularity, thermostats were turned up, lights were on all the time, and littering increased. But not all news was bad.

Most states bordering the Upper Mississippi River enacted zoning laws to protect the floodplain, making non-water-dependent developments difficult to locate there. Farm economics dictated that some levee and drainage districts would be sold back to the government for fish and wildlife habitat. The 1986 enactment of the Waterways Trust Fund, whereby new expansion must be cost-shared by the industry, promised to limit navigation enhancement projects. As a result, Mississippi River resources will apparently be increasingly difficult to exploit without providing adequate mitigation (Wiener et al. 1998). However, critics respond that there never has been environmental mitigation provided for Upper Mississippi River navigation projects and that the EMP can't be considered as mitigation.

During the last four years of the twentieth century, the Clinton–Gore administration showed great promise but fell short in the face of a Republican Congress and scandal. William Clinton was supposed to be our environmental president, but during his watch the research arm of the USFWS was transferred to the USGS, increasing the complexity of the bureaucratic labyrinth. Incredible as it seemed to lay people, few of the new USGS employees had any training in geology. The transfer also increased the void that exists between managers who need information and researchers who create it.

By the year 2000, the environmental backlash that had been building slowly for decades among citizens and industries who felt they had been stifled, belittled, and insulted by regulatory agencies showed up in the ballot box.

By the vote of a single Supreme Court justice, George W. Bush and Richard Cheney, both oilmen, became president and vice president of the United States. They proposed to increase offshore oil drilling, especially in the Gulf of Mexico, and to drill in Alaska's Arctic National Wildlife Refuge. They also promised hundreds of new power-generating plants and a revitalization of the nuclear power industry, saying that conservation and windmills alone were not going to solve the nation's energy shortage.

The fact that a political party with a poor environmental record is presently in the White House and controls both houses of the U.S. Congress may not bode well for environmental management on the Mississippi River. Federal and state management agencies are already suffering drastic budget cuts, hiring freezes, and personnel ceilings.

The past half-century has seen dramatic culture changes as a result of Americans' increased affluence and leisure time. Volunteerism is down. The Izaak Walton League is a good example. Once the river's most influential citizens' environmental organization, the "Ikes" are still powerful nationally, but most local chapters have failed, as have many river-oriented sportsmen's clubs.

In 2002, Upper Mississippi River System wildlife refuge managers were tasked to implement the *Refuge Improvement Act of 1997* and the USFWS directive related to the Act titled "Fulfilling the Promise." Prior to the act each national wildlife refuge had received its "marching orders" from the congressional statute or presidential executive order that created the refuge. But under the 1997 Act, the Secretary of the Interior ordered each of the 530 widely dispersed national refuges to develop a comprehensive conservation planning process leading to a conservation plan. Implementation of the Refuge Improvement Act requires managers to make their plans with full public participation. Critics insist that public use of Mississippi River refuges will be severely curtailed.

The Corps continues to loom as an intimidating agency in the river management scene. Its continuing quest to enlarge the navigation system seems only to be fettered by a national shortage of funds.

We have always had jurisdictional problems in the United States with regard to flowing water systems. The tugs and pulls of the 1960s and 1970s resulted in the transfer of more and more jurisdictional power from the federal government to the states, something that probably should have happened many years earlier. Of course there were inexperienced people in some key positions and inefficiencies in the system. But they were breaking new ground in river management and doing so with one of the greatest rivers in the world! From the rather chaotic beginnings has come a more workable management scheme . . . not perfect, but one that seems to be responding to emerging new pressures and one that has greater focus on state and local control.

23

Can This Be the Same River?

St. Louis to the Mouth of the Ohio and Beyond

"Big Muddy": The Missouri River

The river blocked the explorers' path.

Great masses of trees, their branches interlocked, their great roots spiked and jagged, threatened to smash through the bottom of a canoe or capsize a pirogue as they ducked and bobbed in the swirling, roiling yellow waters pouring in from the west.

It was 1673; Jacques Marquette and Louis Joliet were working their way down the Mississippi from the mouth of the Wisconsin and had come to the Missouri River at the height of the June rise.

It was the white man's first recorded encounter with "Old Muddy"—too thick to drink, too thin to plow, with a tendency to crack when it went around corners, or so claimed Mark Twain.

Joining the Mississippi from the west twenty-three miles upstream from St. Louis, "Old Misery" drains the Great Plains and dramatically changes the character of the Mississippi River below St. Louis.

The Missouri drains 74 percent of the Upper Mississippi River Basin and supplies about 40 percent of the long-term discharge below St. Louis. Its drainage area is more than twice that of the Upper Mississippi River above St. Louis, and its suspended sediment load is more than double that of the Upper Mississippi River.

The Missouri River is 2,340 miles long, making it the longest river in the United States. The Missouri River drainage basin encompasses over 528,000 square miles, including about 9,600 square miles in Canada.

Historically, the Missouri contributed vast quantities of sand and silt from the Rocky Mountains and Great Plains. Most of the Missouri's tributaries enter from the west or southwest, flowing through highly erodable, unglaciated soils adding heavy loads of silt to the river.

At St. Louis, the sediment load of the Mississippi has declined 66 percent from pre-1935 levels. Six major dams, built in the early 1950s, impound 765 miles of the river, trapping much of the river's sediment load. Even so, at St. Louis the Mississippi still receives about 80 percent of its suspended sediment load from the Missouri and only about 20 percent from the Upper Mississippi.

Like the Mississippi, the Missouri has been profoundly shaped by the hand of man. In addition to the reaches impounded behind huge dams, 746 miles of the river have been channelized, leaving only 827 miles of "free flowing" stream. But even the free-flowing sections are controlled by reservoir releases from the dams upstream.

St. Louis to Cairo, Illinois

Below St. Louis, the character of the Mississippi suddenly changes. The undammed, unimpounded Middle River is narrower, deeper, swifter, and scarier. With no impoundments to maintain a seasonally uniform high water level, most wing dams are visible at low flow, rising like threatening shoals above the water surface.

Barge traffic is heavier and tows may be three times as large as they are above St. Louis. The wakes of ten-thousand-horsepower towboats accentuate each other, rebounding off shorelines armored with stony riprap and concrete.

When the Missouri River is running high, the Mississippi runs like chocolate, loaded with driftwood—sometimes whole trees. All of this, and the stigma of pollution, precludes most recreational activity. Compared with the northern impounded reaches, recreational boaters are few and far between.

About ninety-nine river miles downstream from St. Louis, the Upper Mississippi River enters Thebes Gap, a seven-mile-long, narrow gorge the Mississippi cut through Shawneetown Ridge about nine thousand years ago. Thebes Gap extends from Gray's Point, Missouri (RM 46), to Commerce, Missouri (RM 39.5), and separates Missouri's Commerce Hills from the highlands of southern Illinois. The gap is named for the town of Thebes, Illinois, where it is less than three thousand feet wide. Today, limestone bluffs as high as 390 feet form the gap's walls. The gap is like the stem of an inverted funnel, with the constricted Mississippi River suddenly widening as it exits the gap and enters the Mississippi Embayment.

The Mississippi Embayment began as a subsiding basin during Paleozoic time, more than 250 million years ago, and continued to receive sediments into the Cretaceous and Tertiary eras. During these millions of years it was a shallow, northward-extending arm of the Gulf of Mexico. If Cape Girardeau, Missouri, had existed then, it could have been a seaport.

After Tertiary time, about five million years ago, the Gulf of Mexico retreated and the ancestral Mississippi and Ohio Rivers removed much of the accumulated sediment but left upland remnants, such as Crowley's Ridge. The Embayment is still an unstable area and lies in the center of the New Madrid earthquake zone (Unklesbay and Vineyard 1992).

The Lower Mississippi

By definition, the Lower Mississippi River begins at Cairo, Illinois, at the confluence of the Ohio and Mississippi Rivers, and runs out into the Gulf of Mexico about ninety miles beyond New Orleans. The Ohio usually runs clearer than the Mississippi, and their waters don't mix readily. The clear Ohio River water hugs the east shore of the Mississippi for several miles in sharp contrast to the silty waters of the Mississippi.

On shore the contrast between the Upper and Lower Rivers is equally sharp. By the time the Mississippi reaches the mouth of the Ohio River at Cairo its floodplain is about fifty miles wide, but the broad Mississippi floodplain has been cut off from the river that nourishes it (Fremling et al. 1989).

Stereotypically, most people envision the countryside flanking the Lower Mississippi River as scenes from *Gone with the Wind*—the broad, lazy river flanked by antebellum mansions, live oaks festooned with Spanish moss, and great fields of cotton. As a youngster, I know that I envisioned happy black folks dancing, with banjoes strumming in the background.

Even where such a scene might exist, from a small boat any view of the countryside is blocked by the combined height of a river bank rising twenty feet or more, topped by a levee that may rise an additional forty feet above the floodplain. To view the intensively agricultural countryside, tourists must first climb the riverbank and then scale the levee, or ride on the upper deck of a big commercial boat. Towboat captains have the best views of all, but even theirs are curtailed in summer by the high humidity that usually makes the air hazy.

Despite the nostalgia associated with it, the Lower Mississippi has historically been unappreciated and feared—a menace in floodtime, a place to get rid of things, the *cloaca maxima* of the United States. Levees have uncoupled the Lower River from its floodplain, rendering it essentially an armored Corps of Engineers drainage ditch, maintained for navigation and flood control. There is no highway running southward along the Lower Mississippi. Few bridges cross the river, and most are closed to pedestrians. The best views of the Lower Mississippi River are from the high loess bluffs at Vicksburg and Natchez, Mississippi.

Through the great delta there are broad sandy reaches, a shifting channel, and a sparsity of launch ramps and marinas.

Historically, the level of environmental awareness along the Lower Mississippi was abysmal, but it increased rapidly in the last three decades of the twentieth century.

The Lower River has been channelized, armored, and shortened 143 miles, but it remains undammed; its natural floodplain has been decreased about 90 percent by levee construction begun at New Orleans in 1727. The navigation channel from Cairo, Illinois, to the Gulf of Mexico has a minimum depth of forty feet and a minimum width of five hundred feet.

The Lower Mississippi Valley is protected by over seventeen hundred miles of high levees that run along both sides of the Mississippi and up the lower reaches of its tributaries. The west bank levee runs in an unbroken line from Venice, Louisiana, northward to Cape Girardeau, Missouri, except where it joins tributary levees or higher ground. On the east side, levees alternate with high bluffs from Pointe a la Hache, Louisiana, to Cairo, Illinois. Below Baton Rouge, Louisiana, where the levees are most susceptible to damage by river currents and wave wash, levee faces are paved with concrete, asphalt, or rock riprap (Fremling et al. 1989).

Within Louisiana, the flood levees that normally protect the valley and its cities are augmented by a complex series of projects that divert Mississippi River floodwaters into the Gulf of Mexico southwestward via the Atchafalaya River or eastward via the Bonnet Carre Floodway and Lake Ponchartrain, thus detouring as much as two-thirds of flood flows around Baton Rouge and New Orleans.

For residents of "The Big Easy," this is a good thing. Because the Mississippi is constricted by levees as it passes New Orleans, its level has been raised, and the entire city lies below the Mississippi's average annual high-water levels. Most of the city of New Orleans lies several feet below sea level, is surrounded by levees and de-watered by a massive pumping system.

For New Orleans, called the "Sinking City" because the land beneath it is subsiding, the situation is becoming ever more precarious. New Orleans, the country's preeminent river city, is threatened by floods, rising sea level, extreme tides caused by hurricanes, and by the threat of the Mississippi changing its course to take a shorter, steeper path to the sea via the Atchafalaya River, the Mississippi's largest distributary.

Where the river merges with the Gulf, Louisiana's coastal marshes are main nursery areas for fish and other wildlife. Ironically, while marshes along the Upper Mississippi are being choked by too much sediment, the loss of sediment to upstream dams and channel structures has led to protective coastal marshes and barrier islands being washed away by waves and tides faster than they are replenished by sediment from the less muddy Mississippi.

24

Epilogue

The River Yesterday, Today, and Tomorrow

Let's Rejoice

In early September 2002, I was awed by a flock of about 150 majestic white pelicans performing a lazy, synchronized aerial ballet, their black-tipped wings spreading up to nine and a half feet as they soared on thermals over downtown Winona. Thirty-five years ago, these huge birds were a rarity on the Upper Mississippi, but today they are common, along with bald eagles, double-crested cormorants, ring-billed gulls, and bird-watchers.

Those who love the Mississippi River have reasons to rejoice at some of the remarkable things that happened in the last three decades of the twentieth century. The Twin Cities separated their sanitary and storm sewers, stopping raw sewage overflow into the river. St. Louis isn't dumping its ground garbage into the river any more. Today, every city along the river has a sewage treatment plant, complete with at least secondary treatment; many have tertiary treatment. The input of toxic industrial chemicals to the river has virtually ceased.

The results of point-source cleanup can be seen everywhere. The river is noticeably cleaner. The condoms, tampons, foam, paper, and slaughterhouse grease that bobbed on its surface and washed up on its shores are gone. Blooms of nuisance algae are less severe. Walleyes, saugers, and smallmouth bass have returned to the Twin Cities corridor after a century-long absence. *Hexagenia* mayflies have repopulated areas where they hadn't been seen for a hundred years. SCUBA surveys during 2000 and 2001 in the eighty-three-mile corridor through the Twin Cities downstream to Red Wing found twenty-seven species of clams (mussels) where virtually none had existed in the 1970s (Kelner 2002). In the 1970s, however, few divers would have even thought of entering the polluted waters of the corridor.

Cities are turning their attention to the river after thumbing (or holding) their noses at it and turning their backs on it for nearly a century. They're revitalizing their waterfronts with new parks, walkways, marinas, gambling casinos, museums, restaurants, shopping malls, and condominiums. People are swimming and water-skiing in areas where it would have been unthinkable thirty years ago. Paddle-wheel cruises are increasingly popular. Bands of volunteers patrol the river banks and islands, picking up recent litter and hauling off the tons of junk dumped there during the last century.

Therein lies the conundrum. The river is cleaner, safer, and more inviting, but we know there are serious problems and that grave decisions must be made regarding the river's future.

The Big Picture

The Upper Mississippi River is a threatened ecosystem because of what humans have done within it. Man's navigation and flood control structures have locked the Mississippi's main channel in place, arresting natural geologic processes. As a result, biological diversity and abundance are declining and natural communities continue to simplify. Aquatic habitats are being lost to sedimentation. The quality and integrity of riverine resources have been degraded, and public trust in technical "experts" has withered, as has trust in the Corps of Engineers.

In many cases, recent changes in the river evoke a mixed response.

For instance, towboats are a vital part of the river scene. What could be more beautiful than a glistening, white towboat churning downriver with fifteen corn-laden barges in tow, illuminated by the late afternoon sun, contrasted against Wisconsin's towering green bluffs and an ominous backdrop of black storm clouds? Childlike, I enjoy waving to the captain and crew from my fishing boat. The throb of the towboat's mighty diesel engines is pleasing to me, perhaps because I've been lulled to sleep by it many times. Towboats make the river unique, but do we need more tows and bigger locks?

The very nature of the Nine-Foot Channel Project placed immediate and long-term limitations on management of the river for non-navigation purposes. However, the Corps of Engineers continues in its effort to extend the river's commercial activity by building larger locks—going so far as to "cook the books" in its cost-benefit studies to justify lock expansion. Despite the Corps' best (worst?) efforts, most scientists and technicians believe that increasing commercial traffic would hasten the death of riverine lakes and diminish the ecological integrity of the river.

Some of the world's most productive soils make up the farmlands of the Upper Mississippi River Basin—the legacy of ancient seas, glaciers, and thousands of years of slow nurturing in diverse prairie environments. For the first 150 years of his tenure on the land, the white man's farming practices devastated the watershed and the river. Cultivation of small grains, wheat, oats, rye, and barley led to horrific soil erosion, soil depletion, and flooding in an era when a farmer was judged by the straightness of the furrows he plowed.

Contour farming and strip-cropping were introduced in the late 1930s, especially in the Unglaciated Area, as measures to control soil erosion and flooding. Dairying and diversified farming dominated the landscape before World War II, and row crop agriculture, mainly corn and soybeans, was limited but expanded greatly after the war.

A dramatic change in the look of the land began when increasing numbers of farmers, recognizing that it was poor business and poor conservation, discontinued grazing woodlands and steep slopes starting in the 1950s. Today, large-scale burning is no longer permitted, except as a prairie restoration tool in some state and federal preserves. The impact of not grazing or burning is most evident on the steep, south-facing slopes characterized by dryness and temperature extremes. Prairie vegetation on most "goat prairies" has been rapidly displaced by dense stands of red cedar (juniper) that are beautiful in the eyes of most people. Most of the red cedar stands are virgin forests, having invaded areas that had been treeless for thousands of years due to climate, grazing, and burning. Unimagined just fifty years ago, the cedar thickets have become dominant features of the landscape, contrasting with the surrounding deciduous forests and providing valuable winter cover for wildlife.

Throughout the Upper Mississippi Basin, the natural flow of prairie into woodland has been replaced by a sharply fragmented landscape, the result of clearing, plowing, grazing, mowing, draining of wetlands, fire exclusion, logging, the introduction of exotics such as fungal tree diseases (including white pine blister rust, chestnut blight, and Dutch elm disease), spreading urbanization, and larger, specialized agricultural operations that practice monoculture of corn or soybeans.

As we enter the twenty-first century, humans are again reshaping the landscape as the small, diversified, family farms of yesteryear are rapidly being replaced by huge corporate farms and feedlots of hogs and cattle. Narrow, but efficient, contour strips are being eliminated or widened to accommodate massive new farm machinery. Furthermore, recent technological advances have made it possible for farmers to have corn and soybean fields that grow up virtually weedless without so much as turning a clod with cultivator or hoe.

The impact of a changing human agriculture has been felt not only on the land but in the river itself. Increasingly, in the last half of the twentieth century, the agricultural soil-nutrient mix washed into the river carried with it a high-tech chemical stew of fertilizers, insecticides, and herbicides. Sediments and pollutants not irretrievably trapped by the impoundments of the Nine-Foot Channel Project were sluiced through the armored, channelized Lower Mississippi into the Gulf of Mexico.

Sedimentation: The Greatest Threat to the Upper Mississippi River

Human development has profoundly altered the Upper Mississippi River and its floodplain. The natural river and its tributaries wandered freely across their floodplains for thousands of years, cutting new channels, creating new backwaters, and rejuvenating themselves by alternately flooding and drying out. New floodplain lakes and ponds were created as old channels were abandoned. Flooding, erosion, and sedimentation were powerful natural processes that shaped and maintained the floodplain and its biotic communities. Humans changed the hydrological regime of the river, altering these processes (Fremling and Claflin 1984; Grubaugh and Anderson 1988; Sparks 1992).

The European American invasion of the Mississippi River Basin caused environmental changes that were analogous to a great climatic change. The pioneers came into a river environment where aggradation had been underway for over ten thousand years. By baring the land and increasing sediment input, they accelerated the rate of aggradation, as did their engineering work on the river itself. The free-flowing river was "hardened" with rock structures that collected sediment and prevented the river from meandering. The slack water, in-channel lakes created by dam building in the 1930s permanently flooded most of the Upper Mississippi's glacial floodplain; sedimentation and eutrophication (nutrient enrichment) began as soon as the dams were closed. The Nine-Foot Channel dams of the 1930s sealed the fate of the river for the foreseeable future.

Scientists now know that rivers in their natural state have the ability to traffic energy and nutrients through their biological systems at somewhat predictable rates and routes. Sediments are periodically flushed and rearranged during periods of high water, particularly when the floodplain is submerged. During periods of low flow, sediments are regularly exposed to the atmosphere for prolonged periods of time, allowing the organic portion of the sediment to decompose in the presence of oxygen, turning into water and carbon dioxide, and effectively disappearing. This decreases the volume of the sediment and reduces its oxygen demand. Untamed rivers usually reach an equilibrium with regard to sediment input from their tributary streams and sediment discharge at their mouths (Fremling and Claflin 1984).

But when rivers like the Upper Mississippi are dammed, the processes that normally occur in flowing water environments are changed to reflect those that occur in lake environments.

While rivers are virtually immortal, lakes are mortal. Lakes are born, then pass through the stages of youth, middle age, old age, senescence, and death as they inexorably fill with sediments or the products of enrichment. Lakes within agricultural and other fertile watersheds tend to age faster. They live fast and die young. Like lakes, rivers tend to be as fertile as their watersheds.

Eutrophication, as this enhanced aging process is known, is rampant in the impoundments of the Upper Mississippi River. Organic materials and nutrients accumulate in excessive quantities, particularly in nonchannel areas where flushing of sediments during flood stages is reduced. Water levels become stabilized, and low flow conditions occur less frequently. This retards the rate of oxidation of sediments rich in organic matter.

As sediments accumulate in the downstream reaches of an impoundment (pool), they reduce the pool's slope and cause meandering processes to either be slowed or stopped altogether. Dam closure results in increased sediment-trapping efficiency and corresponding increases in rates of eutrophication. Increased cycling of nutrients within the reservoirs results in the conversion of lake-type habitats to marsh-type wetlands, and ultimately to floodplain forest.

Any obstruction in the river that lessens its ability to transport sediments will promote sediment deposition and aggradation. The wing dams and closing dams of the Four-and-One-Half-Foot and Six-Foot Channel Projects inhibited meandering of the Upper Mississippi and increased the rate of aggradation of its floodplain. The rate was further increased by the big Nine-Foot Channel dams.

Following the closure of a dam, sedimentation begins. Usually, it will continue until the sediment level throughout most of the pool reaches the crest of the dam's spillway. The river bed will be raised upstream to the point at which the water surface of the pool intersects the original bed. It may extend even farther upstream if the river is slowed during floodtime by floodplain forests. During extreme floods, sand bars five feet high or higher may be deposited within floodplain forests a half-mile or more from the main river channel. Water storage capacity is decreasing in all pools due to sedimentation. Because the slack water navigation pools are excellent sediment traps, their useful life is severely curtailed.

When the Nine-Foot Channel impoundments were created, they also impounded the lower reaches of tributaries. The Mississippi's hydraulic damming action reduced tributary gradients, causing their beds to be raised. Reduced current velocity resulted in deposition of sediments, causing formation of deltas and new, productive wetlands and floodplain forests in the lower reaches of many tributaries.

Today, most tributaries, especially Wisconsin's Chippewa River, flow through extensive deposits of glacial sand and gravel, poised and ready to wash into the Mississippi. With the notable exception of the Illinois River, most of the Mississippi's tributaries have steeper gradients than the master stream, and they deliver sediments faster than the Mississippi can remove them, thus speeding the aggradation that began at the end of the last Ice Age about ten thousand years ago.

Most of the tributaries are still choked with sand and silt that accumulated in their valleys during the era of maximum soil erosion from 1880 until about 1950. Detrimental effects of sedimentation in the Upper Mississippi River were recognized as early as the 1930s; soil conservation practices improved, but wetland drainage and stream channelization increased.

Soil erosion and sediment transport are inevitable, but we have sped up the process by removing the vegetative cover of the land, dislodging soil with our livestock and machinery, and increasing the rate of water runoff through wetland drainage and urbanization. Likewise, we have temporarily (in a geological time frame) delayed sediment transport to the Gulf of Mexico with our reservoirs and other sediment-capturing devices like highways, railroads, and levees. All of man's river "improvement" projects such as dams, roadways, and bridges accelerate the rate of aggradation because they increase the roughness of the floodplain. Because of all these factors, virtually the entire valley floor has been blanketed by sediments from the postsettlement period.

Casual observers seldom notice that sedimentation of the Upper Mississippi's shallow reservoirs is a deviation-accelerating mechanism. The rate of filling may be imperceptible during the early years of impoundment, but subsequent sediment deposition speeds the accretion of additional sediment. Once a sediment deposit rises to the water surface, things happen fast. Willows, cottonwoods, and silver maples colonize it, creating a thicket that slows the velocity of the river during floodtime, causing more sediment to drop out, creating additional footholds for developing vegetation that will trap more sediment—and so on with increasing rapidity.

Sedimentation is among the most critical ecological problems of the Upper Mississippi River, but accurate predictions are hard to make because sedimentation studies are complex, expensive, and are usually limited to relatively small sample areas. The writing, however, is on the wall. Anecdotally, longtime river rats have already seen many of their prime fishing and hunting areas degrade or disappear since 1940. Today, numerous channels that accommodated houseboats in the 1960s can scarcely handle small fishing boats. More hunters and trappers are using airboats as well as Louisiana-style "digger boats," which can run in extremely shallow water and even tear open new channels through soft sediments and semiterrestrial habitats, creating new problems and speeding the demise of marshes.

A different set of problems arises in windswept river reaches, usually in mid- or lower pool areas. Waves and currents redistribute sediments, eroding shallow areas and filling deeper areas, thus simplifying bottom topography. As islands erode and disappear, the wind has a longer fetch, causing waves to build, resuspending soft sediments, increasing turbidity, decreasing light penetration, and limiting aquatic plant growth. When redeposited, loose bottom sediments provide inadequate footing for rooted aquatic plants that may be torn up by wave action or ripped out by ice in the spring. As a result, in most pools the general trend is away from a rich mosaic of habitats and toward monotony.

Recent experience teaches that the Mississippi River and the estuarine areas at its mouth cannot be studied and managed as discrete biological, geological, and political units. The Mississippi system must be recognized, treated, and appreciated as an integrated, interdependent national resource.

When this fact is ignored, human activities and geologic processes combine to cause massive problems.

For example, while the Upper Mississippi is plagued by sediments, the Louisiana Delta is starved for them. The Upper Mississippi Basin's historic sediment contribution to the Lower Mississippi River has been reduced by improved farming practices in the last half century, but especially by dam construction on most major tributaries—especially on the Missouri River. Since the early 1960s, over half of Big Muddy's sediment load has been trapped by six huge dams.

On the Lower Mississippi we have uncoupled the floodplain from the river with levees that extend about ninety miles beyond New Orleans, thus aggravating the problem. What sediments that are carried as far as the delta are conducted out into the Gulf of Mexico, over the continental shelf, and into the depths.

The loss of sediment trapped on the Upper River and channelization on the Lower River has resulted in Louisiana's coastal wetlands and barrier islands being lost to the sea at the rate of up to thirty-seven square miles per year, with over fifteen hundred square miles lost in the last sixty years. "Even vacating the issues regarding bureaucracy and overlapping jurisdictions, it is imperative that someone or some body of individuals looks at river management from a total basin perspective" (Thomas Claflin, personal communiqué September 2002).

The Corn-Soybean Dilemma

The Upper Mississippi River Basin is the most productive agricultural region in the world, a fact that has resulted in multiple disruptions of the river environment.

In modern agriculture, forage production—growing alfalfa and grasses to feed livestock, primarily dairy cows—most closely mimics the natural prairie environment. Producing hay is much better than growing corn and soybeans for preventing soil erosion, but across the Midwest hay lands are being replaced by fields of corn and soybeans as increasing numbers of dairy farmers go out of business. In eleven southeast Minnesota counties corn and soybean acreage increased from 87 percent of tilled acres in 1975 to 96 percent in 1999.

Unfortunately for the environment, corn–soybean rotation methods produce more soil erosion, allow faster water runoff, and send more nitrogen to surface and groundwater than other forms of agriculture.

In the two-crop rotation system, soybean residues decompose too quickly, leaving the soil vulnerable to erosion, yet in much of the rolling blufflands adjacent to the Mississippi River, 160-acre fields of "Roundup-ready" soybeans blanket erosion-prone slopes—without a single weed or stalk of volunteer corn in sight. After the harvest, those fields will be disasters with the first torrential rain.

Today, nitrogen pollution of the Mississippi River is directly linked to the basin's nearly universal corn–soybean rotation and the nation's highest application of nitrogen fertilizers. A northeast Iowa farmer recently won a yield contest with a harvest of 397 bushels of corn per acre, but he used nearly four hundred pounds of nitrogen fertilizer per acre to achieve the impressive yield.

In pre-agricultural times, naturally-occurring nitrate was denitrified in wetlands and ponds, or taken up by prairie vegetation, but post–World War II farming practices created a cycle of "chemical dependency" that will not, in the long term, be sustainable.

Before 1940, farmers relied on natural rotations of crops and animal manures to build up soil nutrients, but the advent of chemical fertilizers allowed farmers to abandon rotations in favor of monocultures—growing only one crop for long periods of time.

Farms, by necessity, grew larger because chemicals were expensive. Larger farms required larger machinery, which compacted the soil, adding to its degradation, and making it less able to retain water and nutrients. The soil (and its chemical nutrients) washed away, ultimately contaminating the Mississippi River.

Most farmers are dedicated to environmental health; they would initiate conservation programs on their land if there were rewards to offset the additional costs and risks they must assume. But the larger the farms grew, the more corn and soybeans were produced. This lowered profit margins, and farms had to be bigger to show a profit

(Crosbie 2001). To show that profit, farmers became ever more dependent on federal subsidies.

From an environmental standpoint, farm subsidy programs work like a slow poison, discouraging diversification of agriculture. Farmers "farming the government," have little choice but to plant more and more of their acres to "program crops" (more corn and soybeans) and wait for crop subsidy checks.

Corn and soybean farming is becoming less sustainable economically, environmentally, ecologically, and sociologically, but it has been subsidized and sustained by the federal government. Federal loan deficiency payments are made to eligible farmers when posted prices for corn and beans are lower than government established loan rates. This support encourages planting row crops in every nook and cranny of fields because payments are based on the number of bushels produced.

The bulk of subsidy payments goes to the biggest farmers, speeding the demise of the family farm and devastating the watershed.

Today, as farms continue to expand to make raising corn and soybeans profitable, fewer farms and farm families occupy the rural landscape. This, in turn, means student numbers in schools dwindle, church membership shrinks, and small towns wither. Large producers often bypass local businesses to purchase supplies and equipment from larger regional outlets, where prices may be lower due to bigger inventory. As more production contracts are developed between agribusiness and the farmer, the farmer gradually assumes the role of custom operator or indentured servant—and loses the freedom to manage.

To add insult to injury, this monstrous disruption of both the natural and human environments has failed to yield an economic reward. Real net farm cash income is at its lowest since the Great Depression of the 1930s, causing the current farm recession that began in 1998. Furthermore, if corn-soybean farms were pushed to their upper yield levels, not only would there likely be a grain glut but prices would plunge. Massive export increases would be needed to handle the surplus—increasing the pressure to enlarge the Mississippi River System lock and dam system.

Ultimately, we need a slow greening of the landscape with crops other than corn and soybeans.

We Need a Reappraisal of Agricultural Priorities in the Upper Mississippi River Basin

Corn and soybeans require a well-drained warm soil for optimum growth. The basin's till and loess soils are highly productive but have poor internal drainage. As a result, millions of acres are now drained through perforated plastic pipe laid beneath the seasonal water table. Beginning in the late 1880s, federal and state governments paid for wetland drainage to facilitate crop production. By 1913, about six million acres of Minnesota's original wetlands had been drained. Today, about ten million acres have been drained.

Unfortunately for the river environment, tile drainage short-circuits the natural drainage pattern, flushing nitrate out of the soil before it is denitrified or leached to the water table. Collected in small, open ditches, the nitrate-enriched water runs to larger ditches, which lead to tributaries and ultimately to the Mississippi and the Gulf of Mexico, where it is largely responsible for creating the anoxic zone ("dead zone") that has grown larger than the state of New Jersey.

The River Is a Reflection of Its Evolving Agricultural, Urbanized Watershed

Drainage from farm fields will surely be a major environmental issue in the twenty-first century. The single biggest problem facing the Mississippi and other basin rivers is that we don't have a well-designed strategy to combat polluted runoff. The Minnesota River, tributary to the Mississippi, is a good example. To be sure, some of the soil and pollutants that foul the Minnesota come from urban sources: faulty sewage-treatment plants, construction sites, golf courses, parking lots, city streets, and lawns. But, the Minnesota's basin is 92 percent agricultural, and the quality of its river water depends overwhelmingly on agricultural practices that load it with manure, pesticides, and fertilizers that then endanger the river and those who depend on it. The average annual flow of the Minnesota has more than doubled during the past thirty years and is higher now than at any time in the past century. Some of the increased flow has been due to above average rainfall during the past three decades, but much of it has been nitrate-laden seepage, drained off tiled croplands.

Furthermore, the problems caused by urban runoff are beginning to be addressed. As stipulated by the 1986 Clean Water Act and its amendments, American cities with a population of at least ten thousand must monitor their storm water for contaminants—including nitrogen and phosphorus. Storm water that exceeds standards must be treated, usually by ponding, so that aquatic plants can absorb the fertilizer elements. Such monitoring and engineering are straightforward because storm sewers are point sources of pollution. But meanwhile, fertilizer and herbicide runoff from farm fields is virtually uncontrolled because the fields are nonpoint sources that can't be handily monitored.

Does the River Really Need Bigger Locks?

Most of the locks and dams on the Upper Mississippi River have surpassed their fifty-year-design life. Their life spans have been prolonged by continuous maintenance and recent rehabilitation, but it seems inevitable that sooner or later they will have to be completely overhauled or replaced.

Proponents of the towing industry maintain that an efficient national transportation system must include well-maintained and improved waterways, lifelines to the world and to the people who need our products and ship us theirs. Barge lines maintain that improved locks will mean greater profits for Midwest farmers.

The lock and dam system was designed to handle tows up to six hundred feet long. Today's fifteen-barge tows are twelve hundred feet long and must be split in two to navigate most locks, a process that can take from an hour and a half to over two hours, compared to thirty minutes for a single lockage. The short locks are bottlenecks, causing traffic jams that create long delays, especially at Locks 20 through 25, among the busiest in the system.

And time is money. The average seven-hour delay at Lock 22 adds at least $2,800 to the shipping cost of an average tow. With river traffic increasing, delays will surely get longer. The obvious solution is to make the locks bigger, but the cost would be extraordinary. For example, the

Agricultural drainage systems not only aggravate pollution problems in times of ample rainfall, they intensify the impact of a series of dry years.

Because water is being diverted by tile lines and ditches, it is not seeping deeper into the earth to recharge the groundwater supplies that provide a slow, but steady, year-round flow into the Mississippi and its tributaries. If a disastrous drought like the one that caused the Dust Bowl of the 1930s should hit, the only water in some tributaries may be what is coming out of sewage plants, septic systems, and big feed lots.

Should We Expand the Lock and Dam System?

Commodity groups and the barge industry have endorsed the Upper Mississippi River Navigation Project (UMRNP) and its expansion of the locks and dams on the Mississippi and Illinois Rivers, projects

---

reconstruction and expansion of Lock and Dam 26 near St. Louis cost over one billion dollars. Increased commercial traffic would have serious environmental consequences.

The Corp's "Upper Mississippi River–Illinois Waterway Navigation Study" that had begun in 1993 was scrapped in 2000 after charges were substantiated that the Corps had fabricated economic justification for lock expansion. A revised plan was offered in 2004.

Specifically, the proposal called for replacing six-hundred-foot locks with twelve-hundred-foot locks at Locks 20, 21, 22, 24, and 25 on the Mississippi River and the La Grange and Peoria locks on the Illinois River. Locks 19, 26, and 27 are already twelve hundred feet long (and there is no Lock 23). Thus, fifteen-barge tows could operate, without double lockages, on the Upper Mississippi as far upstream as Lock and Dam 18, just north of Burlington, Iowa (McLeod 2004).

If funded, environmental restoration would employ specific restoration tools including creating fish passages at dams, building islands, dredging backwaters and side channels, employing drawdowns to stimulate growth of aquatic plants, and reconnecting the river to its floodplain. It should be noted that the present federal Environmental Management Program received sixteen million dollars in 2003 and that it has never been funded at the maximum annual level of thirty million dollars authorized by Congress (Balcom 2004).

that would cost about one billion dollars, half of which would be funded by taxpayers. Advocates contend that increased barge capacity will lower transportation costs, translating into better crop prices for farmers and economic benefits to the Upper Midwest economy through export of more commodities down the Mississippi, especially corn and soybeans. They argue that water transportation is the most environmentally friendly, energy-efficient mode of agricultural transportation.

This may have been true in the past, but recent evidence shows that railroads have become increasingly efficient, equaling and perhaps even surpassing the efficiency of barge transportation. Trucks, the least efficient haulers, serve as feeders for both water and rail transportation. Thus, truck distance traveled may actually be the determiner of which is more energy efficient, truck–rail or truck–barge. Deregulation of the rail and truck industries has promoted research and development to improve their efficiencies, but the barge industry hasn't kept pace.

Some suggest that instead of barging grain downstream on the Upper Mississippi, those cargoes could be sent southward by rail, bypassing locks, to St. Louis and offloaded to tows on the Lower Mississippi River where water transport is very cost effective.

Critics of river transportation maintain that the barge lines have developed towboats so powerful that they tow more barges than can fit through existing six-hundred-foot locks. This sounds like an ominous new development, but it isn't. There have been diesel-driven towboats and fifteen-barge tows on the Upper Mississippi River for over a half century, but the towing industry's technology hasn't changed appreciably in that time. Also, corn and soybeans that have traditionally been barged down the river and shipped to Europe are increasingly bound for Asia, making rail transportation to West Coast ports a more energy-efficient option (Institute for Agriculture and Trade Policy 2001).

Nationally, the barge industry is the most subsidized form of transport. The taxes it pays on diesel fuel covers only about 10 percent of the $647 million that the Corps spends annually building, operating, and maintaining locks, dams, and navigation channels. The rest is paid by taxpayer dollars. Critics charge that it would be a waste of money to expand the infrastructure on a waterway system that is becoming obsolete and cost inefficient (Institute for Agriculture and Trade Policy 2001). "The end result is the expenditure of massive amounts of dollars to support a very small sector of the national transportation system at a very high cost to the taxpayer. Yet at the same time, these fixed costs are never utilized in the benefit: cost calculations to describe the true cost to the public of the use of the resource. For this reason, waterway transport services are in demand by shippers simply because the rates do not reflect the actual costs of the use of the river. To imagine that barge companies pay a significant percentage of the cost of maintaining the waterway is sheer nonsense" (Thomas Claflin, personal communiqué September 2002).

The Upper Mississippi River–Illinois Waterway System Navigation Study

Shippers have believed for years that the Midwest was losing export markets because the locks are half the size they need to be. With the intent of quantifying that belief, the Corps began an eight-year, sixty million dollar study in 1994 to evaluate the impacts of building a series of new, larger locks on the Upper Mississippi and Illinois Waterway Systems.

Preliminary analysis indicated that navigation improvements may be justified for Mississippi River Locks and Dams 11 through 25 between the years 2000 and 2050, but with increased costs to maintain a navigable channel for larger, more abundant tows. A scandal ensued when a senior Corps economist disclosed that top Corps commanders had "cooked the books" in an effort to justify the $1.5 billion project. The study was halted briefly but was begun again in December 2001, and was expected to proceed for three or four more years. Critics charged that the books wouldn't be cooked again because "there wouldn't be any books."

The resumed study was restructured to incorporate recommendations made by the National Academy of Sciences. It included a new feature called "scenario-based analysis" that does not attempt to make a single forecast of the river system's economic and environmental needs. Instead, it defines a variety of scenarios and evaluates the benefits of economic and environmental alternatives under each one. If an alternative shows up in several different scenarios, it might be concluded that it would be a good thing to do.

The five scenarios cover a broad range of possible trends including variations in global trade, land utilization, demand for agricultural products, crop yields, trade barriers, willingness of China and India to become food importers, acceptance of genetically engineered crops, impacts of climate change on yields, domestic consumption of ethanol-based fuels, the shift in developing countries toward a meat-based diet, and population growth. Each scenario translates into a level of demand for U.S. exports and associated barge traffic, and possible recommendations for navigation improvements and environmental enhancements.

All of this is subject to the capriciousness of world politics. We older folks remember the grain embargo on the Soviet Union during the Cold War when Midwest grain elevators were filled to capacity, and surplus grain was stored in hundreds of barges moored along the Mississippi and Illinois Rivers.

Environmentalist critics of the Corps approach insist river management schemes have a much broader base than the economic advantage of the moment. They point to the extensive damage caused by floods in the last half century to highlight the fact that the Corps has continued to manage the Mississippi in a manner that sacrifices the multiple purposes of river management—including flood control, environmental management, and recreation—to the sole purpose of commercial navigation.

Great floods, like the disastrous one of 1993, would have been devastating to human development on the historic floodplain under any river management scheme, but they have been made worse by drainage of wetlands, channelization of the river, and the proliferation of river-constricting levees. Impoundment of the river has increased sedimentation and elevation of the floodplain and channel beds, further exacerbating flooding.

High levels of erosion and siltation, nitrogen pollution of surface and groundwaters, hypoxia in the Gulf of Mexico, diminished biodiversity, and global climate change are symptoms that indicate present systems are out of control. Expanding locks and dams and increasing feed grain production would makes these local, regional, and global problems worse (Keeney and Muller 2000).

The most severe critics insist that the Corps is maintaining an environmentally destructive, hydrologically insane, agriculturally nonsensical position on the river, supporting an antiquated form of transportation. They ask, "What national system of any kind can you put out of commission three months each year and during big floods?" They maintain that the Corps straightened the river and placed levees not so much to protect farmland as to keep the river in place for navigation purposes.

Environmentalists oppose any new authorizations until certain changes are implemented within the Corps. Reform acts, now pending, would, among other things, require that all Corps project recommendations be subjected to independent peer review before being authorized. The bills would also ensure full mitigation for adverse environmental impacts of a project and secure 50 percent of the mitigation before start of construction (Werner 2002).

But one must remember that environmentalists tend to think tactically, while the Corps thinks strategically. Environmentalists may win battles, but the Corps wins wars.

The Corps' job is to earn economic benefits for American taxpayers. It is the world's premier civil engineering organization. Depending on the congressional mandates that direct it, the Corps can expertly and thoroughly do wonderful or terrible things.

Farmers Are Caught in the Middle

In its analysis of possible economic outcomes from the proposed Upper Mississippi River Navigation Project (UMRNP), the Institute for Agriculture and Trade Policy determined that, as a way to help farmers, the project is fundamentally flawed (Levins, Rice, and Sawin 2000). Because of their weak bargaining position with grain handlers, input suppliers, and non-farm landlords, the best farmers can hope for are small, short-lived improvements in net income per acre.

The UMRNP attempts to address the symptom of low farm prices but does not deal with its cause, mainly ever-expanding yields and increasing power in the hands of non-farm corporations. The river project subsidizes corn, not farmers. Its benefits will go mainly to the corporations that transport and sell grain on world markets. If there are increased profits to farmers through higher prices for their grain, they will be passed on to input suppliers and landlords. The only farmers to benefit will be those that sell the most corn, continuing the long-term trend toward fewer, larger farms.

Exports of U.S. Grains Are on the Decline

In their Upper Mississippi River–Illinois Waterway Navigation System Feasibility Study, the Corps forecasted rapidly rising grain exports at the same time that corn exports have been in a twenty-year downtrend, and wheat exports have fallen dramatically (Baumel 2001). As late as 1994 and 1995, China imported substantial amounts of corn, but in 1996 China exported 160 million bushels of corn, rising to four hundred million bushels in 1999. Presently, China is one of the world's largest corn exporters. In recent years, Taiwan, South Korea, and Japan have accounted for over half of U.S. corn exports, feeding the corn mainly to hogs and poultry.

The United States enjoyed its peak grain exports in 1980. Since then, world grain supplies continue to increase and U.S. grain exports decline. Reasons for decreasing grain exports include: the rapid shift, in recent years, toward exporting grains in the forms of meat, poultry, eggs, and special human-food-quality grains; increases in the value of the dollar in the mid- to late 1990s; inability of Taiwan, South Korea, and Japan to dispose of large amounts of animal waste on limited land areas; and incentive programs within the three countries to become self-sufficient in food production. The severe recent decline in corn and soybean exports to the European Union countries may be due, in part, to European resistance to genetically modified grains, commonly called "Frankenfoods" (Baumel 2001).

There is a trend in many countries for consumers to increasingly switch from grain-fed beef (which require seven or eight pounds of grain to make a pound of beef) to farm-produced seafood and poultry (which require two pounds or less of a grain-animal mix per pound of gain). The trend toward consumers preferring grass-fed beef and mutton is also important (Baumel 2001). Because geese mainly consume grass (and also control weeds), they require even less grain than chickens and may be increasingly important as a replacement for chickens and hogs in China.

Meanwhile, grain production has increased sharply in countries like Brazil and Bolivia where costs of production, including land and labor, are lower than in the United States. In Brazil, soybean production has been sharply upward since 1992 (Baumel 2001). Is it possible that someday Brazilian soybeans could be barged upstream on the Mississippi?

Would an Enhanced Lock and Dam System Enable Us to Feed the World?

One of the arguments for expanding the lock and dam system is that it would enable grain-belt farmers to feed a hungry world—implying that it is the poor of the world, those in most danger of starving, who are being fed. On the contrary, the vast majority of grains shipped on the Mississippi are corn and soybeans, shipped to those who can best afford them, not to those in most need. For example, for every ton of corn exported in 1996 to a poor country with serious malnutrition problems, as designated by the United Nation's Food and Agriculture Organization (FAO), 260 tons were exported to wealthy countries within the Organization for Economic Cooperation and Development (OECD). No soybeans were exported to the poorest countries in 1996, but 17.8 million metric tons went to OECD countries including Japan and the entire European Union, which, like the United States, have well-developed livestock economies. In this country, 76 percent of corn goes for animal feed. Other wealthy countries also use corn more for feeding livestock than for people. In the United States more corn is used to make alcoholic beverages than is exported to feed the world's twenty-five most undernourished countries combined (Muller and Levins 1999).

Soybeans are not often consumed as whole beans but are crushed to yield two products. One is high-quality protein meal for livestock, the other is vegetable oil. The United States exports about a third of its soybeans and crushes the rest. As with corn, exported soybeans go to the richest countries, not to the poorest countries who most need them. The river project will do virtually nothing to alleviate world hunger (Muller and Levins 1999).

To see the ultimate dysfunctionality of an export-based farm economy one need look no farther than Iowa—the country's quintessential farm state. Because corn and soybeans (most of which are not consumed directly by humans) make up 95 percent of Iowa's crops, Iowa imports more than 80 percent of its food (Crosbie 2001). The state that "feeds the world" can't feed itself.

Fred Kirschenmann, director of the Leopold Center for Sustainable Agriculture at Iowa State University says, "Perhaps it is time to reexamine the Native American vision of feeding the village first. Certainly farmers must know by now that making exports their top marketing priority for the last three decades has not served them well" (Crosbie 2001, 16).

What Can Be Done?

Our present agricultural system is dependent on a grain economy with low economic returns and high nutrient and sediment runoff. Moreover, it has moved toward the industrial model, increasingly decoupling the farmer from the land. Farmers share the global agricultural economy with powerful, transnational grain giants, suppliers, and non-farm landlords. Individual farmers are squeezed between powerful agribusiness interests. On one side, they sell their grain to a few very large buyers like Cargill and Archer Daniels Midland. On the other side, they buy supplies and equipment from huge corporations like DuPont, Monsanto, and John Deere.

There are alternatives that will give us lower production costs, lower environmental impacts, and higher farm incomes without extensive government intervention. Replacing or reducing the need for nitrogen in row crops, mainly corn and soybeans, is a key component of policies that would address these problems.

Land stewardship suffers when farms grow too large and the operations' owners are not its farmers. We need farm families living in local ecosystems with the knowledge of those local ecologies passed from one generation to the next (Crosbie 2001).

Novel and rediscovered agricultural practices could have beneficial impacts on the entire basin, the Mississippi, and the Gulf of Mexico, but the main beneficiaries would be rural communities and smaller family farmers (Keeney and Muller 2000).

Traditionally, farmers plowed after harvest each fall with moldboard plows that sliced deep into the soil, turning the soil over, and burying corn stalks and other crop residue. This allowed the black soil to warm up and dry out faster in the spring, but it left it exposed to the ravages of winter winds, snowmelt, and spring rains. This accelerated erosion and increased the input of sediments, nutrients, and pollutants to the Mississippi and its tributaries. To their credit, more farmers are now helping our rivers by leaving crop residues on their fields after harvest. Also, relatively new techniques like ridge tilling, strip tilling, and chisel plowing leave the soil less vulnerable to erosion without sacrificing yields.

We must expand upon these techniques and move to employ agricultural practices that can sustain the world's demands today without compromising its needs tomorrow. Sustainable practices must be environmentally sound, economically viable, and socially responsive (Crosbie 2001).

Growing numbers of farmers are changing tactics, believing passionately that not only is it more environmentally successful to grow crops organically (without use of pesticides, herbicides, or chemical fertilizers), it is also more successful economically.

Iowa farmers have increased organic crops sixfold since 1996. Nationally, the organic industry has grown 900 percent from 1995 and continues to grow 20 percent annually. The main reason farmers don't farm organically is a lack of information and support. Since a farmer using manure to fertilize his field isn't turning a dollar for Archer Daniels, and achieving pest control through crop rotation provides slim profit for Monsanto, there is little corporate support for organic agriculture. Furthermore, industrial agriculture has been a heavy influence in schools of agriculture at major universities. As a result, nationally, university extension programs have been cited as an impediment to organic production (Crosbie 2001).

Nevertheless, organic agriculture is a technically sophisticated endeavor, using a variety of techniques rooted in both tradition and technology to achieve levels of productivity that compare favorably with their petrochemical-based counterparts. For example:

- Using ridge tillage in their Iowa beef–hog operation, a farm family rotates plantings of corn and soybeans with oats, eliminating high-cost chemical inputs. Their livestock manure and biosolids from a nearby urban sewage treatment plant provide all of the fertilizer for the farm (Crosbie 2001).

- Using an age-old practice that was finely honed during the 1930s soil conservation days, but quickly abandoned when nitrogen fertilizers came on the scene, nitrogen-fixing legumes could be used in place of fertilizers. Rotations including several years of legumes would lessen weed pressures, break insect and soil-borne pathogen cycles, and provide diverse agriculture more suited to smaller land holdings and family farms. Other benefits would include less soil erosion and lowered atmospheric carbon dioxide levels due to carbon sequestering in soils. Atmospheric nitrous oxide, another important greenhouse gas, is derived from the oxidation of ammonium. Its concentrations would be reduced, further improving the world's climate-change issues (Keeney and Muller 2000).
- Soybeans are nitrogen-fixing legumes, but modern high-yield varieties use more soil nitrogen than they fix. Soybeans would be included in the rotation, just as now, probably following two years of corn preceded by three to four years of a legume such as alfalfa. Other crops would include sunflower and canola.

  Rates of land application of swine, dairy, and beef manures would be based on phosphorus needs, although it may cause problems for feed lot operations that now dispose of manure on legumes (which don't need it) as well as on corn. Operations that do not have enough land available for manure disposal would be required to have secondary treatment for it, including nitrogen removal. Methane gas generated during treatment could be sold or used for power, lessening methane's impacts on global warming. Swine could be pastured or raised in hoop houses to provide an environmentally friendly alternative to large indoor operations.

  Instead of being solely dedicated to corn and soybeans, erodable lands could be converted to perennial energy crops like switchgrass in tandem with locally owned and operated ethanol and electricity plants. The grassland system would have to be integrated with nitrogen-fixing legumes to avoid the use of nitrogen fertilizers (Keeney and Muller 2000).
- Some farmers are converting corn acres to Indiangrass and big bluestem, historic prairie species. Seeds are harvested from the tops of the grasses, which may be as high as seven feet, and sold to prairie restoration companies for about ten dollars per pound. The durable grass stems remain year round, holding the soil in place. They survive the weight of heavy winter snows, providing winter cover and nesting cover for birds (Herwig 2002).

Others are converting farmland to ranches where native grasses are grazed by buffalo harvested during paid hunts. Packaged buffalo meat, hides, skulls, and related products are sold at the ranch store located near the ranch house that offers lodging and meals.

Should the Big Dams Be Removed?

The fate of the Upper Mississippi River was sealed for the foreseeable future when the Nine-Foot Channel dams were closed, but we should not cast blame for the action since it was done in a time very different from now. The Nine-Foot Channel Project would never have been started in today's more ecologically aware conditions. Just writing the massive, complex environmental impact statements for such a project would be prohibitive. Not many baby boomers or Generation-Xers have the first-hand knowledge or view of recent history to understand fully the extent of the poverty of those Depression years.

The exciting Nine-Foot Channel Project promised immediate employment, future jobs, and a stimulus to the regional economy. It did all of these things and created an ecological "boom time" that has been relatively short lived. The Nine-Foot Channel Project and other public works projects certainly helped alleviate some of the problems of the Depression era. Where do we go from here?

The argument over dam abandonment boils down to pitting the interests of boaters, swimmers, water-skiers, and most fishermen against those of a relative few environmental extremists who want the river returned to its "natural state" even if they don't understand what that means. Once created, big dams are difficult to abandon.

Environmentalists are dreaming if they think that if the dams were removed the Mississippi would revert back to its natural, 1930s course in their lifetimes. The river's basin isn't the same basin that it was before the dams were put in—and going back is not a viable option.

The Mississippi and its tributaries have had over sixty years to adapt to the navigation pools—and to store sediments. Removing the big dams would cause the river to scour its bed deeper. In turn, tributaries that have stored massive amounts of sediments in their lower reaches would respond by cutting their beds commensurately, belching their stored sediments into the Mississippi, speeding the rise of the floodplain, terminating most recreational boating, and causing other problems that cannot even be imagined.

If the dam gates were rolled up permanently and the sills of spillways were lowered, it would be possible during low water years to wade across the Mississippi in many areas. Most marinas and boat launches would be left high and dry. Big (and even moderate-sized) pleasure boats would be eliminated from the river, dealing an extreme economic hardship to river cities. Boat dealerships, marinas, repair shops, sports shops, bait stores, gas stations, motels, restaurants, liquor stores, bars, and many other river-related businesses would be severely crippled. Many would go belly-up. Adverse ripple effects would be felt over a wide area.

Why Don't We Know More about Riverine Ecosystems?

Of Earth's water resources, only the oceans and a few other large rivers exceed the Mississippi in terms of scale and complex ecology. Compared to the nation's lakes, the Mississippi is poorly understood, mainly because of the logistical difficulties of working on it. The equipment and effort required to study just the river's diverse fish populations is many times greater than in a small stream or lake. Presently there are insufficient data and understanding to support major policy decisions being made in the name of river restoration and conservation.

River resource managers, scientists, and environmental advocates have grown increasingly frustrated with the inability of existing institutions and government programs to effectively address the decline in the quality and diversity of the Upper Mississippi River ecosystem. The Corps of Engineers (or the U.S. Fish and Wildlife Service, or the U.S. Environmental Protection Agency) could be authorized by Congress to develop an ecosystem management plan for the Upper Mississippi, but it would be necessary to clarify for congressional committees why the Corps' present Environmental Management Program authorization, which is generally viewed as a fairly comprehensive environmental program, is insufficient for purposes of ecosystem planning and management. Creation of a new authority could lead to redundancy, increased bureaucratic complexity, new turf wars, and increased public bewilderment.

The current situation in large river science is similar to that of oceanography before World War II. A major investment in large river research is needed, comparable to investments made over the years in ocean sciences. Moreover, multi-investigator, multi-institutional programs should be developed that involve academic researchers (Sparks 2002).

However, we must remember that the great strides made in oceanographic research after World War II were fueled largely by the Cold War when rival fleets of nuclear submarines roamed the oceans, and the nation's very survival could hinge on its knowledge of the ocean depths. At present there is no analogous crisis for the word's rivers.

In the 1970s we wondered if the nation could afford to implement the Clean Water Act and carry on its space race and Cold War with the Soviet Union. Today, in the wake of terrorist destruction of New York's World Trade Center in 2001, we wonder if we can embark on holistic, ecological management programs for the nation's rivers and also fund burgeoning programs for national security.

Are We Due for Drought?

The Great Plains and the Upper Mississippi River Basin have suffered repeated droughts for thousands of years, but the last seven hundred years have, in fact, been unusually wet. Studies of lake sediments reveal that in the past, extreme dry spells not only persisted for centuries at a time but occurred much more frequently than they do today. No one knows what caused the cycles of droughts in the past, but they may increase in frequency because humans are apparently altering the climate with greenhouse gases and doing things to the climate that have never been done before.

Computer models indicate that central North America is likely to become warmer and drier, probably causing the coniferous forests of the Mississippi Headwaters to be replaced by hardwood forests. Present tallgrass prairies may be replaced by mid- or short-grass prairies. Agriculture in America's "breadbasket" will be forced to adapt to the climate change, with southern Minnesota farmers growing very different crops, perhaps cotton and peanuts. Wetlands and lakes will lose water, and many will dry up (Tester 1995). These natural and human-induced changes will obviously impact the Mississippi River in ways we have yet to consider.

In one scenario, the Mississippi has been seen as a potential water source to satisfy areas of the Great Plains that are sucking the great Ogallala aquifer dry. We shouldn't be surprised to one day debate if water from the Mississippi should be piped southwest to water lawns and golf courses in Dallas and Oklahoma City.

In Conclusion

In considering the river's future, all that is absolutely certain is that it will be different from the present. But based on environmental and social trends I feel it is fairly safe to predict that within a half century or so:

- Much of the Upper Mississippi will look much like it did before it was impounded, with islands, bars, and other landforms in about the same place as in pre-impoundment days, but with pool areas elevated at least to dam height.
- A corridor of bottomland will remain inviolate because it is in federal ownership, but the bluff tops and bluff faces will be plastered with houses and condominiums placed cheek to jowl "California style." Because zoning and building codes are lax in most of the valley, such development could detract from its natural beauty.

Looking beyond the fairly immediate future, we are probably less able to imagine today what the river will look like in two hundred years than Zebulon Pike was when he explored it in 1805.

In the meantime we must not succumb to the "Chicken Little syndrome" created by doomsayers who predict that the biological world is going to implode and that we humans will soon be extinct. Such talk makes people feel powerless, causing them to believe that the situation is hopeless, to become apathetic and thus encourage them to rationalize all kinds of destructive behavior.

We must realize that the plant and animal communities that we've become accustomed to aren't suddenly going to become extinct. They are dynamic and are metamorphosing constantly. Today's communities may disappear, but they will be succeeded by others. Open water areas may become marshes, which in turn may

From a low-flying aircraft we see the city of Winona and its landmark "Sugar Loaf" with the Mississippi River and Wisconsin bluffs in the background. When Zebulon Pike stood atop Sugar Loaf in 1805 and described the panorama unveiled before him (chapter 1), he couldn't have imagined what it would look like today. Winona was founded in 1851, built in the Mississippi floodplain on a treeless expanse of prairie. Sugar Loaf was quarried until 1887, providing dolomitic limestone for Winona's sidewalks, trim for buildings, and lime for cement. Today, only a vestige of the original bluff remains, but it is still affectionately called Sugar Loaf. The body of water in the foreground is Lake Winona, an ancient Mississippi River channel that was navigated by steamboats during floodtime as late as 1852. Deltas of several Minnesota streams segmented the old channel, isolating it from the river. The causeway that divides the lake into two basins was built of dredge spoil from the lake in 1913 as was most of the lake's parkland. Because most of the city lies lower than the 1965 "hundred-year" flood level, it is protected by over nine miles of levees and is dewatered in floodtime by huge pumps. During floodtime, the city is virtually an island (photo by Robert J. Hurt, Architectural Environments, courtesy of Winona Area Convention and Visitors Bureau).

become floodplain forests that would be host to different associations of birds, mammals, reptiles, insects, and sundry other organisms. The future river will be a bit less diverse, a bit more monotonous, and a bit grayer, but it will still be here. The transformed riverine environment may not be what we want, but it may be all that we deserve.

Hemmed in by bluff lands, Old Man River will probably occupy his present valley from the Twin Cities to St. Louis at least until the next glacial epoch—unless he is forced to change his course by humans or an earthquake. The Nine-Foot Channel dams are speeding the rise of the river's bed, but in geologic time the brief dam era will be a minor sputter, a hiccough, in the river's tenure. To the Old Man, humans may someday only be a dim memory of a minor annoyance, gnats that were brushed away while he was busily and impatiently trying to do his important job of transporting erosion products to the sea.

Somehow, I find comfort in that.

Bibliography

Index

Bibliography

For those interested in the Headwaters, there are pocket-sized, detailed maps describing a 610-mile Mississippi River canoe route that runs from the river's source at Lake Itasca as far south as the Iowa–Minnesota border. The maps show campsites, accesses, rest areas, dams, portages, rapids, sources of drinking water, and river miles. They are sufficiently detailed to allow side trips into the beautiful backwaters downstream from the Twin Cities. The maps are available free from the Minnesota Department of Natural Resources, Division of Parks and Recreation, Rivers Section, Centennial Building, St. Paul, MN 55155.

Large books of detailed maps of the Upper Mississippi (from Minneapolis to the mouth of the Ohio River at Cairo, Illinois) can be purchased from the U.S. Army Corps of Engineers, U.S. Post Office and Custom House, St. Paul, MN. Maps of the Lower Mississippi River (Cairo to the Gulf of Mexico) can be purchased from the USACE in Memphis, Tennessee. The maps are indispensable to travelers of the Mississippi's main stem because they show locations of navigation hazards and navigation aids, river mile numbers, locks and dams, and other important physical features.

A&E Television Networks. 1998. *Sultana: The Mississippi's Titanic.* VHS Documentary. 50 min. Cat. No. AAE40378. New Video Group, New York.

Alexander, E. Calvin, Jr. 1985. "Karst in the Upper Mississippi Valley." In *Pleistocene geology and evolution of the Upper Mississippi Valley: A working conference,* 3–4. Winona State University, MN.

Alexander, R. B., R. A. Smith, and G. E. Schwarz. 1995. "The regional transport of point and nonpoint source nitrogen to the Gulf of Mexico." In *Proceedings of the Hypoxia Management Conference, Gulf of Mexico Program, New Orleans, Louisiana, December 5–6, 1995, Kenner Louisiana,* 127–32.

Ambrose, Stephen E. 1996. *Undaunted courage: Meriwether Lewis, Thomas Jefferson and the opening of the American West.* Simon & Schuster, New York.

Ambrose, Stephen E., and Douglas G. Brinkley. 2002. *The Mississippi and the making of a nation.* National Geographic Society, Washington, D.C.

American Rivers, Inc. 1999. *In harm's way: A report on floods and floodplains.* American Rivers Publications, Washington, D.C.
Anderson, Richard C. 1985. "Preglacial drainage in the Upper Mississippi Valley region." In *Pleistocene geology and evolution of the Upper Mississippi Valley: A working conference,* 9–10. Winona State University, MN.
Anderson, Wayne I. 1983. *Geology of Iowa: Over two billion years of change.* Iowa State University Press, Ames.
Anfinson, John O. 1996. *Henry Bosse's views of the Upper Mississippi River.* U.S. Army Corps of Engineers, St. Paul District.
Anonymous. 1829. *The hat makers manual containing a full description of hat making in all its branches: also copious directions for dyeing, blocking, finishing, and water proofing; together with a list of hat manufacturers and trades connected therewith in the metropolis; mode of obtaining employment.* Cowie and Strange, London.
Arnold, J. L. 1988. *The evolution of the 1936 Flood Control Act.* U.S. Army Corps of Engineers, Office of History, Fort Belvoir, VA.
Arzigian, Constance M., R. F. Boszhardt, H. P. Halverson, and J. Theler. 1994. "Animal remains in the Gunderson Site: An Oneota village and cemetery in La Crosse, Wisconsin." *Journal of the Iowa Archaeological Society* 41: 38–53, and Appendix B.
Arzigian, Constance M., Robert F. Boszhardt, James L. Theler, Roland L. Rodell, and Michael J. Scott. 1989. "Human adaptation in the Upper Mississippi Valley: A study of the Pammel Creek Oneota Site (47Lc61) La Crosse, Wisconsin." *The Wisconsin Archeologist* 70 (1–2).
Austin, George S. 1972. "Paleozoic lithostratigraphy of southeastern Minnesota." In *Geology of Minnesota: A centennial volume,* ed. P. K. Sims and G. B. Morey, 459–73. Minnesota Geological Survey, St. Paul.
Balcom, Trudy. 2004. "The green chapter of the navigation study." *Big River* 12 (July–August): 15, 45, 51.
Baldwin, Leland D. 1941. *The keelboat age on western waters.* University of Pittsburgh Press, Pittsburgh.
Barry, John M. 1997a. *Rising Tide: The great Mississippi flood of 1927 and how it changed America.* Simon & Schuster, New York.
———. 1997b. "Eades vs. the Mississippi and the Army." *Invention and Technology* 13 (1): 12–21.
Baumel, Phillip C. 2001. *How U.S. grain export projections from large scale agricultural sector models compare with reality.* Institute for Agriculture and Trade Policy, Minneapolis.
Beauvais, S. L., J. G. Wiener, and G. J. Atchison. 1995. "Cadmium and mercury in sediment and burrowing mayfly nymphs (*Hexagenia*) in the Upper Mississippi River, USA." *Arch. Environ. Contam. Toxicol.* 28: 178–83.

Belt, C. B. Jr. 1975. "The 1975 flood and man's constriction of the Mississippi River." *Science* 189 (4204): 681–84.

Bettis, E. Arthur III, and G. R. Hallberg. 1985a. "The Savanna (Zwingle) terrace and 'red clays' in the Upper Mississippi River Valley: stratigraphy and chronology." In *Pleistocene geology and evolution of the Upper Mississippi Valley: A working conference,* 41–43. Winona State University, MN.

———. 1985b. "Quaternary alluvial stratigraphy and chronology of Roberts Creek Basin, northeastern Iowa." In *Pleistocene geology and evolution of the Upper Mississippi Valley: A working conference,* 44–45. Winona State University, MN.

Billington, R. A., and M. Ridge. 1982. *Westward expansion.* Macmillan, New York.

Bolland, T. 1980. *A classification of wing and closing dams on the Upper Mississippi River bordering Iowa.* Report to the Fish and Wildlife Management Work Group of GREAT II. Iowa Conservation Commission, Des Moines.

Botkin, Daniel B. 1990. *Discordant harmonies. A new ecology for the twenty-first century.* Oxford University Press, New York.

Bray, Edmund C. 1985. *Ancient valleys, modern rivers: What the glaciers did.* Science Museum of Minnesota, St. Paul.

Breining, G., and L. Watson. 1977. *A gathering of waters, a guide to Minnesota's rivers.* Minnesota Department of Natural Resources, St. Paul.

Bunnell, Lafayette H. 1897. *Winona and its environs on the Mississippi in ancient and modern days.* Jones and Kroeger, Winona, MN.

Capano, Daniel E. 2003. "Chicago's war with water." *Invention and Technology* 18 (4): 50–58.

Carlander, Harriet B. 1954. *History of fish and fishing in the Upper Mississippi River.* Upper Mississippi River Conservation Committee, Rock Island, IL.

Carlander, Kenneth D., C. A. Carlson, V. Gooch, and T. L. Wenke. 1967. "Populations of *Hexagenia* mayfly naiads in Pool 19, Mississippi River, 1959–1963." *Ecology* 48 (5): 873–78.

Carter, Hodding. 1942. *The rivers of America: Lower Mississippi.* Farrar and Rinehart Inc., New York.

Catlin, George. 1841. *Letters and notes on the manners and customs, and condition of the North American Indians.* Vol. 2. Reprinted 1965, Ross and Haines, Inc., Minneapolis.

Chapman, Carl H., and Eleanor F. Chapman. 1983. *Indians and archaeology of Missouri.* Rev. ed. University of Missouri Press, Columbia.

Childs, Marquis W. 1982. *Mighty Mississippi: Biography of a river.* Ticknor and Fields, New Haven and New York.

Clay Floyd M. 1986. *A Century on the Mississippi: A history of the Memphis District: U.S. Army Corps of Engineers 1876–1981.* U.S. Army Corps of Engineers, Memphis District.
Clemens, S. L. [Mark Twain]. 1985. *Life on the Mississippi.* Bantam Books, New York.
Coffin, B. A. 1988. *The natural vegetation of Minnesota at the time of the Public Land Survey: Biological Report No. 1.* Minnesota Department of Natural Resources, St. Paul.
Cooper, W. S. 1935. *The history of the Upper Mississippi River in late Wisconsin and postglacial time.* Minnesota Geological Survey Bulletin 26, St. Paul.
Coues, E. 1965. *The Expeditions of Zebulon Pike to the headwaters of the Mississippi River, through Louisiana Territory, and in New Spain, during the years 1805–6, –7.* Vol. 1. Ross and Haines, Minneapolis.
Crawford, Dave. 1996. *The geology of Interstate State Park.* Minnesota Department of Natural Resources, St. Paul.
Crosbie, Karol. 2001. "Sustainable agriculture: A journey." *Iowa State University Visions* Jan/Feb: 12–23.
Curtis, John T. 1959. *The vegetation of Wisconsin: An ordination of plant communities.* University of Wisconsin Press, Madison.
Dahl, T. E. 1990. *Wetlands: Losses in the United States, 1780s to 1980s.* U.S. Department of the Interior, Fish and Wildlife Service, Washington, D.C.
Dahl, T. E., and C. E. Johnson. 1991. *Status and trends of wetlands in the coterminous United States, mid-1970s to mid-1980s.* U.S. Department of the Interior, Fish and Wildlife Service, Washington, D.C.
Davis, N. D. 1982. *The father of waters: A Mississippi River chronicle.* Sierra Club Books, San Francisco.
Devore, Brian J. 2001. "Dead zone puzzle." *Minnesota Conservation Volunteer* 64 (377): 11–21.
Diaz, David. 1996. "Under pressure." *Invention and Technology* 11 (4): 53–63.
Dobbs, Clark A., and Howard Mooers. 1991. *A Phase I archaeological and geomorphological study of Lake Pepin and the upper reaches of Navigation Pool 4, Upper Mississippi River (Pierce and Pepin Counties, Wisconsin; Goodhue and Wabasha Counties, Minnesota).* Report of Investigations No. 44, Institute for Minnesota Archaeology, Minneapolis.
Dobney, Frederick J. 1978. *River engineers on the Middle Mississippi: A history of the St. Louis District.* U.S. Army Corps of Engineers. Superintendent of Documents, U.S. Government Printing Office, Washington, D.C.
Ebling, W. H., C. D. Caparoon, E. C. Wilcox, and C. W. Estes. 1948. *A century of Wisconsin agriculture.* Wisconsin Crop and Livestock Department Service Bulletin No. 290.
Eddy, Samuel, and James C. Underhill. 1974. *Northern fishes: With special*

reference to the Upper Mississippi Valley. University of Minnesota Press, Minneapolis.

Eggers, Steve D., and Donald M. Reed. 1987. *Wetland plants and plant communities of Minnesota and Wisconsin.* U.S. Army Corps of Engineers, St. Paul District.

Eliot, T. S. 1943. *Four quartets:* "The dry salvages." Harcourt, Brace and Company, New York.

Ellet, C., Jr. 1852. *Report on the overflows of the delta of the Mississippi.* The War Department, Washington, D.C.

Ellis, W.S. 1993. "The Mississippi River under siege." In *Water: The power, promise, and turmoil of North America's fresh water.* Special Edition, *National Geographic Magazine.* National Geographic Society, Washington, D.C.

Elson, J. A. 1983. "Lake Agassiz—Discovery and a century of research." In *Glacial Lake Agassiz,* ed. J. T. Teller and L. Clayton, 21–41. Geological Association of Canada, Special Paper 26, St. John's, Newfoundland.

Esling, S. P., and G. R. Hallberg. 1985. "Development of the Mississippi River from late Illinoian through early Holocene as recorded in terrace remnants along the Lower Iowa and Cedar Rivers, southeast Iowa." In *Pleistocene geology and evolution of the Upper Mississippi Valley: A working conference,* 55–57. Winona State University, MN.

Faber, Scott. 1995. *The real choices report: The failure of America's flood control policies.* American Rivers, Washington, D.C.

———. 2003a. "Removing locks would restore river's falls." *Mississippi Monitor* 4 (October): 1, 11.

———. 2003b. "Special section: Citizen's guide to the Navigation Study." *Mississippi Monitor* 4 (October): 6–7.

Fertey, Andre. 1970. *The journals of Joseph N. Nicollet: A scientist on the Mississippi Headwaters with notes on Indian life, 1836–37.* Martha C. Bray, ed., Minnesota Historical Society Press, St. Paul.

Fiedel, Stuart J. 1999. "Older than we thought: Implications of corrected dates for Paleoindians." *American Antiquity* 64 (1): 95–115.

Fisk, Harold N. 1944. *Geological investigation of the alluvial valley of the Lower Mississippi River.* U.S. Army Corps of Engineers, Mississippi River Commission, Vicksburg.

Fowler, Melvin L. 1974. *Cahokia: Ancient capitol of the Midwest.* Addison-Wesley Publishing Co., Reading, MA.

Fremling, Calvin R. 1960a. *Biology of a large mayfly, Hexagenia bilineata (Say), of the Upper Mississippi River.* Iowa State University Agriculture and Home Economics Experiment Station Bulletin 482: 842–52.

———. 1960b. "Biology and possible control of nuisance caddisflies of the Upper Mississippi River." *Iowa State University Agriculture and Home Economics Experiment Station Bulletin* 483: 853–79.

——. 1968. "Documentation of a mass emergence of *Hexagenia* mayflies from the Upper Mississippi River." *Transactions of the American Fisheries Society* 97: 278–80.
——. 1978. "Biology and functional anatomy of the freshwater drum, *Aplodinotus grunniens* Rafinesque." Laboratory Manual. Nasco International, Fort Atkinson, WI.
——. 1980. "*Aplodinotus grunniens* Rafinesque, freshwater drum." In *Atlas of North American Fishes,* ed. D.S. Lee et al., 756. North Carolina State Museum of Natural History, Raleigh.
——. 1987. "Human impacts on Mississippi River ecology." In *Proceedings of the Tenth National Coastal Society Conference, New Orleans, Louisiana,* 235–40.
——. 1989. "*Hexagenia* mayflies: Biological monitors of water quality in the Upper Mississippi River." *Journal of the Minnesota Academy of Science* 55: 139–43.
Fremling, Calvin R., and T. O. Claflin. 1984. "Ecological history of the Upper Mississippi River." In *Contaminants in the Upper Mississippi River,* ed. J. G. Wiener, R. V. Anderson, and D. R. McConville, 5–24. Butterworth Publishers, Boston, MA.
Fremling, Calvin R., Donald V. Gray, and Dennis N. Nielsen. 1973. *Environmental impact study of Pools 4, 5, 5A and 6 of the northern section of the Upper Mississippi River Valley. Phase III.* Report in 4 volumes. North Star Research Institute, Minneapolis.
Fremling, Calvin R., and D. K. Johnson. 1990. "Recurrence of *Hexagenia* mayflies demonstrates improved water quality in Pool 2 and Lake Pepin, Upper Mississippi River." In *Mayflies and stoneflies, Life Histories and Biology, Proceedings of the Fifth International Ephemeroptera Conference,* ed. I. C. Campbell, 243–48. Kluwer Academic Publishers, Norwell, MA.
Fremling, Calvin R., Dennis N. Nielsen, David R. McConville, Rory N. Vose, and Ray Faber. 1979. *The feasibility and environmental effects of opening side channels in five areas of the Mississippi River (West Newton Chute, Fountain City Bay, Sam Gordy's Slough, Kruger Slough and Island 42).* Vol. 1, 2. U.S. Fish and Wildlife Service, Twin Cities.
Fremling, Calvin R., Dennis N. Nielsen, Rory N. Vose, and David R. McConville. 1976. *The Weaver Bottoms: A field model for the rehabilitation of backwater areas of the Upper Mississippi River by modification of standard channel maintenance practices.* U.S. Army Corps of Engineers, St. Paul District.
Fremling, Calvin R., J. L. Rasmussen, R. E. Sparks, S. P. Cobb, C. F. Bryan, and T. O. Claflin. 1989. "Mississippi River fisheries: A case history." In *Proceedings of the International Large River Symposium,* ed. D. P. Dodge, 309–51. Canadian Special Publication of Fisheries and Aquatic Sciences.

Fry, J. C., H. B. Willman, and R. F. Black. 1965. "Outline of glacial geology of Illinois and Wisconsin." In *The Quaternary of the United States,* ed. H. E. Wright Jr. and D. G. Frey, 43–61. Princeton University Press, Princeton, NJ.

Fugina, Frank J. 1945. *Lore and lure of the Upper Mississippi River.* Published by author, Winona, MN.

Galloway, Gerald E. 2000. "Three centuries of river management along the Mississippi River: Engineering and hydrological aspects." In *New Approaches to River Management,* ed. A. J. M. Smits, P. H. Nienhuis, and R. S. E. W. Leuven, 51–64. Backhuys Publishers, Leiden, Netherlands.

Gore, Rick. 1997. "The most ancient Americans." *National Geographic* 192 (4): 92–99.

Graham, R. W., C. V. Haynes, D. L. Johnson, and M. Kay. 1981. "Kimmswick: A Clovis-mastodon association in eastern Missouri." *Science* 213: 1115–17.

Grahame, Kenneth. 1966. *The wind in the willows.* Illustrated by Tasha Tudor. The World Publishing Company, Cleveland and New York.

Great River Environmental Action Team. 1980. *A public trust: An executive summary of GREAT I.* U.S. Army Corps of Engineers, Twin Cities District.

Greenbank, John. 1946. "Effects of midwinter drawdowns of the Upper Mississippi River on aquatic wildlife." In *Upper Mississippi River Conservation Committee, Investigational Reports,* ed. K. D. Keenlyne, 17–24. Upper Mississippi River Conservation Committee, Rock Island, IL.

Grubaugh, J. W. 1988. "Long-term effects of navigation dams on a segment of the Upper Mississippi River." *Regulated Rivers: Research and Management* 4: 97–104.

Grubaugh, Jack W., and Richard V. Anderson. 1988. "Spatial and temporal availability of floodplain habitat: Long-term changes in Pool 19, Mississippi River." *The American Midland Naturalist* 119 (2): 402–10.

Grubaugh, J. W., R. V. Anderson, D. M. Day. K. S. Lubinski, and R. E. Sparks. 1986. "Production and fate of organic material from *Sagittaria latifolia* and *Nelumbo lutea* on Pool 19, Mississippi River." *Journal of Freshwater Ecology* 3 (4): 477–84.

Grunwald, Michael. 2000. "Engineers of power: An agency of unchecked clout." *Washington Post.* September 10: AO1–AO3 (first of a five-part series, September 10–14).

Haapoja, Margaret A. 1997. "Renaissance for white pine." *The Minnesota Volunteer.* May–June: 10–19.

Haines, Francis. 1975. *The buffalo: The story of American bison and their hunters from prehistoric times to the present.* Thomas Y. Crowell Company, New York.

Hajic, Edwin R. 1985. "Terminal Pleistocene events in the Mississippi Valley near St. Louis as inferred from Illinois Valley geology." In *Pleistocene geology and evolution of the Upper Mississippi Valley: A working conference*, 49–52. Winona State University, MN.

Hall, Steve. 1987. *Fort Snelling, colossus of the wilderness*. Minnesota historic sites pamphlet series, no. 20. Minnesota Historical Society Press, St. Paul.

Hall, Thomas F., William T. Penfound. 1942. "The biology of the American lotus, *Nelumbo lutea* (Willd.) Pers." *The American Midland Naturalist* 31 (3): 744–58.

Hallberg, George R., E. A. Bettis III, and J. C. Prior. 1984. "Geologic overview of the Paleozoic Plateau Region of Northeastern Iowa." *Proceedings of the Iowa Academy of Science* 91 (1): 5–11.

Hallberg, George R., and E. A. Bettis III. 1985a. "Overview of landscape evolution on northeastern Iowa. I: Pre-Wisconsinan." In *Pleistocene geology and evolution of the Upper Mississippi Valley: A working conference*, 33–36. Winona State University, MN.

———. 1985b. "Overview of landscape evolution in northeastern Iowa. II: Wisconsinan." In *Pleistocene geology and evolution of the Upper Mississippi Valley: A working conference*, 36–40. Winona State University, MN.

Hallberg, G. R., B. J. Witzke, E. A. Bettis III, and G. A. Ludvigson. 1985."Observations on the evolution and age of the bedrock surface in eastern Iowa." In *Pleistocene geology and evolution of the Upper Mississippi Valley: A working conference*, 15–19. Winona State University, MN.

Harris, C. M. 1989. "The improbable success of John Fitch." *American Heritage of Invention and Technology* 4 (3):24–28, 30–31.

Hartsough, Mildred L. 1934. *From canoe to steel barge on the Upper Mississippi*. University of Minnesota Press for the Upper Mississippi Waterway Association, Minneapolis.

Havighurst, Walter. 1964. *Voices on the river: The story of the Mississippi waterways*. Macmillan, New York.

Hawkeye Pearl Button Company. undated. *Pearls of the Mississippi*. 34-page trade booklet. Record Printing Co., Muscatine, IA.

Henderson, W. C. 1931. "Memorandum for hearing to be held with the Army Engineers at Wabasha, Minnesota on February 26, 1931." 3 mimeographed pages. U.S. Bureau of Biological Survey.

Herwig, Mark. 2002. "Farms turning wild." *Minnesota Conservation Volunteer* 65 (384): 30–39.

Hesse, L. W., J. C. Schmulbach, J. M. Carr, K. D. Keenlyne, D. G. Unkenholz, J. W. Robinson, and G. E. Mestl. 1989. "Missouri River fishery resources in relation to past, present, and future stresses." In *Proceedings of the International Large River Symposium*, ed. D. P. Dodge,

352–71. Canadian Special Publication of Fisheries and Aquatic Sciences 106.
Hey, Donald L., and Nancy S. Philippi. 1997. "Reinventing a flood-control strategy." In *Proceedings of the Scientific Assessment and Strategy Team workshop on hydrology, ecology, and hydraulics,* ed. G. E. Freeman and A. G. Frazier, vol. 5 of *Science for floodplain management into the 21st century,* ed. J. A. Kelmelis. U.S. Government Printing Office, Washington, D.C.
Hill, Alberta K. 1961. "Out with the fleet on the Upper Mississippi, 1898–1917." *Minnesota history: The North Star State in words and pictures* 37 (7): 283–97.
Hill, Libby. 2000. *The Chicago River: A natural and unnatural history.* Lake Claremont Press, Chicago.
Hobbs, Howard C. 1985. "Quaternary history of southeastern Minnesota." In *Pleistocene geology and evolution of the Upper Mississippi Valley: A working conference,* 11–14. Winona State University, MN.
Holling, C. H. 1951. *Minn of the Mississippi.* Houghton Mifflin, Boston.
Hunter, Louis C. 1969. *Steamboats on the western rivers: An economic and technological history.* Octagon Books, New York.
Hutton, James. *Theory of the Earth.* 1795. Vol. 2. Edinburgh.
Institute for Agriculture and Trade Policy. 2001. *Myth: Barges are the most fuel efficient mode of transportation for agricultural commodities.* Agriculture Myth Series. Institute for Agriculture and Trade Policy, Minneapolis.
Iowa Department of Natural Resources. 1997. *Mines of Spain State Recreation Area.* Pamphlet.
———. 1997. *Pikes Peak State Park.* Pamphlet.
Johnson, Elden. 1988. *The prehistoric peoples of Minnesota.* Rev. 3d ed. Minnesota Prehistoric Archaeology Series, No. 3. Minnesota Historical Society Press, St. Paul.
Johnson, Frederick L. 1990. *The Sea Wing disaster.* Goodhue County Historical Society, Red Wing, MN.
Kane, Lucile M. 1987. *The Falls of St. Anthony: The waterfall that built Minneapolis.* Minnesota Historical Society Press, St. Paul.
Kane, Lucile M., June D. Holmquist, and Carolyn Gilman, eds. 1978. *The northern expeditions of Stephen H. Long: The journals of 1817 and 1823 and related documents.* Minnesota Historical Society Press, St. Paul.
Keating, B. 1971. *The mighty Mississippi.* National Geographic Society, Washington, D.C.
Keeney, Dennis, and Mark Muller. 2000. *Nitrogen and the Upper Mississippi River.* Institute for Agriculture and Trade Policy, Minneapolis.
Kelner, Dan. 2002. "New mussels in the Old Miss." *Minnesota Conservation Volunteer* 65 (384): 18–29.

Knox, James C. 1985. "Geologic history of valley incision in the driftless area." In *Pleistocene geology and evolution of the Upper Mississippi Valley: A working conference,* 5–8. Winona State University, MN.

Knox, James C., P. J. Bartlein, K. K. Hirschboek, and R. J. Muchenhim. 1975. *The response of floods and sediment fields to climatic variation and land use in the Upper Mississippi Valley.* University of Wisconsin–Madison, Institute for Environmental Studies, Report 52. Madison.

Koch, Donald L., Jean C. Prior, and Samuel J. Tuthill. 1973. *Geology of Pikes Peak State Park, Clayton County, Iowa.* Iowa Geological Survey, Iowa City.

Kohlmeyer, F.W. 1972. *Timber roots: The Laird Norton story, 1885–1905.* Winona County Historical Society, Winona, MN.

LaBerge, Gene L. 1994. *Geology of the Lake Superior region.* Geoscience Press, Phoenix, AZ.

Latrobe, C. J. 1833. *The rambler in North America.* London.

Leopold, Aldo. 1966. *A Sand County almanac.* Oxford University Press, New York.

Leopold, Luna B. 1994. *A view of the river.* Harvard University Press, Cambridge, MA.

Levins, Richard A., Philip W. Rice, and Elizabeth R. Swain. 2000. *Will it really help farmers? The Upper Mississippi River Navigation Project.* Institute for Agriculture and Trade Policy, Minneapolis.

Lively, R. S., and E. C. Alexander Jr. 1985. "Karst and the Pleistocene history of the Upper Mississippi River Valley." In *Pleistocene geology and evolution of the Upper Mississippi Valley: A working conference,* 31–32. Winona State University, MN.

Lubinski Kenneth S., A. van Vooren, G. Farabee, J. Janecek, and S. D. Jackson. 1986. "Common carp in the Upper Mississippi River." *Hydrobiologia* 136: 141–54.

Lutgens, Frederick K., and Edward J. Tarbuck. 2000. *Essentials of geology.* 7th ed. Prentice Hall, Upper Saddle River, NJ.

MacCleery, Doug. 1996. "When is a landscape natural?" *Minnesota Conservation Volunteer* (September–October): 42–52.

MacGregor, Molly, and Peter L. Card. 1995. *Mississippi Headwaters guidebook. A guidebook to the natural, cultural, scenic, scientific and recreational values of the Mississippi River's first 400 miles.* Mississippi Headwaters Board, Walker, MN.

Madson, John. 1982. *Where the sky began: Land of the tallgrass prairie.* Houghton Mifflin Company, Boston.

———. 1985. *Up on the river.* Nick Lyons Books/Schocken Books, New York.

Marsden, Roger D., and Fred F. Shafer. 1924. *The overflowed lands on the Mississippi River between St. Paul and Rock Island, and the practicability of reclaiming them for agriculture.* U.S. Department of Agriculture. Bureau of Public Roads, Washington, D.C.

Martin, Lawrence. 1965. *The physical geology of Wisconsin.* 3rd ed. University of Wisconsin Press, Madison.

Masteller, E. C., and E. C. Obert. 2000. "Excitement along the shores of Lake Erie—*Hexagenia*—echoes from the past." *Great Lakes Research Review* 5 (1): 25–36.

Matsch, Charles L. 1976. *North America and the great ice age.* McGraw-Hill, New York.

Mauck, W. L., and L. E. Olson. 1977. "Polychlorinated biphenyls in adult mayflies (Hexagenia bilineata) from the Upper Mississippi River." *Bulletin of Environmental Contaminants and Toxicology* 17: 387–90.

Mc Leod, Reggie. 2004. "Marketing lock expansion." *Big River* 12 (July-August): 15, 49.

McKnight Foundation. 1996. *The Mississippi River in the Upper Midwest: Its economy, ecology, and management.* The McKnight Foundation, Minneapolis.

Meade, Robert H., ed. 1995. *Contaminants in the Mississippi River, 1987–92.* U.S. Geological Survey Circular 1133. Denver, CO.

Merrick, G. B. 1987. *Old times on the Mississippi: The recollections of a steamboat pilot from 1854 to 1863.* Minnesota Historical Society Press, St. Paul.

Merritt, Raymond H. 1979. *Creativity, conflict and controversy: A history of the St. Paul District.* U.S. Army Corps of Engineers, St. Paul.

———. 1984. *The Corps, the environment, and the Upper Mississippi River Basin.* U.S. Army Corps of Engineers, Historical Division. EP 870: 1–19.

Metcalf, C. L., and W. P. Flint. 1939. *Destructive and useful insects: Their habits and control.* McGraw-Hill, New York and London.

MICRA. 1993. *River crossings* 2 (5).

Mississippi River Commission, U.S. Army Corps of Engineers. 1983. *Flood control and navigation maps of the Mississippi river.* Chart 1. General map: *Middle and Upper Mississippi River.*

Mossler, J. H. 1999. *Geology of the Root River State Trail area, southeastern Minnesota.* Minnesota Geological Survey, Educational Series 10, St. Paul.

Muller, Mark, and Richard Levins. 1999. *Feeding the world? The Upper Mississippi River Navigation Project.* Institute for Agriculture and Trade Policy, Minneapolis.

Nace, R. L. 1970. "World hydrology: Status and prospects." International Association for the Science of Hydrology. Publ. 92: 1–10, Symposium of Reading (England).

National Geographic Society. 1984. *Great rivers of the world.* National Geographic Society, Washington, D.C.

National Park Service. 1997. *Effigy Mounds: Official map and guide.* Government Printing Office, 417-648/60053, Washington, D.C.

Nelson, George. 2002. *My first years in the fur trade: The journals of 1802–1804.* Ed. Laura Peers and Theresa Schenck. Minnesota State Historical Society Press, St. Paul.

Nelson, John C., Lynne DeHann, and Larry Robinson. 1998. "Presettlement and contemporary vegetation patterns along two navigation reaches of the Upper Mississippi River." In *Perspectives on the land-use history of North America: A context for understanding our changing environment,* ed. T. D. Sisk. U.S. Geological Survey, Biological Resources Division, Biological Science Report USGS/BRD/BSR-1998-0003 (revised September 1999).

Nelson, John C., Anjela Redmond, and Richard E. Sparks. 1994. "Impacts of settlement on floodplain vegetation at the confluence of the Illinois and Mississippi Rivers." *Transactions of the Illinois Academy of Science* 87 (3 and 4): 117–33.

Nelson, John C., and Richard E. Sparks. 1998. *Forest compositional change at the confluence of the Illinois and Mississippi Rivers.* Transactions of the Illinois State Academy of Science 91 (1 and 2): 33–46.

Neuzil, Mark. 2001. *Views on the Mississippi: The photographs of Henry Peter Bosse.* Foreword by Merry A. Foresta. University of Minnesota Press, Minneapolis.

Nielsen, D. N., R. N. Vose, C. R. Fremling, and D. R. McConville. 1978. *Phase I study of the Weaver-Belvidere area, Upper Mississippi River.* Final report to U.S. Fish and Wildlife Service, St. Paul.

Nygren, Doug. 2003. "Biologists replaced: The Missouri River saga continues. *River Crossings* 12 (November/December): 1–4.

Ockerson, J. A. 1898. "Dredges and dredging on the Mississippi River." *Transactions, American Society of Civil Engineers.* 40: 215–310.

Office of Communications, U.S. Department of the Interior. 1976. *A passing in Cincinnati, September 1, 1914.* Superintendent of Documents, Washington, D.C.

Ogg, Frederic A. 1904. *The opening of the Mississippi: A struggle for supremacy in the American interior.* Greenwood Press Publishers, New York.

Ojakangas, Richard W., and Charles L. Matsch. 1982. *Minnesota's geology.* University of Minnesota Press, Minneapolis.

Olsen, Bruce M. 1985. "The bedrock geology of southern Minnesota and its relationship to the history of the Mississippi River Valley." In *Pleistocene geology and evolution of the Upper Mississippi Valley: A working conference,* 21–24. Winona State University, MN.

Olson, K. N., and M. P. Meyer. 1976a. *Vegetation, land, and water surface changes in the upper navigable portion of the Mississippi River Basin over the period 1939–1973.* University of Minnesota, Institute of Agriculture, Forestry, and Home Economics, Remote Sensing Laboratory, Research Report No. 76-4.

———. 1976b. *Vegetation, land, and water surface changes in the upper navigable portion of the Mississippi River Basin over the period 1929–1973.* University of Minnesota, Institute of Agriculture, Forestry, and Home Economics, Remote Sensing Laboratory, Research Report No. 76-5.

Parfit, Michael. 2000. "Hunt for the first Americans." *National Geographic* 198 (6): 41–67. Map Supplement: *Peopling of the Americas.*

Patrick, Ruth. 1998. *Rivers of the United States: The Mississippi River and its tributaries north of St. Louis.* John Wiley, New York.

Paull, Rachel K., and Richard A. Paull. 1980. *Geology of Wisconsin and Upper Michigan including parts of adjacent states.* Kendall/Hunt, Dubuque, IA.

Petersen, W. J. 1965. "Mississippi River floods." *The Palimpsest/Iowa Heritage Illustrated* 46 (7): 305–64.

———. 1968. *Steamboating on the Upper Mississippi.* State Historical Society of Iowa, Iowa City.

Phillips, Paul. C. 1961. *The fur trade.* Vols. 1, 2. Concluding chapters by J. W. Smurr. University of Oklahoma Press, Norman.

Pielou, E. C. 1991. *After the ice age. The return of life to glaciated North America.* University of Chicago Press, Chicago.

Rabun, Joanne T. 1997. *Death on the dark river: The story of the Sultana disaster in 1865.* The SultanaWebring. http://www.rootsweb.com/~genepool/sultana.htm.

Rasmussen, J. L. 1994. "Management of the Upper Mississippi: A case history." In *The rivers handbook,* ed. P. Calow and G. E. Petts, 2:441–63. Blackwell Scientific Publications, Boston.

Rasmussen, J. L., and J. Milligan. 1994. *The river floodway concept: A reasonable and common sense alternative for flood control.* Department of the Interior, U.S. Fish and Wildlife Service, Columbia, MO.

Rasmussen, J. L., John M. Pitlo Jr., and Mark T. Steingraeber. 2004. "Upper Mississippi River resource management issues." In *Upper Mississippi River Conservation Committee Fisheries Compendium,* 3rd ed, ed. John Pitlo and Jerry Rasmussen, 50–96. Upper Mississippi River Conservation Committee, Rock Island, IL.

Redfern, Ron. 1983. *The making of a continent.* Times Books, New York.

Reps, John W. 1998. *Bird's eye views: Historic lithographs of North American cities.* Princeton Architectural Press, New York.

Reimer, Ron. 2001. "Sailing the River." *Big River* 9 (July–August): 18–21.

Robinson, Ann, and Robbin Marks. 1994. *Restoring the big river: A clean water act blueprint for the Mississippi.* Izaak Walton League of America, Minneapolis; Natural Resources Defense Council, Washington, D.C.
Russel, Charles. 1928. *A-rafting on the Mississip'.* Century Co., New York.
Sansome, Constance J. 1983. *Minnesota underfoot. A field guide to the state's outstanding geologic features.* Voyageur Press, Stillwater, MN.
Sardeson, F. W. 1916. *Minneapolis-St. Paul folio.* U.S. Geological Survey, folio 201, Washington, D.C.
Saucier, Roger T. 1994. *Geomorphology and quaternary geologic history of the Lower Mississippi Valley.* Vols. 1, 2. Mississippi River Commission, Vicksburg, MS.
Scarpino, P.V. 1985. *Great River: An environmental history of the Upper Mississippi, 1880–1950.* University of Missouri Press, Columbia.
Schmitt, C. J., and C. M Bunck. 1995. "Persistent environmental contaminants in fish and wildlife." In Our living resources: A report to the nation on the distribution, abundance, and health of U.S. plants, animals, and ecosystems, ed. E. T. LaRoe, G. S. Farris, C. E. Puckett, P. D. Doran, and M. J. Mac, 20–23. U.S. Department of the Interior, National Biological Service, Washington, D.C.
Schoenhard, Carl B. Jr. 1988. *Galena: Hidden treasure.* Published by author.
Schwartz, George M., and George A. Thiel. 1963. *Minnesota's rocks and waters: A geological story.* University of Minnesota Press, Minneapolis.
Shaw, Margaret F. 1929. "A microchemical study of the fruit coat of *Nelumbo lutea.*" *American Journal of Botany* 16 (5):259–75.
Shipley, Sara, Eric Heisler, and Christopher Carey. 2003. "A flood of development." *St. Louis Post-Dispatch.* Five-part series, July 27–31.
Simons, D. B., P. F. Lagasse, Y. H. Chen, and S. A. Schumm. 1975. *The river environment: A reference document.* U.S. Department of the Interior, Fish and Wildlife Service, Twin Cities.
Sioli H. 1975. "Tropical river: The Amazon." In *River ecology,* ed. B. A. Whitton, 461–88. University of California Press, Berkeley.
Sloan, Robert E. 1985. "Pleistocene fluvial geomorphology of southeastern Minnesota." In *Pleistocene geology and evolution of the Upper Mississippi Valley: A working conference,* 27–30. Winona State University, MN.
———. 1987. "Tectonics, biostratigraphy and lithostratigraphy of the middle and late Ordovician of the Upper Mississippi Valley." In *Middle and late Ordovician lithostratigraphy and biostratigraphy of the Upper Mississippi valley,* ed. R. E. Sloan, 7–20. Minnesota Geological Survey, Report of Investigations 35, St. Paul.
———. 1998a. "Living fossils in the big river." *Big River* 6 (May): 1–3.
———. 1998b. "Rock walls along the river." *Big River* 6 (November): 1–3.
———. 1998c. "Where does all that sand come from?" *Big River* 6 (August): 1, 4.

——. 1998d. "Why is the river where it is?" *Big River* 6 (February): 1–3.
——. 1999. "Karst country—water flowing underground." *Big River* 7 (April): 1, 4–5.
Sparks, Richard E. 1992. "Can we change the future by predicting it?" In *Proceedings of the 48th Annual Meeting of the Upper Mississippi River Conservation Committee, March 10–12, Red Wing, Minnesota.*
——. 1995. "Need for ecosystem management of large rivers and their floodplains." *BioScience* 45 (3): 168–82.
——. 2002. "National need for expansion of large river research: Lessons from oceanography." In *Program of the Annual Meeting of Mississippi River Research Consortium, University of Illinois, April 25–26, 2002*, 1–4.
Staff of the Department of Geology, University of Minnesota. 1956. "Geology of the Minneapolis-St. Paul region." Prepared for the Minneapolis meeting of the Geological Society of America, Mineralogical Society of America, and the Paleontological Society.
Stanley, Steven M. 1986. *Earth and life through time.* W. H. Freeman and Company, New York.
Steele, Ray C. 1933. "The Upper Mississippi River Wild Life and Fish Refuge." *Minnesota Waltonian* 4 (9): 4, 14.
Stuart, George E. 1972. "Who were the 'mound builders'?" *National Geographic* 142 (6): 782–801.
Stuiver, Minze. 1971. "Evidence for the variation of atmospheric C-14 content in the late Quaternary." In *Late Cenozoic Glacial Ages,* ed. K. K. Turekian, 57–70. Yale University Press, New Haven, CT.
Tester, John R. 1995. *Minnesota's natural heritage: An ecological perspective.* University of Minnesota Press, Minneapolis.
Theler, James L. 1987a. "Prehistoric freshwater mussel assemblages of the Mississippi River in southwestern Wisconsin." *The Nautilus* 101 (3): 143–50.
——. 1987b. *Woodland tradition economic strategies: Animal resource utilization in southwestern Wisconsin and northeast Iowa.* Report 17. Office of the State Archaeologist, University of Iowa, Iowa City.
——. 1999. "Animal remains recovered at native American archaeological sites in the Upper Mississippi River Valley." Unpublished manuscript.
Thiel, George A. 1944. *The geology and underground waters of southern Minnesota.* Minnesota Geological Survey Bulletin 31, University of Minnesota Press, Minneapolis.
Thomson, John. 1868. *A treatise on hat-making and felting, including a full exposition of the singular properties of fur, wool and hair.* Henry Carey Baird, Industrial Publisher, Philadelphia.
Thornbury, William D. 1965. *Regional geomorphology of the United States.* Wiley, New York.

Thwaites, Reuben G., ed. 1905. *Early western travels 1748–1846. Part I of James's account of S. H. Long's expedition, 1819–1820.* Arthur H. Clark Company, Cleveland, OH.

Time-Life Books. 1975. *The rivermen.* Text by Paul O'Neil. Time-Life Books, New York.

Trimble, Stanley W., 1983. "A sediment budget for Coon Creek basin in the Driftless Area, Wisconsin, 1853–1977." *American Journal of Science* 283: 454–74.

——. 1995. "The cow as a geomorphic agent—a critical review." *Geomorphology* 13: 233–53.

Trimble, Stanley W., and S. W. Lund. 1982. *Soil conservation and reduction of erosion and sedimentation in the Coon Creek Basin, Wisconsin.* U.S. Geological Survey Professional Paper 1234. Washington, D.C.

Trowbridge, Arthur C. 1959. "The Mississippi in glacial times." *The Palimpsest/Iowa Heritage Illustrated* 40 (7): 257–88.

Tweet, Roald D. 1978. *Taming the Des Moines Rapids: The background of Lock 19.* U.S. Army Corps of Engineers, Rock Island District.

——. 1980. *A History of navigation improvements on the Rock Island Rapids (The background of Locks and Dam 15).* U.S. Army Corps of Engineers, Rock Island District.

——. 1983. *History of transportation on the Upper Mississippi and Illinois Rivers. National waterways study.* U.S. Army Engineer Water Resources Support Center, Washington, D.C.

——. 1984. *A history of the Rock Island District U.S. Army Corps of Engineers 1866–1983.* U.S. Army Engineer District, Rock Island, IL.

Unklesbay, A. G., and Jerry D. Vineyard. 1992. *Missouri geology: Three billion years of volcanoes, seas, sediments, and erosion.* University of Missouri Press, Columbia.

Upper Mississippi River Conservation Committee. 1993a. *The Upper Mississippi River: America's river faces its greatest threat: A call for action from the Upper Mississippi River Conservation Committee.* Upper Mississippi River Conservation Committee, Rock Island, IL.

——. 1993b. *Facing the threat: An ecosystem management strategy for the Upper Mississippi River.* Upper Mississippi River Conservation Committee, Rock Island, IL.

——. 2000. *A River that works and a working river: A strategy for the natural resources of the Upper Mississippi River System.* Upper Mississippi River Conservation Committee, Rock Island, IL.

U.S. Army Corps of Engineers. 1973. *Phase I Report. Mississippi River year-round navigation.* North Central Division, Chicago.

——. 1993. *Economic impacts of recreation on the Upper Mississippi River System: Economic impacts report.* U.S. Army Corps of Engineers, St. Paul District.

——. 1994. *Upper Mississippi River–Illinois Waterway System navigation study: Baseline initial project management plan.* U.S. Army Corps of Engineers, St. Paul District.

U.S. Bureau of Fisheries. 1931. *Report* by Captain Culler at hearing with Army Engineers at Wabasha, MN. February 26.

U.S. Fish and Wildlife Service. 2001. "Landmark fine assessed in caviar trade, fraud case." *Fish and Wildlife News* (April–June): 6.

U.S. Geological Survey. 1999. *Ecological status and trends of the Upper Mississippi River System 1998: A report of the Long Term Resource Monitoring Program.* U.S. Geological Survey, Upper Midwest Environmental Sciences Center, La Crosse, WI.

van der Schalie, H., and A. van der Schalie. 1950. "The mussels of the Mississippi River." *American Midland Naturalist* 44: 448–66.

Vinton, E. L. 1943. "Healing the wounds that man has made." In *Coon Valley in pictures,* ed. M. N. Daffinrud. Vernon County Historical Society, Viroqua, WI.

Visser, Margaret. 1988. *Much depends on dinner: The extraordinary history and mythology, allure and obsessions, perils and taboos of an ordinary meal.* Collier Books/Macmillan, New York.

Warren, G. K. 1874. "An essay concerning important physical features exhibited in the valley of the Minnesota River and upon their signification." Government Printing Office, Washington, D.C.

Waters, Thomas. F. 1977. *The streams and rivers of Minnesota.* University of Minnesota Press, Minneapolis.

Weflen, Kathleen. 2001. "Big, old trees." *Minnesota Conservation Volunteer* 64 (379): 2–3.

Werner, Paul. 2002. "Behind the scenes: A glimpse at the Water Resources Development Act debate." *The Waterways Journal* 116 (8): 16–17.

Wheeler, Mark. 1996. "The vertical forest." *Discover* (February): 76–82.

Wheelwright, Jeff. 2001. "How a genetically modified corn called Starlink that wasn't intended for humans got into your food supply." *Discover* (March): 35–43.

Wiebe, A. H. 1927. "Biological survey of the Upper Mississippi River, with special reference to pollution." *Bulletin of the U.S. Bureau of Fisheries* 43 (2):137–67.

Wiener, James G., C. R. Fremling, C. E. Korschgen, K. P. Kenow, E. M. Kirsch, S. J. Rogers, Y. Yin, and J. S. Sauer. 1998. "Mississippi River." In *Status and trends of the nation's biological resources,* ed. M. J. Mac, P. A. Opler, C. E. Puckett Haecker, and P. D. Doran, 1: 351–84. U.S. Department of the Interior, U.S. Geological Survey, Reston, VA.

Wiggers, Raymond. 1997. *Geology underfoot in Illinois.* Mountain Press Publishing Co., Missoula, MT.

Winchell, N. H., and Warren Upham. 1884. *The geology of Minnesota.* Vol. 1

of the final report. 1872–1882. The Minnesota Geological and Natural History Survey of Minnesota. Johnson, Smith, and Harrison, State Printers, Minneapolis.

———. 1888. *The geology of Minnesota.* Vol. 2 of the final report. 1882–1885. The Minnesota Geological and Natural History Survey of Minnesota. Pioneer Press Co., State Printers, St. Paul.

Wisconsin Department of Natural Resources. 1994. *Fishing and boating on the Mississippi River: Wisconsin, a great state to fish.* Publication FM-745-94.

Wright, H. E. Jr. 1972a. "Quaternary history of Minnesota." In *Geology of Minnesota: A centennial volume,* ed. P. K. Sims and G. B. Morey, 515–47. Minnesota Geological Survey, St. Paul.

———. 1972b. "Physiography of Minnesota." In *Geology of Minnesota: A centennial volume,* ed. P. K. Sims and G. B. Morey, 561–78. Minnesota Geological Survey, St. Paul.

———. 1985. "History of the Mississippi River in Minnesota below St. Paul." In *Pleistocene geology and evolution of the Upper Mississippi Valley: A working conference,* 1–2. Winona State University, MN.

———. 1989. "Origin and developmental history of Minnesota Lakes." *Journal of the Minnesota Academy of Science* 55 (1): 26–31.

Wright, H. E., and R. V. Ruhe. 1965. "Glaciation of Minnesota and Iowa." In *The Quaternary of the United States,* ed. H. E. Wright Jr. and D. G. Frey, 29–41. Princeton University Press, Princeton, NJ.

Wright, Karen. 1999. "First Americans." *Discover* 20 (2): 52–63.

Young, Beloine W., and Melvin L. Fowler. 2000. *Cahokia, the great American metropolis.* University of Illinois Press, Urbana and Chicago.

Zumberge, J. H. 1952. *The lakes of Minnesota: Their origin and classification.* Minnesota Geological Survey Bulletin 35, St. Paul.

Index

CPSIA information can be obtained
at www.ICGtesting.com
Printed in the USA
LVHW081546101218
599911LV00010B/222/P